The RF in RFID

T0297617

The RF in RFID

The RF in RFID
UHF RFID in Practice

Daniel M. Dobkin

AMSTERDAM • BOSTON • HEIDELBERG • LONDON
NEW YORK • OXFORD PARIS • SAN DIEGO
SAN FRANCISCO • SINGAPORE • SYDNEY • TOKYO
Newnes is an imprint of Elsevier

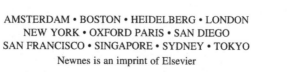

Newnes is an imprint of Elsevier
The Boulevard, Langford Lane, Kidlington, Oxford OX5 1GB, UK
225 Wyman Street, Waltham, MA 02451, USA

Second edition 2013

Copyright © 2013 Elsevier Inc. All rights reserved

No part of this publication may be reproduced, stored in a retrieval system or transmitted in any form or by any means electronic, mechanical, photocopying, recording or otherwise without the prior written permission of the publisher

Permissions may be sought directly from Elsevier's Science & Technology Rights Department in Oxford, UK: phone (+44) (0) 1865 843830; fax (+44) (0) 1865 853333; email: permissions@elsevier.com. Alternatively you can submit your request online by visiting the Elsevier web site at http://elsevier.com/locate/permissions, and selecting obtaining permission to use Elsevier material

Notice

No responsibility is assumed by the publisher for any injury and/or damage to persons or property as a matter of products liability, negligence or otherwise, or from any use or operation of any methods, products, instructions or ideas contained in the material herein. Because of rapid advances in the medical sciences, in particular, independent verification of diagnoses and drug dosages should be made

British Library Cataloguing-in-Publication Data
A catalogue record for this book is available from the British Library

Library of Congress Cataloging-in-Publication Data
A catalog record for this book is available from the Library of Congress

ISBN: 978-0-12-394583-9

For information on all Newnes publications
visit our web site at www.newnespress.com

Printed and bound in United States of America

13 14 15 16 17   10 9 8 7 6 5 4 3 2 1

Working together to grow
libraries in developing countries

www.elsevier.com | www.bookaid.org | www.sabre.org

ELSEVIER   BOOK AID
            International   Sabre Foundation

Contents

CHAPTER 1

Introduction

1.1 What, When, and Where, Wirelessly

To a quantum mechanic the whole universe is one godawful big interacting wavefunction — but to the rest of us, it's a world full of separate and distinguishable objects that hurt us when we kick them. At a few months of age, human children recognize objects, expect them to be permanent and move continuously, and display surprise when they aren't or don't. We associate visual, tactile, and in some cases audible and olfactory sensations with identifiable physical things. We're hardwired to understand our environment as being composed of separable things with specific properties and locations. We understand the world in terms of what was where when. So one can forgive us for being disappointed that the computers and networks that form so large a part of our lives, and often seem so intelligent in other respects (at least on a good day), are clueless when it comes to perceiving and recognizing all these discrete physical objects we so easily detect and categorize. Why do we have to laboriously inform a computer database, by typing or mousing or tapping a screen, that a perfectly recognizable object has arrived at our doorstep? Why is so much human intervention needed for such a simple task?

It is to correct this deficiency of networked sensibilities that the field of *automated identification* (auto-ID) has arisen. Auto-ID includes any means of automating the task of identifying a physical object. To date, by far the most common means of doing so is to print a special machine-decipherable *bar code* on an object, and then image or scan the code using an optical transducer to extract an identifying number. One-dimensional bar codes (so named because information is obtained in traversing the pattern in a single direction, not because such patterns are in fact absent width and height) are easily deciphered and, in the form of the *Universal Product Code* (UPC) and its more modern descendents, nearly ubiquitous in the commercial world. Two-dimensional bar codes are also available, and pack more information into the same space. *Optical character recognition* (OCR) can be used to acquire information from conventional human-readable text, at the cost of an increase in computing requirements and decreased reliability. However, all optical methods of identifying an object have some deficiencies. Most fundamentally, the sensing device must be able to see the identifying mark: optical techniques require a clear *line of sight*. Not only objects, but dirt, paint, ink, and other objectionably opaque but relentlessly commonplace substances can distort or deface bar codes and other optical marks, obscuring the information optical auto-ID techniques require. Mechanical damage to the marks or labels degrades their readability. To store more data requires more space, or the use of finer markings visible from a shorter distance. Finally, data stored in printed marks on a surface is not readily modified or extended, save perhaps by wholesale replacement. While optical techniques for object identification are versatile and inexpensive, it is clear that in many cases another approach may be helpful.

To remedy some of the deficiencies of optical ID, we can turn to an alternative technique, *radio-frequency identification* (RFID). RFID is the use of radio communications to identify a physical object. RFID is really not one but a suite of identification technologies, because of the differing characteristics of the radio waves of varying frequency employed, and the differing approaches to operating the sensors that serve to identify individual objects. RFID has existed for more than half a century, but its widespread application has had to wait for inexpensive integrated circuits to enable small, low-cost *transponders* (the parts of the system that get attached to an object to be identified, more commonly known as RFID *tags*) to be fabricated. Over the last three decades, as the capability of integrated circuits has doubled and the cost per function halved about every two years, religiously attending to Gordon Moore's famous law, new RFID applications have become economically feasible. In particular, since the mid-90's, a great deal of effort has been focused on the application of RFID in the manufacture and distribution of goods: *supply chain* management, where until recently the bar code reigned supreme. To serve the needs of manufacturing, distribution, and shipment functions, RFID tags must be very inexpensive, compact, mechanically robust, and readable from at least a meter or two away. As we will examine in more detail in chapter 2, this combination of requirements has led to the choice of *ultra-high-frequency* (UHF) radio waves and *passive* RFID tags as the approach of choice for many supply chain applications, and it is UHF RFID technology that is the main topic of this book.

1.2 Why Would You Read This Book?

The purpose of **The RF in RFID** is to provide users of UHF RFID with an understanding of how identification information gets from a tag to a reader and in some cases back to the tag. We will use that understanding to see how the system of tags, readers, and antennas goes together, and analyze the capabilities and limitations resulting from the choices of tag, reader, antenna, and protocol. This book is for people who want to know why things RFID are the way they are and what (if anything) can be done about it, and perhaps be entertained upon occasion along the way.

As we will see as we proceed, the replacement or supplement of bar codes with RFID tags may give rise to a substantial increase in the amount of information available about objects being made, shipped, or sold, and thus create a need for improved software solutions to enable useful integration of this new knowledge into the existing infrastructure for managing such transactions. The reader must, alas, turn elsewhere for advice and insight on software integration issues: this book is focused on tags, readers, and their interactions. For folks familiar with the OSI reference model for communications systems, **The RF in RFID** is a book about the *physical layer* of a UHF RFID system, with some digressions into the *data link layer*, but no higher.

You don't need a prior acquaintance with radio technology to read the book (although it doesn't hurt), but familiarity with basic electrical engineering concepts of current, voltage, power, frequency, capacitance, and inductance is very helpful. A general familiarity with algebraic manipulation, and the concepts of an integral and derivative, will be needed to follow the derivations of key formulas; a brief review of a few more specialized mathematical tools that are widely used in electrical engineering is provided within the Appendices in the interests of completeness.

1.3 What Comes Next?

The structure of the remainder of this book is depicted in Figure 1.1.

Chapter 2 is a general introduction to RFID, including a bit of history, some terminology, and an examination of the various flavors of RFID and their characteristics and uses. Chapter 3 introduces the reader to the basics of radio technology: transmission, modulation, bandwidth, signal voltage and power. Chapter 4 describes how the specific radios used in

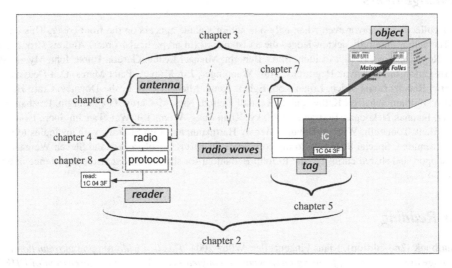

Figure 1.1
Overview of this book.

UHF RFID readers work. Chapter 5 delves into the operation of passive UHF RFID tags. Chapter 6 examines reader antennas: how they work and how they are characterized and described. Chapter 7 extends this discussion to the peculiar requirements of passive tag antennas. Chapter 8 reviews the tag-reader protocols employed in UHF RFID. Chapter 9 returns to some old applications in the light of issues discussed in the book and introduces current and future applications of RFID. A brief Afterword rounds out the main text of the book. Appendices cover some supplemental information, including the radio regulatory world in which manufacturers and users of RFID systems must operate, and some electrical engineering background useful for those from other fields.

In the interests of clarity, detailed citations are not provided within the text. However, each chapter contains a Further Reading section directing the still-curious reader to additional materials related to the topics covered therein. Each chapter (except this one) also contains exercises to provide an opportunity to exercise concepts introduced in the text; answers may be found at the author's web site, www.enigmatic-consulting.com.

Readers familiar with the first edition of this book will find fewer minor errors, often courtesy of folks like themselves who brought said mistakes to my attention. New materials include a discussion of link budgets and wake-up challenges for active and battery-assisted tags, some recent progress in nonradiating antennas, an improved discussion of the T-match for tag antennas and some examples of near-metal antenna design, and coverage of battery-assisted and sensor provisions of ISO 18000-6, all in addition to the whole chapter on Applications. References to ETSI EN 302 208 have also been updated.

Acknowledgements

A book is a collective endeavor even when only one author's name appears on the front cover. This one is no different. The author gratefully acknowledges the assistance of (in no particular order) Andrew Crook, Richard Woodburn, Roger Stewart, Douglas Litten, Barry Benight, Michael Leahy, Gordon Hurst, John Myers, Stephen Colby, Chris Parkinson, Dewayne Hendricks, Titus Wandinger, Jim Mravca, Peter Mares, Dan Deavours, Tali Freed, Gabriel Rebeiz, Louis Sirico, Lilian Koh, Egbert Kong, Michael Lim, Leslie Downey, Craig Harmon, Gene Donlan, William Schaffer, Kathy Radke, Kendall Kelsen, Nick McCurdy, Greg Durgin, Prashant Upreti, Jim Buckner, Faranak Nekoogar, Bertrand Teplitxky, Kuan Sung, Wong Tak Wai, Tan Jin Soon, Brian Ogata, Sanjiv Dua, Hank Tomarelli, William Devore, Merrily Hartmann, and Bob Stewart, with apologies to those inadvertently omitted. Special thanks go to my colleagues Dan Kurtz, Nathan Iyer and Steven Weigand, for consistent support and shared curiosity, and to John Bellantoni for sharing his extensive experience in every aspect of radio design.

Further Reading

RFID Handbook (2nd edition), Klaus Finkenzeller, Wiley 2004. *This is a wide-ranging introduction to RFID technology and applications, focusing on inductively-coupled systems but with a discussion of UHF RFID. The technical material is challenging for someone unfamiliar with the field (though perhaps it will be more accessible after you finish* **The RF in RFID**). *An updated third edition of this book was released in 2010.*

RFID Field Guide, Manish Bhuptani and Shahram Moradpour, Sun Microsystems Press 2005. *Marketing- and ROI-focused; a useful complement to the current volume and Finkenzeller's book.*

RFID Sourcebook, Sandib Lahiri, IBM Press, 2006. *A nice guide for folks who need to implement supply-chain RFID systems, with guidelines, rules of thumb, and checklists for the various aspects of the project.*

A great number of diverse web sites touch upon RFID-related matters. Some useful ones are:

RFID Journal, www.rfidjournal.com

Association for Automatic Identification and Mobility (AIM Global), www.aimglobal.org

EPCglobal Inc., www.epcglobalinc.org

RFID Network, www.rfid.net

RFID Online Solutions, www.rfidonlinesolutions.com

RFID Tribe, www.rfidtribe.org

RFID Revolution, www.rfidrevolution.com

History and Practice of RFID

2.1 It All Started with IFF

By the 1930's, the primitive biplanes of fabric and wood that had populated the skies above the battlefields of World War I had become all-metal monoplanes capable of carrying thousands of kilograms of explosives and traveling at hundreds of km per hour: by the time observers could visually identify an incoming flight, it was too late to respond. Detection of airplanes beyond visual range was the task of microwave radar, also under rapid development in the 30's, but mere detection of the presence of aircraft begged the key question: whose side were they on? It was exactly this inability to identify aircraft that enabled the mistaken assignment of incoming Japanese aircraft to an unrelated US bomber flight and so ensured surprise at Pearl Harbor in 1941. The problem of identifying as well as detecting potentially hostile aircraft challenged all combatants during World War II.

The Luftwaffe, the German air force, solved this problem initially using an ingeniously simple maneuver[1]. During engagements with German pilots at the beginning of the war, the British noted that squadrons of fighters would suddenly and simultaneously execute a roll for no apparent reason. This curious behavior was eventually correlated with the interception of radio signals from the ground. It became apparent that the Luftwaffe pilots, when they received indication that they were being illuminated by their radar, would roll in order to change the backscattered signal reflected from their airplanes (Figure 2.1). The consequent modulation of the blips on the radar screen allowed the German radar operators to identify these blips as friendly targets. This is the first known example (at least to the author) of the use of a *passive backscatter* radio link for identification, a major topic of the remainder of this book. *Passive* refers to the lack of a radio transmitter on the object being identified; the signal used to communicate is a radio signal transmitted by the radar station and *scattered* back to it by the object to be identified (in this case an airplane).

As a means of separating friend from foe, rolling an airplane was of limited utility: any aircraft can be rolled, and no specific identifying information is provided. That is, the

---

[1] Unfortunately, at the moment this is an unverified Internet report, for which I have been unable to find an archival source — but it is such a fun story I had to include it anyway! An authoritative citation for (or debunking of) the story would be appreciated. —DMD

attitude change varies signal
backscattered to radar

Figure 2.1
The use of backscattered radiation to communicate with a radar operator (not to scale!).

system has problems with *security* and the size of the *ID space* (1 bit in this case). More capable means of establishing the identity of radar targets were the subject of active investigation during the 1930's. The United States and Britain tested simple IFF systems using an active beacon on the airplane (the XAE and Mark I, respectively) in 1937/1938. The Mark III system, widely used by Britain, the US, and the Soviet Union during the war, employed a mechanically tunable receiver and transmitter with six possible identifying codes (that is, the ID space had grown to 2.5 bits). By the mid-1950's, the radar transponder still in general use in aviation today had arisen. Modern transponders are *interrogated* by a pair of pulses at 1030 MHz, in the *ultra-high frequency* (UHF) band about which we will have a lot more to say shortly. The transponder replies at 1090 MHz with 12 pulses each containing one bit of information, providing an ID space of 4096 possible codes. A mode C transponder is connected to the aircraft altimeter and also returns the current altitude of the aircraft. A mode-S transponder also allows messages to be sent to the transponder and displayed for the pilot. Finally, the typical distance between the aircraft and the radar is on the order of 1 to a few kilometers. Since it takes light about 3 μs to travel 1 kilometer, the radar reflection from a target is substantially delayed relative to the transmitted pulse, and that delay can be used to estimate the distance of the object.

An aircraft transponder thus provides a number of functions of considerable relevance to all our discussions in this book:

- identification of an object using a radio signal without visual contact or clear line of sight: *radio-frequency identification*

- an ID space big enough to allow unique[2] identification of the object
- linkage to a sensor to provide information about the state of the object identified (in this case the altitude above ground)
- location of each object identified (angle and distance from the antenna)
- transmission of relevant information from the interrogator to the transponder

These functions encompass the basic requirements of most RFID systems today: RFID has been around for a long time. However, for many years wider application of these ideas beyond aircraft IFF was limited by the cost and size of the equipment required. The early military transponders barely fit into the confined cabins of fighter airplanes, and even modern general aviation transponders cost US\$1,000–5,000. In order to employ radio signals to identify smaller, less-expensive objects than airplanes, it was necessary to reduce the size, complexity, and cost of the mechanism providing the identification.

2.2 Making It Cheap

The most important underlying dynamic enabling the wide use of RFID has been the unprecedented increase in capability and decrease in cost of electronics ever since the invention of the transistor, most particularly the scaling of integrated logic circuitry according to Gordon Moore's famous law ever since the 1960's. This trend has impacted every aspect of modern life, including the use of radios (and other technologies) for identification.

However, RFID also has some specific requirements not normally encountered in other radio communications systems. These requirements are essentially economic in nature: any method of identifying an object must cost less than knowledge of the identity of that object is worth. Such constraints are of minimal consequence when identifying an airplane that costs anywhere from \$100,000 to \$100,000,000 and where an error in identification or location could result in a midair collision costing billions of dollars. The same insensitivity to cost is less appropriate when the object to be identified is an automobile or rail car, and inconceivable when the object is a case of disposable diapers.

Thus, RFID applications often require radios that are extraordinarily cheap and simple. In addition to simply reducing the cost of the electronics, there are several special pieces of technology that enable very inexpensive identifying tags to be produced:

- ***no transmitter:*** we've already examined above the idea of skipping the transmitter by varying some property of the signal reflected back from an object. Since transmitters

---

[2] Unique in this context, anyway. Obviously there are more than 4096 aircraft in the world, but rarely 4096 aircraft within range of one radar. Note that aircraft flying under ***visual flight rules*** instead of positive air traffic control all use the same code, 1200.

are large, complex, and power-hungry, particularly at high frequencies, skipping a transmitter is very useful.

- *no battery:* batteries, as any parent of small children can tell you, are expensive, maintenance-intensive, and easily tripped over. A tag that either doesn't need power or can find it somewhere else is a good thing.
- *simple circuitry*: the less the tag has to think about, the better. Reducing the intelligence in the tag usually involves a tradeoff with security and the size of the available ID space.
- *standing out of the noise*: a tag that uses the transmitted signal to communicate creates a problem for the reader, that of hearing the tiny tag signal in the big reader signal.

Let's examine how the tricks that enable an RFID system to satisfy these requirements have come about. As we noted above, the use of backscattered radiation to send one bit to a radar operator was demonstrated in the late 30's, but a single bit is not a very flexible payload. The earliest reported description of passive communications I have so far found is a patent assigned to Evenor Brard, issued by the US patent office in 1930 (US 1,744,036). This basically describes an inductively coupled transmitter and receiver; the receiver being equipped with a switch allowing a human operator to vary its tuning and thus return a signal to the transmitter by varying the inductive load. Very little detail is provided, and it is not clear how the system was to be used.

A more ambitious investigation of the use of backscattered radiation to communicate substantial information was reported by Harry Stockman in 1948; one of the configurations he explored is shown schematically in Figure 2.2. A conventional microphone and speaker coil were used to modulate the position of a receiving antenna according to the sound received by the microphone. This positional modulation in turn affected the signal reflected back to the transmitter, where the sound could be demodulated and reproduced despite the fact that there is no radio transmitter associated with the microphone. This is the first example of a *backscatter radio link* conveying substantial amounts of information back to the transmitter.

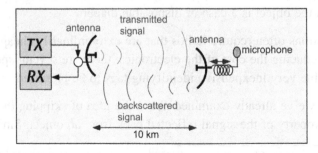

Figure 2.2
Use of backscattered radiation to communicate; after Stockman, Proc. I.R.E., October 1948.

In the early 1950's, passive backscatter links of this type were investigated with, among others, the object of creating an inexpensive wireless telephone system (the cellular phone being at that time very far in the future!), as exemplified in a 1960 patent due to Harris (US 2,927,321). Another patent of the same era (Figure 2.3) shows the use of a signal received on an antenna and rectified by a diode, with the DC used to power a transistor oscillator to produce an identifying signal at a second, unrelated frequency: an early RFID application, although in this case the frequencies envisioned are rather low and the "tag" antennas would be quite ungainly by modern standards!

In many applications of this type, unlike Mr. Stockman's 10-km link, ranges of less than a meter are quite sufficient. When the distance is a few tens of centimeters, it may no longer be necessary to launch a wave and reflect it: the transmitter and receiver can be *inductively coupled*, so that the load presented to the transmitter by the receiver can be intentionally varied to produce a signal at the transmitter, again without the need for active signal generation by the receiver. Inductively coupled systems can operate at much lower frequencies than those used in radar, typically tens of kiloHertz to around 10 MHz. This combination of simplicity, low frequency, and short range enables one to produce compact, low-cost transponders and practical, reasonably inexpensive interrogators. The simplest inductively-coupled transponders, like the Luftwaffe aircraft, transmit only one bit signaling their presence. A compact, low-cost transponder of this type may be constructed using a strip of magnetically sensitive metal mechanically resonant at the frequency at which the interrogator operates, so that when the transponder is placed near a reader antenna, it vibrates and extracts energy from the antenna. (A second strip, whose magnetization can be changed, is placed adjacent to the first, and used to detune the transponder when, for example, the associated object has been purchased.) Such transponders were developed in the 1960's for the purpose of preventing theft of retail goods, and are still in wide use.

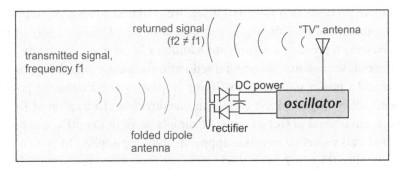

Figure 2.3
Passive retransmitting identification system using oscillator driven by DC power harvested from incoming RF; after Crump, US 2,943,189, filed 1956, granted 1960.

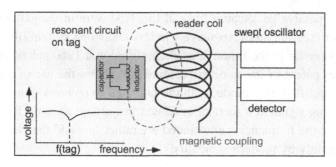

Figure 2.4
Resonant circuit coupled to a reader coil acts as an identifying tag; after Walton,
US patent 3,752,960.

A transponder with a larger potential ID space, but still inexpensive and compact and with no need for a battery, may be achieved by using a ***resonant circuit***, composed of a capacitor and inductor which together determine a unique frequency at which a large current flows through the transponder. The resonant frequency can be readily varied by adjusting the values of the components. Charles Walton, among others, patented several types of inductive identifying transponders in the early 1970's; an example is depicted in Figure 2.4. The reader sweeps over a range of frequencies containing the resonant frequency of the tag, and detects a relatively abrupt change in the voltage across the coil when the tag resonance is encountered. Such a system can convey more than one bit, since the resonant frequency of the tag can be used as an identifying means, and one tag may display multiple resonances, allowing for a reasonably large ID space.

Tags of this type were used in one of the first major commercial implementations of RFID technology by the Schlage Lock Company around 1972–73. Several million electronic keys, using multiple resonators in the 3–32 MHz region, were produced.

To fabricate a more sophisticated identifying tag, we'd like to add some circuitry to perform logical operations, but this requires electrical power. The usual way of providing such power, a battery, represents a significant addition of cost and complexity in many applications. Instead, we can use the transmitted radio-frequency signal both as a means of communication and a power source, by rectifying it: passing the alternating-polarity signal through a diode, which only conducts current in one direction. Extraction of DC from RF had already been envisioned in (for example) Crump's work in the 50's, and by the early 70's Cardullo and Parks showed how this approach could be applied to power a more modern bit of circuitry (Figure 2.5). A diode and capacitor (to store some of the resulting current and so smooth out the fluctuations in the output voltage) are used to extract power from the received signal; this power is then used to process the signal and produce a reply. (The diode in Figures 2.5 and 2.6 are used to indicate that rectification is performed. The

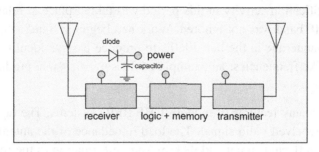

Figure 2.5
Extraction of DC power from RF power; after Cardullo and Parks, US patent 3,713,148.

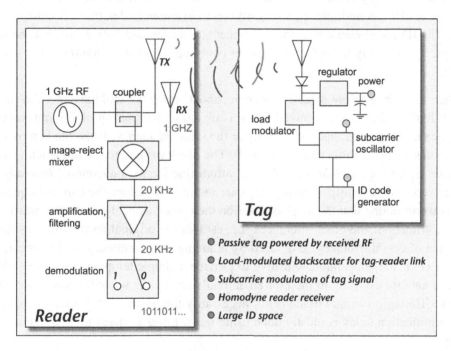

Figure 2.6
An early UHF passive tag system; after Koelle et. al. Proc. 1975, IEEE p. 1260.

actual configuration employed needs to provide a DC path to the ground, which an antenna may or may not do depending on its configuration. We will dive into the details in chapters 5 and 7.) In practice a regulating function is also necessary, to account for the wide variations in signal strength likely to be encountered when the distance between the transponder and the reader antenna varies.

Though much identification activity in this period was taking place at relatively low frequencies, the UHF band was not ignored. Work had begun at Sandia National Laboratories in Albuquerque in the late 1960's to produce passive identifying systems operating at radar-like frequencies; an example of the sort of system produced is shown in Figure 2.6.

In this work we see many features of modern UHF RFID systems. The tag is powered by rectification of the received radio signal. The load impedance of the antenna is changed in order to modify the reflected signal and thereby send information to the reader. The tag uses a *subcarrier* modulation scheme; that is, the tag antenna impedance is switched at 20 KHz, and the degree of modulation of the antenna impedance is small for a binary '0' and large for a binary '1'. Subcarrier modulation is slow because many switching cycles are used to send one binary bit, but it is simple to implement and relatively robust to noise, because the reader has multiple signal transitions to examine to determine the nature of each bit. The ID code generator in this implementation provided only 3 bits, but the size of the ID space was mainly limited by the power consumption of the circuitry rather than the approach.

The reader uses a *homodyne* detection scheme, about which we will have more to say in chapter 4. In this scheme, the received signal is mixed with a portion of the transmitted signal, since it is expected that both will be at the same frequency (if the tag is moving slowly, so that Doppler shifts can be ignored). The result of the mixing operation is the information signal itself (the *baseband*), in contradistinction to conventional *heterodyne* radio receivers, where multiple mixing steps are required to reduce the carrier frequency to zero. It is worth noting that the authors describe their work as being directed towards animal identification, and other references suggest that the original motivation was the preservation of cowhide otherwise defaced by branding. The frequency of the subcarrier was purposely made temperature-sensitive to provide a simple temperature sensor, which the authors anticipated using to monitor an animal's health. As we will learn in section 5, sticking a UHF tag on animal tissue is not necessarily the best use of this technology; animal identification today is mostly done using much lower frequencies and inductive coupling.

The system described by Koelle and coworkers provided read ranges of a few meters using a 1 GHz, 4-watt signal. Koelle noted that much longer ranges could be obtained if the tag had its own source of power. Note also that the electronics available at this time didn't allow for writing new information to the tag, which doesn't do much other than broadcast its ID continuously when powered up.

Higher microwave frequencies were also explored in the early 70's. For example, work done by Klensch and coworkers at RCA Laboratories involved a very simple tag containing a high-frequency diode and a rather clever antenna, printed onto a piece of metal just about the same size as an automobile license plate, which of course this technology was designed to supplement. The tag was illuminated by a microwave transmitter operating at around 8–9 GHz. When a radio signal at any frequency f is applied to a diode, a significant amount of power is obtained at a frequency of $2f$, the **second harmonic** of the original signal. By varying the bias voltage on the diode, the amount of second-harmonic power generated could be varied. The resulting signal, at around 17 GHz, was captured in a receiver and demodulated to identify the tag. Because the return signal is at a different frequency from the transmitted signal, it is relatively easy to detect. However, other technologies have supplanted this approach for automated identification of vehicles.

2.3 Making and Selling: Tracking Big Stuff

A recurring theme in the business of RFID is the basic economics of identification: you can't spend more identifying an object than the object is worth. As a consequence, applications generally involve either very expensive objects or very cheap tag technologies.

By the early 1970's, the rapidly decreasing cost of electronic components had lead to the demonstration of more sophisticated identifying systems using inductive coupling, UHF backscatter radios, and microwave radios, as described in section 2 above. One of the obvious applications of such non-contact identifying technologies was traffic management, because automobiles and trucks are expensive, mobile, and uniquely identified with an owner or owners. The application of RFID techniques to traffic control at this time was nicely summarized in an article by William Arnold, published in Electronics in September of 1973. After reviewing some of the major pilot projects, Mr. Arnold provides a remarkably prescient list of the applications envisioned for RFID: toll collection, toll parking, dynamic traffic control, vehicle identification, trucking location, and surveillance.

The implementation of these visions was largely constrained by barriers of cost, regulation, and institutional risk, though early observers were also already concerned about driver privacy. The main solution to the obstacle of cost was time: during the 1970's and 1980's semiconductor manufacturing improved dramatically and the cost of virtually all electronic goods fell commensurately.

One of the first major adopters of RFID technology in this area was the rail industry in the United States. This far-flung industry was faced with the problem of keeping track of tens of thousands of rail cars across tens of thousands of kilometers of track, made more challenging by the fact that companies often swapped cars in order to avoid paying to move empty cars. In the 1970's, the railways tried to make use of a circular bar code identifier, but optical identification in the rainy, muddy, snowy outdoors didn't work very well. (In fact, it was the railcar identification problem that Mr. Cardullo relates as triggering his earliest work in RFID.) Railroads represent a particularly favorable problem for a passive

RFID solution: trains travel on tracks, so the qualitative location of any rail car can be inferred if its presence at a limited number of choke points is known. Further, the location of the cars at these choke points is well-controlled by the tracks, so only one reader is needed for each track. The industry is mature and follows uniform practices across the continental US. Finally, rail cars are fairly expensive assets, so the cost of the transponders was not a significant barrier to adoption. By the late 1980's, the rail industry established a standard for railcar identification (AAR S-918), based on a backscatter transponder operating in the *US industrial, scientific, and medical* (ISM) band at 902–928 MHz. The standard envisions both active and passive tags, but in practice passive tags are used because their short read range of around 3 meters is desirable to avoid counting cars on neighboring tracks. By 1994, essentially all rail cars in the United States were equipped with S-918 compliant transponders (Figure 2.7).

Progress has been slower in automobile identification. A few narrowly-focused implementations in the early 1980's, such as the Coronado Bridge in San Diego, California, demonstrated the feasibility of UHF radio traffic management. Around the same time, Philips Electronics developed an inductive system known as V-com/V-tag, requiring antennas embedded in the roadways, and mainly useful to track buses or other large vehicles following repetitive routes. In order to encourage adoption in the US, Amtech Corp. provided most of the funding for implementation of automated tolling on the North Dallas Turnpike in 1989. This effort was successful enough to get the ball rolling in the US. The EZpass interagency group was formed in 1990 to promote automated tolling in the Eastern US. Open-highway automated tolling was implemented in Oklahoma in 1991, and the title 21 standard was promulgated for use in California and the Western US in 1992. Use of the transponders is not legally required but is encouraged by special toll lanes and faster passage through lines; adoption is reported to be several million vehicles at the time of this writing, a large number but still a small percentage of the total.

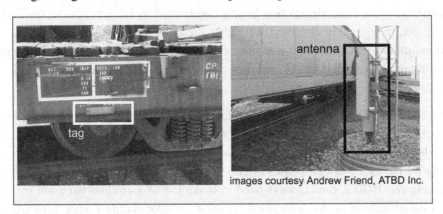

Figure 2.7
Example of typical passive tag and reader antenna for identification of railcars.

Probably the most advanced implementation of RFID for traffic management can be found in the city-state of Singapore, where urban crowding and congestion are magnified by the geographical constraints of the tiny state. In the 1970's and 1980's a paper licensing scheme was in use to provide rights to drive in the downtown areas, but this was inflexible and expensive to administer. In 1991 the government of Singapore initiated a program to automate tracking and payment. After several years of development, they settled on a system using powered backscatter transponders operating at 2.45 GHz, developed primarily by Philips Electronics and Mitsubishi Heavy Industries. Smart payment cards are used to allow anonymous transactions and preserve driver privacy. Clearly-marked gantries above roadways leading into the downtown areas signify when the system is active; payments are deducted from the smart card as the car passes below the gantry, and the license plates of violators are automatically photographed. The system demonstrated a very low missed-read rate of a few per million cars in development testing. The system is active during heavy traffic periods each day, using payments to discourage excessive travel in the most congested areas. The government had given some early consideration to instantaneous pricing to account for actual congestion conditions, but this scheme was rejected as too complex for practical implementation. Adoption was encouraged by initial government financing of transponder installation. The transponders are also widely used for payment of parking fees, just as Mr. Arnold had foreseen two decades previously. Implementation of this system was clearly assisted by unique circumstances: a single government with undiluted jurisdiction and relatively broad powers, a well-defined traffic problem in a small geographic area, and high vehicle costs due to existing government practices, as well as consideration of privacy concerns early in the project.

As noted in section 2, the work at Sandia Laboratories was partially directed towards livestock management. Cows are high-value assets and may be allowed to range over large areas to find feed, and the cattle industry in the US is highly fragmented, with many ranchers in a specific geographical region providing cattle to a smaller number of auction locations and slaughterhouses. Tracking of ownership of the animals is of commercial importance. In addition, particularly since the appearance of human-transmissible bovine spongiform encephalopathy ('mad cow' disease), it has been of considerable interest to be able to trace individual animals so that tainted meats can be specifically recalled from human use. Traditional means of branding animals wastes hide and are not readily automated. In these applications, use of frequencies around 100 KHz and inductive coupling, described in more detail later in this chapter, was initially done, because at these frequencies the water in the tissues of an animal has little effect on the radio operation (although cows apparently don't like to line up, so the short range of these transponders is a challenge). Work in the 1980's led to a standards effort initiated at the International Organization for Standardization (ISO) in 1991, resulting in approval of the ISO 11784 and 11785 standards in 1996, recently updated as ISO 14223. Transponders can be attached to an animal's ear or hide, injected under the skin, or in the case of cows swallowed and embedded within the stomach. The beef industries in Canada and Australia are actively encouraging implementation of RFID. Inexpensive inductive transponders are also now widely available for insertion in pets (and humans, apparently stylish in at least one Spanish resort).

Invented around 1956, the use of multi-mode shipping containers revolutionized worldwide logistics over the next three decades. Standardized shipping containers now carry goods on trucks, trains, or ships. A typical container costs a few thousand dollars and carries tens of thousands of dollars worth of goods. Around 18 million containers are currently in use throughout the world. Identifying and locating these containers and their contents using paper manifests is awkward, labor-intensive, and wasteful, as the US Department of Defense found during its 1991 actions in response to the Iraqi invasion of Kuwait. In subsequent years the DoD has made extensive use of radio identification and location using battery-powered transponders from Savi Technologies, which operate under the ANSI 371.2 standard at 433 MHz (ISO 18000-7), and identify a container's location and contents, as well providing some tamper protection. A similar solution, operating at 2.4 GHz based on ANSI 371.1, was developed by another company, Wherenet, in the 1990's and serves commercial tracking requirements.

People are important, mobile, and expensive (as any parent knows), so identifying and tracking them can make economic sense. Tracking of people using radio technology has become practical in the 1990's and has been implemented in several special circumstances. Many corporate employees now carry short-range radio badges, typically using inductively-coupled transponders, to allow admission into company facilities. Similar technologies are used in smart payment cards, noted above in connection with the Singapore traffic control project and also seeing increasing use in retail applications. These cards are usually compliant with ISO standards, typically ISO 14443 or ISO 15693. Prisoners can be tracked using active tags (obviously using anti-tamper provisions to avoid tag removal). Children and family members can be tracked in theme parks. Wider implementation of RFID for people tracking runs squarely into issues of privacy, surveillance, and security, which are important but beyond the scope of this book save for a few remarks in the Afterword, and the citations therein.

2.4 Tracking Small Stuff: AutoID and the Web of Things

As noted briefly above, by the 1960's simple bimetallic tags were available for one-bit monitoring functions usable for theft prevention in retail. These tags have no electronics and are inexpensive, and because they carry only 1 bit of information, no data infrastructure is needed to implement them. They are practical to attach to low-value assets, and are widely used. Very simple electronic frequency doublers, again carrying only 1 bit of information, are also available for this application.

More sophisticated tags that simultaneously provided low cost solutions for tagging individual stuff with large ID spaces had to wait for the availability of inexpensive, low-power integrated circuits. By the early 1990's, it had become possible to implement a simple communications protocol and store a reasonably large identifying number on a single integrated circuit. The circuit could be attached to an inductive antenna (a few loops of wire formed by photolithographic techniques on a plastic substrate) to create a low-cost transponder capable of attachment onto an item or box. Several companies developed low-cost inductive identification systems, notably Texas Instruments' Tag-IT and Philips u-code products, and these systems were successful in many *closed-loop* applications (those where the required information is contained within a single organization), ranging from tracking library books to beer kegs. More general applications throughout the retail supply chain were inhibited by the relatively high cost of the tags (around US$1−3) and by the number of incompatible standards and proprietary implementations.

A broader vision for the implementation of RFID was triggered by MIT researcher David Brock in 1998. Brock was working on the seemingly unrelated problem of robotic vision: how does a robot navigate through a room or building and identify and manipulate the objects it finds therein? Brock realized that this task would become a lot easier if the objects were all uniquely identifiable by some other means, such as RFID. To be useful, such a system would need to be widespread and inexpensive. Brock's MIT colleagues Sanjay Sarma, Sonny Siu, and Eric Nygren extended his observation to the idea of an ubiquitous and globally unique *Electronic Product Code* (EPC) to identify every manufactured object, along with a software infrastructure to provide access to information about the object so identified. Assisted by supporters in industry, notably Kevin Ashton of Proctor and Gamble, and Alan Haberman of UCC (the consortium administering bar code identification for retail products), in 1999 MIT launched the Auto-ID Center. Over the next three years, the center added research facilities at the University of Cambridge in the U.K., the University of Adelaide in Australia, Keio University and Fujan University in Japan, and the University of St. Gallen in Switzerland.

Through the activities of the Auto-ID laboratories, Sarma and coworkers promoted several concepts related to ubiquitous RFID. A tag should be as simple (and thus cheap) as possible, foregoing encryption of the data, complex collision resolution protocols, and extra memory beyond the EPC and error correction. A standardized infrastructure should exist, analogous to the Domain Name Service that forms the basis of the World Wide Web, to locate information about an object whose EPC is known. A markup language based on Hypertext Markup Language, the scheme used to define the appearance of web pages, should be defined to describe the characteristics of an object. Specialized software agents, originally known as *Savants*, would act to reduce the undifferentiated mass of data generated by RFID readers to a smaller comprehensible collection of meaningful events in the life of a box. Finally, after some examination of the various alternative approaches, the Auto-ID Labs emphasized RFID tags operating in the 900-MHz regime as the best compromise of cost, read range, and capability, and created a hierarchy of classes of tags with increasing capabilities (but also increasing cost). Two startup companies, Matrics (later acquired by Symbol Technologies, and that in turn by Motorola) and Alien Technology, became involved in the period 2000–2002, mixing Auto-ID Center concepts with their own technology to form the two "first-generation" air-interface standards for Class 0 and Class 1 tags.

To support wide dissemination of the technology and concepts developed within the Auto-ID centers was increasingly requiring the participants to stretch the boundaries of what was appropriate in an academic environment. By 2002 it was apparent that a new organizational basis was needed, and in 2003 EPCglobal Inc. was founded as joint venture of the administrators of the international bar code system, then UCC and EAN (now merged into the GS1 organization). EPCglobal is a non-profit corporation charged with promulgation of supply-chain RFID standards, advancement of public policy, and support for training, marketing, and implementation of RFID.

One of the first major activities taken up by EPCglobal was the resolution of the UHF air-interface conundrum created by the earlier activities of the Auto-ID center. The Class 0 and Class 1 standards mandate the use of EPCs for identification and both use the 860–960 MHz bands, but they are otherwise wholly incompatible in modulation, packet definition, collision resolution, and just about any other protocol property. This was just the situation Sarma had set out to eliminate four years before. It was apparent that a second generation standard needed to be defined that would be reasonably neutral for the existing players in the industry, and flexible enough to cover most applications without introducing mutual incompatibility. A hardware action group was formed early in 2004 to create such a standard, and in a remarkable marathon (at least by the standards of standards setting) was able to reach agreement on a completely new protocol for passive UHF tags by the end of 2004. This Generation 2 protocol, about which the reader will have the opportunity to learn more than they probably want to know in chapter 8, was ratified with minor modifications by the International Organization for Standardization as ISO 18000-6C in 2006, and has become the basis for a globally accepted protocol for conversing with passive RFID tags.

Roughly coincident with the formation of EPCglobal, Wal-Mart, at that time and this the world's largest retailer, brought its activities in the RFID arena to the attention of the world by mandating that its top 100 suppliers would need to provide RFID tags on all cases and pallets delivered to Wal-Mart by January of 2005. The next 100 suppliers would need to follow by January 2006. While some of the expectations for system performance advertised at that time turned out to be unrealistically optimistic, both these mandates were eventually met, and such data as has been made public suggests that the investment has been worthwhile for Wal-Mart, though perhaps less obviously so for its vendors. Wal-Mart's announcement was soon followed by similar mandates from other large retailers: Tesco, Metro, and Target. The US Department of Defense, which as noted above had already made extensive use of RFID in tracking shipping containers, also mandated that high-value cases should be RFID-tagged by 2005, though serious enforcement of those provisions seems to have begun only late in 2006.

The failed attempt at an initial public offering in the US stock markets by Alien Technology in 2006 signaled a period of retrenchment in the RFID market. In the clarification of hindsight, the limitations of the available technology (which will be more apparent to the reader at the completion of this volume) were not at first properly comprehended by the community of users. With the passing of time, wide implementation of the EPC Generation 2 standard for tag-reader communication, and availability of improved hardware, made possible more thoughtful targeted implementation of the technology. By 2010, it was clear that UHF tags are particularly suitable for tracking garments, and wide use of RFID for this purpose began to create the mass market envisioned by the MIT team years before.

The ferment surrounding this association of large amounts of money with radio-frequency identification has created a considerable increase in participation by vendors, integrators, and ordinary folks who might buy the stuff, but has had the unfortunate consequence of obscuring key distinctions between the various technological approaches to identification by radio. We have alluded in passing to these issues above; let us now pause in reflection on the past, and delve into the details of how the veil of objective anonymity is stripped aside by an invisible vibration in a non-existent ether.

2.5 RFID Systems and Terminology

An archetypal RFID system consists of an *interrogator*, more often known as a reader, a *transponder* or tag, and *antennas* to mediate between voltages on wires and waves in air (Figure 2.8). The reader antenna or antennas may be integrated with the reader, or physically separate and connected with a cable; the tag antenna is generally physically integrated with the tag. Most tags have at least one *integrated circuit* (IC), often known as a silicon chip, containing the tag ID and the logic needed to navigate the protocol that guides discussions between the tag and reader. There are tag technologies that do not employ silicon IC's, though we will touch only peripherally on them in this book. The reader may contain a user interface of its own, but more often will be connected to a network or a particular host computer, which interacts with the user to control the reader, and stores and displays the resulting data.

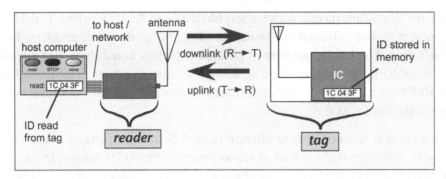

Figure 2.8
Overview of RFID system.

A link in radio parlance is the data-carrying connection that exists between a specific radio transmitter and receiver. Although they occupy the same physical space and generally use the same antennas, engineers often distinguish between the communications channel carrying information from the reader to the tag – the *downlink* or *forward link* – and that carrying information from the tag to the reader – the *uplink* or *reverse link*. In a real application it is not unlikely that multiple tags will be present in the neighborhood of the reader, and also possible that many readers will be located in close proximity to one another.

Readers and tags usually live in a larger world of information storage and handling, from which point of view an RFID reader is just another sensor, sharing that position with bar code scanners, keyboards, touchscreens, and other data collection apparatus (Figure 2.9). In a small organization, this infrastructure might consist of a database running on the local host, or even a spreadsheet that just records the list of unique tag reads. In a large organization or company, operating activities are managed by a much bigger database, often known as an *enterprise resource planning* system (ERP), *manufacturing resource planning* (MRP), or perhaps a *warehouse management system* (WMS), depending on the context. Because an RFID system is likely to distinguish between specific individual objects rather than just between classes of objects, it may generate a lot more data than traditional tracking systems. A whole class of software applications, generically known as RFID *middleware*, is arising to provide a bridge between the ERP database and business processes and the newfangled RFID equipment. Though this matter is not a topic of the present volume, it is important for the reader to realize that the mere collection of data is not equivalent to the production of useful information, and the implementation of an infrastructure of RFID readers and tags should be regarded as the enabler for improvements in information handling and business processes that create improved efficiency, rather than an end it itself.

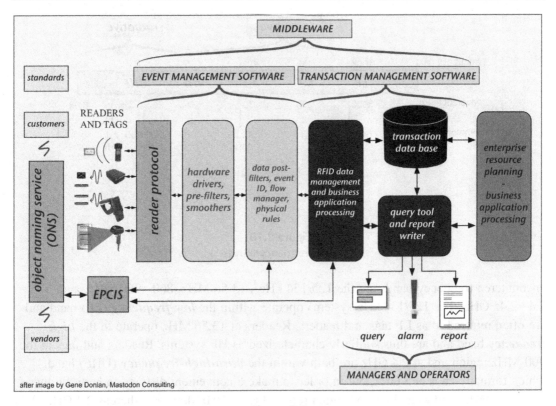

Figure 2.9
RFID as a sensor within an overall software infrastructure.

2.6 Types of RFID

We have alluded to key aspects of RFID systems in our discussion of the history of the technology in sections 1–4 above. RFID systems are crucially distinguished by the frequency of the radio waves the employ, by the means used to provide power to the tags, and by the protocols employed to communicate between tag and reader. The choice of frequency, power source, and protocol has important implications for range, cost, and features available to the user.

2.6.1 Frequency Bands for RFID

RFID systems employ frequencies varying by a factor of 20,000 or more, from around 100 kHz to over 5 GHz (Figure 2.10). Systems rarely operate arbitrarily across this vast swath of spectrum; most of the activity is concentrated in fairly narrow bands that have been made available by regulators for unlicensed industrial activities. The most commonly-

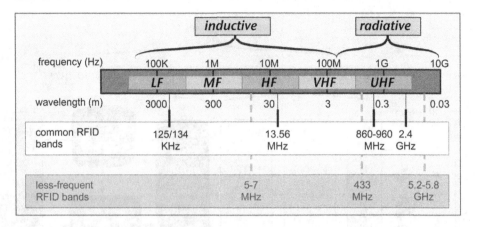

Figure 2.10
RFID frequency bands.

encountered frequency bands are the 125/134 kHz[3], 13.56 MHz, 860−960 MHz, and 2.4−2.45 GHz. The 125/134 kHz systems operate within the *low-frequency* (LF) band and are often referred to as LF tags and readers. Readers at 13.56 MHz operate in the *high-frequency* band and are thus similarly characterized as HF systems. Readers and tags in the 900-MHz region and at 2.4 GHz are both within the *ultra-high-frequency* (UHF) band, which formally ends at 3 GHz, but in order to make a convenient distinction between these two, 900-MHz readers and tags are often referred to as UHF devices, whereas 2.4 GHz systems are known as microwave readers.

Corresponding to the range in frequency is a huge range in wavelength. Recall that electromagnetic waves travel in vacuum at the speed of light (and almost as fast in air), $c = 300,000$ kilometers per second. The wavelength is the distance between successive peaks or trough of the wave, and so is the ratio of the speed of propagation c to the frequency f: mathematically speaking,

$$\lambda = \frac{c}{f} \tag{2.1}$$

Thus a wave with a frequency of 300,000 Hz (300 kHz) will have a wavelength of (300,000 km/s)/(300,000 peaks/sec) = 1 kilometer. The wavelengths in common use in RFID range from about 2000 meters − comparable to a sizable mountain unless you live in Tibet or Chile − to about 12 cm. The antenna sizes used in RFID have no comparable range of variation: an antenna is always about human-sized, with the largest ones around

---

[3] The reference here to two frequencies differing by 9 kHz is related to the means used by some protocols to communicate, in which these two frequencies are used to denote differing binary bits (a *frequency-shift keying* scheme).

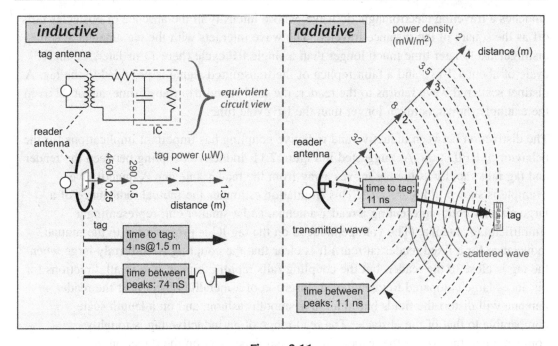

Figure 2.11
Inductive coupling (13.56 MHz, 50 cm diameter antenna) vs. radiative coupling (900 MHz), with associated power and time delays.

1 meter in diameter and the smallest 1–4 cm. As a consequence, RFID systems can also be categorized by whether the wavelength is comparable in size to the antenna, or vastly larger than the antenna.

Systems where the wavelength is much larger than the antenna are typically *inductively coupled*: almost all the available energy from the reader antenna is contained within a region near the reader antenna and comparable to it in size, falling away as the cube of distance or faster as we move away (Figure 2.11). Within this region, communication between tag and reader is effectively instantaneous, since the propagation time to the tag is a small fraction of the time for a complete cycle of the RF voltage. In the figure, even at the distance of 1.5 meters from the antenna (where there is not really enough power to run the tag), the delay is only about 4 ns, about 6% of the RF cycle at 13.56 MHz, which is about 74 ns. Under these circumstances, it is difficult to speak of a separate transmitted and backscattered wave. Instead, we think of changes in the tag antenna as inducing changes in the electrical impedance of the reader antenna: the reader − tag system acts as a magnetic *transformer* providing coupling between the current flowing in the reader and the voltage across the tag.

Systems where the antenna is comparable in size to the wavelength usually employ *radiative coupling* to communicate between the reader and tag. The reader antenna launches a traveling electromagnetic wave, whose intensity in the absence of obstacles falls off as the square of the distance traveled. The wave interacts with the tag antenna at some distinguishably later time much longer than a single RF cycle (here 11 ns later, vs. an RF cycle of about 1.1 ns), and a faint replica of the transmitted signal is provided to the tag. A distinct scattered wave returns to the reader; the total round-trip transit time, about 22 ns in the example shown, is much longer than the RF cycle time.

The distinction between inductive and radiative coupling has important implications for the behavior of RFID tags. As suggested in Figure 2.11, inductive coupling between the reader and tag falls rapidly as the tag moves away from the reader antenna. A quantitative example is shown in Figure 2.12. This simulation estimates the mutual inductance of a large circular coil, representing a reader antenna, and a smaller coil representing a simplified tag antenna. (The voltage induced on the tag IC is proportional to the mutual inductance for a fixed reader current.) It is clear that the coupling is uniformly large when the tag is close to the reader, but the coupling falls rapidly to near zero in all directions for distances large compared to the antenna. Insertion of a metallic object near the reader antenna will distort the fields but in a fairly smooth fashion, and on a length scale comparable to that of the obstacle. The read range of an inductive tag is roughly comparable to the size of the reader antenna, and dependent on the direction of displacement relative to the antenna (and the relative orientation of the tag and reader, not

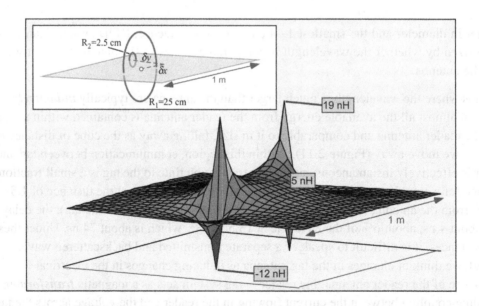

Figure 2.12

Calculated mutual inductance (nanoHenry) of circular coil antennas as the smaller antenna moves in a plane relative to the larger antenna.

shown here). To a good approximation, when an inductive tag is close to the antenna it will be reliably read, and when the tag is far from the antenna it is invisible to the reader.

The situation is very different when radiative coupling is used. Because the power falls slowly with distance, and the wavelength is small compared to typical tag-reader distances, reflections from distant obstacles can propagate back into the region of interest and interfere with the waves launched by the reader antenna, creating a very complex dependence of received power on location of the tag. An example is shown in Figure 2.13, depicting the relative power received by a tag antenna from a reader placed within a simple rectangular room with partially-reflecting walls and floor, and no other obstacles. (The reader antenna is taken to be omnidirectional; see chapter 6.) The received power falls monotonically near the transmitter (reader), but for distances greater than about 1 meter, the propagation environment becomes very complex.

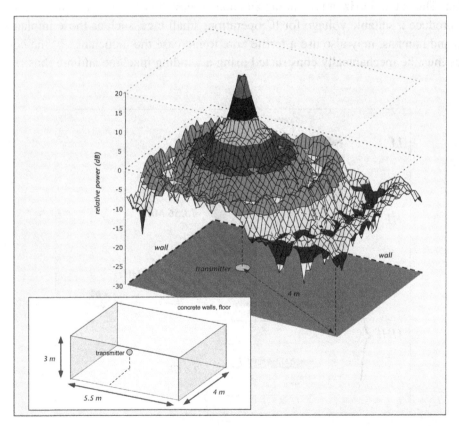

Figure 2.13
Simple model of power density in a room with partially-reflecting walls and floor.

Moving a tag away from the reader antenna by distances on the order of half a wavelength may lead to an increase in received power; in consequence, a radiatively-coupled tag may disappear and then reappear (perhaps several times) to the reader as it travels away from the antenna. Furthermore, this large and complex read zone will overlap with other similarly-unpredictable zones when multiple readers are present in close proximity to one another: interference is more likely with long-range UHF systems than short-range HF or LF systems. The read range of a radiatively-coupled system can be longer than an inductively-coupled system, but at the cost of a much more complex propagation environment, and a discontinuous and unpredictable read zone.

The frequency of operation has a major impact on the type of antennas used. A by-no-means-exhaustive set of examples of the various tag antenna configurations employed at differing frequencies is shown in Figure 2.14.

Inductive-coupled systems use coils for antennas. The voltage induced on a coil is proportional to the number of turns of the coil, the size of the coil, and the frequency of operation. Thus at 125 kHz, a typical tag antenna requires tens to over a hundred turns of wire to produce a suitable voltage for IC operation; small tags, such as those implanted into animals and humans, may also use a ferrite core to increase the inductance of the coil. Such antennas must be mechanically constructed using a winding machine and are thus relatively

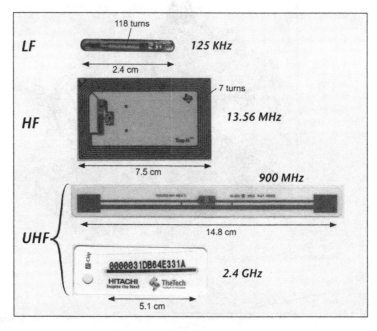

Figure 2.14
Examples of tag antenna configuration designed for different operating frequencies.

expensive. Antennas with fewer turns can be used but with a concomitant reduction in read range. Reader antennas are sized by the range desired, and may also involve 10–20 turns of wire. At 13 MHz, 100 times higher in frequency, a typical credit-card form factor tag requires only 3–6 turns to produce several volts at a reasonable range of a few tens of centimeters. This type of antenna can be fabricated in batch using lithographic techniques. Reader antennas at HF usually use only a single coil.

Coils (loop antennas) a few centimeters in diameter are not very effective antennas at UHF frequencies. Most UHF/microwave tags use variants of a ***dipole*** design, essentially a wire split in the middle, about which much more will be said in chapter 7. A dipole antenna has a natural size of about half the wavelength, roughly 15 cm at 900 MHz, or 6 cm at 2.4 GHz. Smaller dipole-like antennas can be used, but at a sacrifice of either bandwidth or performance or both. Operation at 2.4 GHz allows for compact dipole antennas, but for reasons we will discuss in chapters 6 and 7, 2.4 GHz tag antennas collect less power than a similar 900 MHz tag for the same radiated power, so the read range of 2.4 GHz tags is in general shorter than that of 900 MHz tags. Some UHF tags use small loop antennas and couple inductively despite high frequency operation, but these tags are limited to short-range operation (typically 5–10 cm read range).

When we change the frequency of operation, we also change the way the fields created by the reader antenna interact with materials commonly encountered in use, particularly metallic objects and water (of which people, plants, and animals are mostly constructed). An electromagnetic wave impinging on a conductive object penetrates to an extent known as the ***skin depth***. The skin depth δ depends on the frequency f, the magnetic permeability $\mu = \mu_0$ except for magnetic materials, and the electrical conductivity σ of the object in question:

$$\delta = \sqrt{\frac{1}{\pi \mu_0 \sigma f}} \tag{2.2}$$

Approximate values for the skin depth in differing materials at the most common RFID frequencies are given in Table 2.1. (The values for water and animal tissue are rough estimates, because the frequency dependence of the ionic conductivity has not been accounted for.) It is apparent that at 125 kHz, water and water-containing materials have essentially no effect on RFID operation, and that a thin sheet of metal is readily tolerated, but a thick metallic sheet acts as an effective shield. At 13.56 MHz, penetration into cows or people is substantial but not unlimited, and only thin metal films can be tolerated. (This doesn't mean that 13 MHz tags are insensitive to water. A 13 MHz tag can be surrounded by your hand with no effect, but if the tag comes into contact with your skin, the high dielectric permeability of the tissue changes the capacitance of the coil and greatly reduces readability.) At UHF frequencies, penetration through water is minimal, and all but the thinnest metallic films are obstacles to propagation.

Table 2.1: Skin Depth for Various Common Materials

Material	Skin Depth At:			
	125 kHz	13.56 MHz	900 MHz	2.4 GHz
tap water	8 m	2 m	4 cm	8 mm
animal tissue	2 m	60 cm	2 cm	8 mm
aluminum	0.23 mm	71 μm	2.7 μm	1.6 μm
copper	0.18 mm	55 μm	2.1 μm	1.3 μm

1 μm = 10^{-6} m

Finally, as we will discuss in more detail in chapter 3, the amount of information you can send over a link is constrained by the amount of bandwidth you have available. This fact shows up as a limitation on the speed of data transfer for LF tags. For example, let us look at the popular scheme alluded to briefly above, where transmission at 125 kHz or 135 kHz is used to send binary '0' and '1' symbols. In order to reliable distinguish between the two frequencies, the receiver must be able to count enough waves to tell what frequency is being sent, possibly from a small signal in the presence of noise. The length of a single cycle would have to be determined very accurately to distinguish between two frequencies separated by only 7%. If instead we send (say) 10 cycles, an error of 10% of the cycle time in finding the peak of the last cycle only causes a 1% error in measuring the frequency. So we might use ten RF cycles in each bit to get good noise resistance, but this means that it takes us ten times longer to send a bit. In general, LF tags send data very slowly: data rates around 1 kilobit per second (kbps) are typical, so merely to send an identifying number and error check with 100 bits would take 0.1 second, ignoring the overhead of operating an identification protocol. LF tags are necessarily slow.

HF and UHF tags can operate at much higher speeds, though in this case the data rate is often limited by regulatory restrictions on the amount of bandwidth available. In the US, it is possible for a UHF tag to send data to a reader at hundreds of kbps, and HF tags regularly operate at 10's of kbps. This is fast enough to exchange lots of data and operate complex anti-collision protocols while still identifying tens of tags in less than a second, or one tag in a few milliseconds.

Let us briefly review the consequences of the choice of frequency:

•   LF and HF operation involves inductive coupling and range comparable to antenna size; UHF operation provides range limited by transmit power
•   inductively-coupled read zones are generally small but simple; radiative read zones are larger but complex and often discontinuous, and nearby readers can interfere with each other
•   LF tags use coil antennas with many turns; HF antennas need fewer turns

- UHF tag use simple dipole-like antennas that are easily fabricated, but size tends to be constrained by the wavelength of the radiation
- LF radiation penetrates water and common aqueous materials to a distance much longer than the read range of a typical system; HF penetration into water is comparable to ordinary read ranges; UHF/microwave penetration into water is negligible in comparison to typical read range in air, except for inductively-coupled near-field operation
- LF radiation can penetrate thin layers of conductive metals; HF and UHF radiation are effectively shielded by even quite thin films of metal
- LF tags are limited to low datarate, whereas HF and UHF tags can supply tens or hundreds of kbps

The differing characteristics associated with each frequency band mean that the optimal applications are different for each band.

LF RFID is particularly appropriate for animal and human ID. Tags and readers are quite unaffected by the presence of water, salt or otherwise. Ranges of 1 meter or less are acceptable and often desirable. Tags are relatively costly (a few US dollars), but this is not a major impediment to the identification of expensive livestock, beloved pets, or important people. Tags can be attached to an animal's ear, inserted in the stomach (in the case of a cow), or implanted beneath the skin using a glass-encapsulated transponder such as that depicted in Figure 2.14. It is usually straightforward to arrange for only one tag to be in the read zone at any given time; cows can be moved through a portal with room for only one animal at a time, and people and pets are generally inspected with a short-range handheld reader. People and animals don't move very fast under conditions of interest for this sort of identification, so several seconds are available to read a single tag, and low data rates are not a problem.

LF tags and readers are also popular for access control. Short-range LF readers can be implemented at very low cost, as signal frequencies of 100 KHz present very little challenge to modern digital circuitry. Tags can be implemented in a credit-card form factor, with a several-turn coil antenna, and used as identification badges allowing entry into secured facilities. Near-contact ranges of a few cm are acceptable, and ensure that only one badge is presented to the reader at any given time. LF tags are also used in a key-fob form factor, using a larger number of coil turns to compensate for the small size, to provide unique identification of a driver to an automobile-mounted reader. Again, only one tag is present at a time, and a delay of on the order of 1 second is acceptable. LF tags are useful in robust identification of metal compressed gas cylinders.

HF tags are widely used for non-contact *smart cards*, credit-card-like transponders that contain an integrated circuit and antenna and support secure financial transactions. The rapid increase in tag power as the tag nears a reader means that a slight decrease in read range is sufficient to provide the considerable power needed for cryptographic operations, so HF tags can readily carry out secure communications with a reader. Short range also helps ensure against interception, and inadvertent activation of the cards when they are e.g. contained in a user's wallet or purse. High data rates can support a relatively complex exchange to allow a sophisticated financial transaction to occur. Like LF systems, HF-equipped badges can also be used for access control. HF tags are also increasingly used in RFID-equipped passports and travel documents.

HF tags are also widely employed for asset tracking and supply management. HF tags have ample ID space to support unique identification of a considerable quantity of items. In an asset application, the short read range of HF systems can be a challenge, typically addressed by some combination of large-sized antennas, large form-factor tags, process constraints that force the items to pass sufficiently close to a reader, and handheld or portable readers. The availability of high power at short range means that HF tags can support large memory spaces, up to several thousand bytes, allowing a user to record a substantial amount of unique information on a tag in the field. This capability is very useful when users need to interact with the tags when out of reach of networks or relevant databases.

UHF tags benefit from the potential for long range. The relative simplicity of UHF antenna designs, which involve only a few features with no critical dimensions and no need for crossovers or multiple layers, help reduce the cost of fabrication. On the other hand, at least one component in the tag's circuitry must operate at very high frequencies, which until recently added significantly to the cost of the circuitry. UHF tags are widely used in automobile tolling and rail-car tracking, where ranges of several meters add considerable installation flexibility. They are increasingly used in supply chain management, transport baggage tracking, and asset tracking, where the future potential for very low cost tags is important, and relatively long range adds flexibility in applications (at the cost of some ambiguity in locating the tags that are read). UHF tags equipped with batteries (about which more below) can have ranges of tens or hundreds of meters, and are used for tracking shipping containers and locating expensive individual assets in large facilities.

The distinction between UHF operation at 860−960 MHz and 2.4 GHz is rather more subtle than that between inductive and radiative systems. The worldwide regulatory environment in the 860−960 MHz region is very complex, as RFID at these frequencies competes directly with cellular telephony and other popular and important applications (Figure 2.15), and different countries have made different choices about what can operate where. The 2.4−2.45 GHz band, on the other hand, is available for unlicensed operation in almost every major jurisdiction, but in consequence is also crowded with other devices, so that interference is a major issue. As noted above, 2.4 GHz tags are in general smaller than 900 MHz tags, making them more convenient to use and lower in cost, but reducing read range (around 1−3 meters at 2.4 GHz vs. 2−10 meters at 900 MHz).

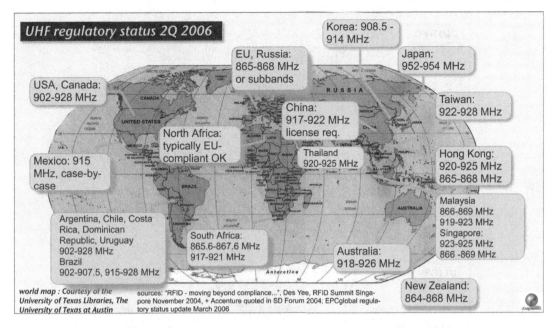

Figure 2.15
Capsule summary of worldwide UHF RFID band allocations.

2.6.2 Passive, Semi-Passive, and Active Tags

We have noted already that it is often advantageous to eliminate the radio transmitter and battery from an RFID tag to save money and space. The presence or absence of these components forms the basis of a second means of classifying RFID systems, by the power source and capabilities of the tags (Figure 2.16).

Passive tags have no independent source of electrical power to drive the circuitry in the tag, and no radio transmitter of their own. Passive tags depend on rectification of the received power from the reader to support operation of their circuitry, and modify their interaction with the transmitted power from the reader in order to send information back from the tag. *Semi-passive* tags, also known as *battery-assisted passive* tags, provide a local battery to power the tag circuitry but still use backscattered communications for the tag-to-reader (uplink) communications. *Active* tags have both a local power source and a conventional transmitter, and are thus configured as conventional bidirectional radio communications devices.

A passive tag is shown in a bit more detail in Figure 2.17. An antenna structure interacts with impinging electromagnetic fields, producing a high-frequency (RF) voltage. The voltage is rectified by a diode (a device which only allows current to flow in one direction) and the resulting signal is smoothed using a storage capacitor to create a more-or-less constant voltage that is then used to power the tag's logic circuitry and memory access.

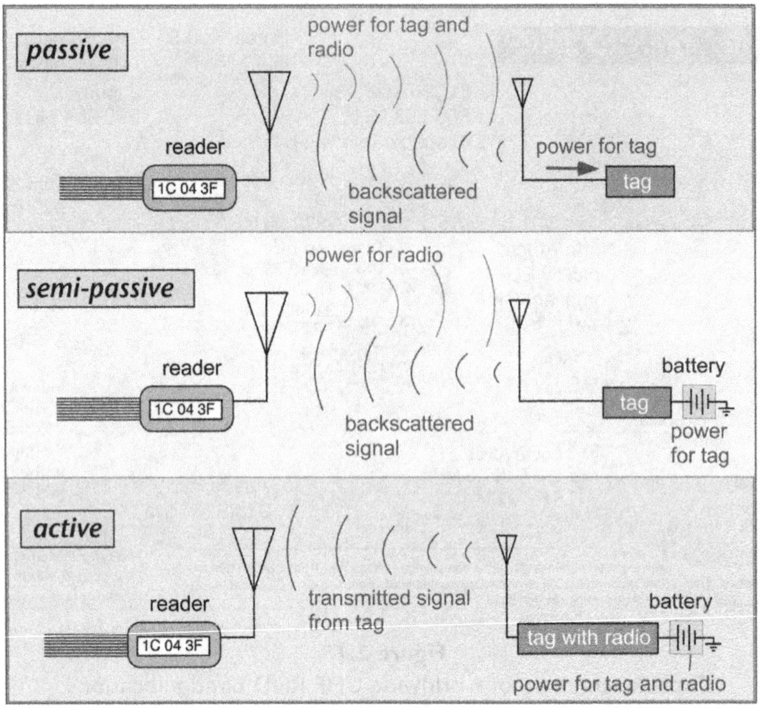

Figure 2.16
Options for tag power/ transmit configuration.

Passive tag memory circuitry is always non-volatile since the tag power is usually off. A similar rectification circuit, using a smaller capacitance value to allow the voltage to vary on the timescale of the reader data, is used to demodulate the information from the reader. This technique is known as ***envelope detection***. Finally, to transmit information back to the reader, the tag changes the electrical characteristics of the antenna structure so as to modify the signal reflected from it, somewhat analogous to tilting a mirror. Here we have shown a field-effect transistor (FET) used as a switch; when the FET is turned on, the antenna is grounded, allowing a large current to flow, and when it is off the antenna floats allowing very little antenna current. Real tags are a bit more sophisticated but use an essentially similar mechanism for modulation. The same conceptual scheme is used for all frequency bands, though the details of implementation differ for LF, HF, and UHF tags.

The tremendous advantage of a passive tag is its simplicity and consequent low cost. Passive tags have no battery, no crystal frequency reference, no synthesizer to create a high-frequency signal, no power amplifier to amplify the synthesizer signal, and no low-noise amplifier to capture the reader signal. These functions are relatively expensive

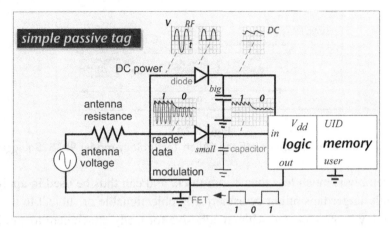

Figure 2.17
Schematic depiction of simple passive RFID tag.

compared to logic circuitry, and in some cases (e.g. a crystal) would require placement of a separate component onto the tag. Their elimination greatly reduces the cost of tag manufacture. Furthermore, because there is no battery, passive tags need no maintenance, and last as long as the materials of which they are composed endure.

In exchange for this low cost, passive tags give up a lot. Read range is limited by the need to power up the circuitry, and so is short relative to the range at which signals from the reader could be detected by the tag; the limitation is particularly acute at UHF, where propagation-limited read range might be large. The tags are also generally dumb. Because they depend on received RF for power, they must be designed to use very little of it. Computational power is minimized to avoid power consumption, so the readers must use very simple protocols to avoid overtaxing the tags, and integration of sensors is limited by the lack of power except when near a reader. Security and privacy are necessarily compromised due to the limited resources available to implement cryptographic algorithms, though this deficiency can be ameliorated for HF tags if short range is acceptable. Passive tags, particularly at UHF frequencies, are unreliable relative to more sophisticated systems: they won't power up at all unless they receive a strong reader signal, and are thus often not seen. Furthermore, the limited computational capability means that many of the techniques used to improve link quality for more capable radios, such as error-correcting codes, interleaving, gain adjustment, and retransmission, are not practical for passive tags.

An example of a typical UHF passive tag is shown in Figure 2.18. The tag is almost wholly composed of the plastic substrate or *inlay* and the antenna structure. The single very small integrated circuit is mounted on a *strap* (which conceals it from direct view in this image).

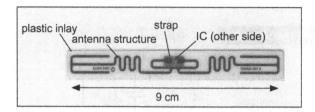

Figure 2.18
Typical commercial passive UHF tag (Alien Technology model 9238 'Squiggle').

The whole assembly is much less than 1 mm thick, and can thus be used in applications where physically larger tags might be esthetically objectionable or subject to mechanical damage. The inlay is often coated with an adhesive for ready attachment to an object, or embedded within a printed adhesive label.

Incorporating a battery to power the tag circuitry produces a semi-passive tag (Figure 2.19). Circuit complexity and peak power consumption can greatly exceed that used in a passive tag, enabling the use of standard commercial IC's rather than solely custom designs. We have shown the use of simple envelope detection just as in the passive tag for acquiring reader data, but with the availability of a battery one can also consider adding high-frequency amplification and other RF functions. The uplink still employs modulation of the antenna load to create a backscattered signal.

Semi-passive tags can achieve ranges in the tens of meters to as much as 100 meters, and are much more reliable than passive tags in the sense that they are much more likely to respond to a valid interrogation. They are often used in automobile tolling applications (where a missed tag translates into missed revenue or an inappropriate citation), and in tracking of airplane parts and other high-value reusable assets. The tradeoff is that they require a battery with concomitant increases in size, cost, and maintenance requirements. Battery life is improved by operating at very low duty cycle, and/or using a detector circuit to keep most of the system off except when a reader signal is probably present. In tolling applications, this limitation can be avoided, at the cost of increased installation complexity, by providing power from the automobile electrical system, a scheme used in the Singapore traffic control (ERP) system described in section 3 above.

An example semi-passive tag is depicted in Figure 2.20. This is an automobile tolling tag, of a type commonly used in the Western United States. The tag is about 9.5 wide and 1.5 cm thick. We have removed the cover in the view on the right of the figure to display the construction of the tag. This tag uses a number of commercial components instead of the single custom integrated circuit in the passive tag of Figure 2.18. This tag uses two separate antennas, one for receiving the signal from the reader and another for transmitting a return signal. The largest component is a lithium battery, which provides about 5 years of ordinary operation. When the

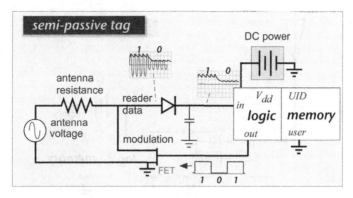

Figure 2.19
Schematic depiction of simple semi-passive tag.

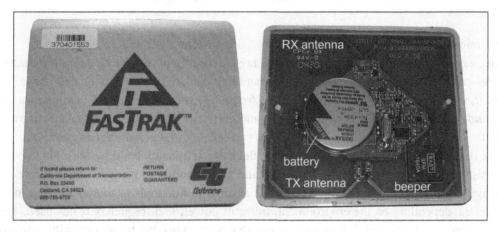

Figure 2.20
Commercial semi-passive tag: external view (left) and internal circuitry (right).

battery is exhausted, the tag must be replaced, representing a significant expense and administrative burden for the tolling authorities. A beeper is used to signal the driver that the exchange with the tolling authority has been successful. These tags are typically mounted on the windshields of cars, and provide about 10 meters of read range, sufficient to allow >95% successful ID acquisition even at full highway speeds. A tag costs about US$20–$30.

Active tags are full-fledged radios, with a battery, receiver, transmitter, and control circuitry (Figure 2.21).

The active tag synthesizes a carrier signal using a local oscillator and crystal reference, so it can communicate within a specific frequency band and can use this capability to

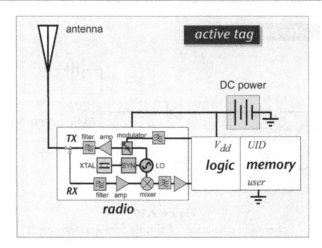

Figure 2.21
Schematic depiction of active tag.

communicate in the presence of other tags by using different frequency channels (*frequency-division multiplexing*). An active tag can use amplitude modulation like passive or semi-passive tags, but it can also transmit and demodulate more sophisticated phase-based modulations (*phase-shift keying*, PSK, *frequency shift keying*, FSK, and *quadrature amplitude modulation*, QAM), which can be more efficient users of available spectrum and provide superior noise robustness. Active tags can use high-rate *code-division multiple access* (CDMA) techniques to allow reuse of the same frequency band by multiple tags. With significant transmit power and filtering and amplification to provide good receive sensitivity, the read range of an active tag is measured in hundreds of meters or even in kilometers, depending on the environment, transmit power, and frequency bands used. Just as important, the increased link margins mean that active tags can be successfully employed in environments where the tag-reader path is significantly obstructed. For example, they are used to mark metal shipping containers, and can be read even when the containers are stacked in close proximity to one another with no line of sight from reader to tag.

Active tags naturally suffer from the additional cost, size, and maintenance requirements of a full-fledged radio. More components imply larger size, or the cost of a custom radio chip design. In addition, an active transponder must be certified as an active radio emitter, and must therefore meet regulatory standards for spectral purity, out-of-band emissions, and frequency accuracy, which are either inapplicable or relatively less stringent for passive and semi-passive tags. The long read range can be detrimental, in the sense that the location of a tag is not well-known simply because it has been read by a reader antenna of known location. (This problem can be addressed by using multiple receivers; by measuring relative time delays to the receivers, the tag can be located within a few meters across several hundred meters.)

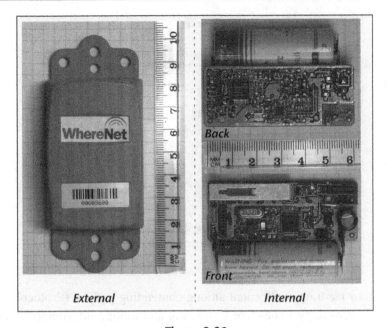

Figure 2.22

Commercial active RFID tag: external and disassembled views (shielding removed); photos from
FCC report.

An example of an active tag is depicted in Figure 2.22. This is a commercial tag employed
in locating large assets, such as shipping containers. These types of tags cost about US$50
each, in quantities of a few thousand, at the time of this writing. The internal view
demonstrates the relative complexity of an active tag (particularly in contrast to the very
simple passive tag of Figure 2.18), and the relatively large battery needed. This particular
tag type produces about 60 mW output power in the 2.4 GHz band, providing about
900 meter outdoor range. The communications protocol, ANSI 371.1, prescribes a
pseudo-noise coded beacon, which in this cases occupies much of the 2.40–2.45 GHz
unlicensed band. The coding permits unique detection of tags in the presence of interferers,
and accurate timing of the beacon. The rate at which beacons are transmitted can be varied
from about two per second to once per hour; at a typical rate of one every four minutes, the
battery is expected to last about 6 years. Triangulation at multiple receive antennas can
locate a tag to within a few meters in an outdoor environment.

2.6.3 Communications Protocols

Every means of communication requires a protocol: an agreement on how information will
be exchanged (Figure 2.23). Protocols must address what sort of signals will be used, what
types of symbols are employed, how they are combined to make meaningful data, and how

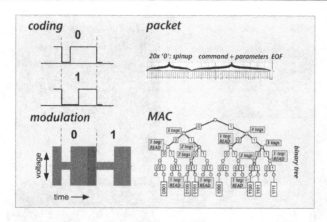

Figure 2.23
Elements of a typical RFID communications protocol.

the communications medium is allocated among contending parties. (Protocols may also need to deal with networking issues like addressing and routing information, which are beyond the scope of our discussion and of most RFID systems.) For example, human speech can be considered to be a complex protocol with the following elements:

- *symbols and coding*: words are constructed using a small set of phonemes standard to each spoken language
- *modulation*: human speech uses an involved mixture of variations in tone and loudness — that is, of differential amplitude and frequency (AM and FM) modulation;
- *packet construction*: words are assembled into statements (packets) according to a language grammar;
- *medium allocation*: in informal speech, people use a variant of *carrier sense multiple access with collision detection* (CSMA-CD): we talk when we feel like it, but if two folks talk at once, they detect the situation and wait a short but random time before trying again. Alternative rule sets are available; when operating under Robert's Rules of Order, a large group of people remain silent, while a designated central authority allocates the medium (the 'floor') to a single speaker based on speech or gestures during a brief contention period.

A key aspect of all communications protocols is that they must be shared: if I speak only Mandarin and you understand only Finnish, conversation will be unprofitable even if both of us are the souls of courtesy, and rigorously adhere to the proper grammar and pronunciation of our respective tongues. With commercial communications devices there are two paths to ensuring agreement on the details of the protocol: either both ends of the link are manufactured by the same vendor and guaranteed to interoperate, or multiple vendors make products to an agreed-upon *communications standard*. Some standards are simply *de facto*, often the adoption by an industry of a particularly successful product from one vendor, but many are codified by multi-vendor industry bodies, quasi-governmental authorities, or national governments. Among important standards-setting bodies in communications today are the *Institute of Electrical and Electronic Engineers* (IEEE), the *American National Standards Institute* (ANSI), the *International Organization for Standardization* (ISO), the *Internet Engineering Task Force* (IETF), and specifically for RFID, EPCglobal Inc.

The action within a standards body typically takes place in committees and working groups, most of the members being volunteers. While a few of these folks participate simply because they are interested in the work, the majority are employees of companies or organizations with a perceived financial stake in the outcome. As a consequence, an important part of standards setting is achieving a resolution that all the participating organizations can tolerate. A superior technology developed by one participant may be disadvantaged in a standards activity because of the concern that competitors may need to pay a royalty to the developing company to use it. For example, in the very popular IEEE 802.11b (WiFi) standard, two coding methods were provided, CCK and PBCC. It has been reported that CCK was included despite its unproven status at the time of the original standardization because PBCC was developed by Texas Instruments, and other companies were reluctant to be wedded to that approach, given TI's predilection for seeking licensing revenues. In practice, almost all 802.11b equipment has implemented CCK rather than PBCC. The ISO18000-6A and −6B standards (about which we shall have more to say in chapter 8) are mutually incompatible despite originating from a nominally common process, because the standards descend from distinct proprietary technologies, and no compromise was reached. EPCglobal class 0 and class 1, as noted before, share the same problem. Standards setting is a complex compromise of competing interests: vendors who may have large investments in specific technologies and associated patent portfolios from which they hope to profit, competing vendors who don't want to pay royalties, users who want the standard to serve their particular application requirements, and favor multiple suppliers to keep prices low and ensure supply, governments seeking harmony with their own activities and other regulated users, and competing governments concerned about perceived hegemony. Standards activities often founder in their attempts to navigate this maze, or surrender and disseminate multiple incompatible solutions from which the market is supposed to identify a winner. On the other hand, a good standard can change the world: today, nearly every computer in existence is equipped with one or more Ethernet (IEEE 802.3) ports, and communicates with almost any other computer on the planet using Transmission Control Protocol (TCP) running over Internet Protocol (IP).

If the organizational challenges weren't serious enough, standards bodies are faced with a multitude of technical alternatives, each choice implying different tradeoffs for different applications scenarios. For example, in most jurisdictions governments regulate the spectrum available for specific uses and put constraints on the operation of radio devices, so readers and tags have only a certain amount of bandwidth available to operate in. One would like to use this bandwidth efficiently to allow as many readers as possible to operate without interfering with one another, but the most bandwidth-efficient methods of modulating signals require resources that are prohibitively expensive and power-hungry for passive tags. Tags can receive and decode amplitude-modulated signals readily, but amplitude modulation involves turning the transmitted power down some of the time, which is bad if this is also the power used to keep the tags running. Therefore, it becomes desirable to use very short 'off' pulses with the reader power on most of the time, but (as we will see in the next chapter) such coding techniques use spectrum very inefficiently and allow fewer readers in the same location. Every choice made in formulating a standard involves a similar balance of competing constraints, usually made in the absence of complete information, since often only prototype devices or perhaps just simulations are available during the standards-setting activities.

The issue of forward and backward compatibility inevitably arises. Very few standards break completely new ground, so one must always evaluate the desirability of making a new standard backward-compatible to an existing standard or at least a subset thereof. Previous standards were generally formulated with more primitive technology and older applications, so backward-compatibility usually involves compromising the potential of the new standard, but leverages existing hardware and software. Forward compatibility is even more challenging: a good standard should have flexibility to allow new applications, very frequently not envisioned by the folks who make the standard. However, burdening a standard with every possible bell and whistle makes the resulting document difficult to understand and implement, and raises the cost of compliant devices. For example, compliant EPCglobal class 1 generation 2 tags must be able to reply to a reader inquiry at data rates as high as 640 kbps and as low as 5 kbps: this capability makes them very flexible but also inevitably raises the cost of manufacture above that of a simpler single datarate requirement.

Finally, ensuring compliance to the standard is always an issue when multiple vendors are present. Brief descriptions of a protocol may be ambiguous, and careful and rigorous standards may be long and difficult to read and comprehend. In practice, the protocols must be implemented by real people designing and building hardware, and writing software to run it, with organizational boundaries and competition standing in the way of information exchange. To ensure that different implementations of a standard can communicate with each other, it is useful to establish compliance tests and a certification process, as well as applicationcharacteristics and conventions for data exchange.

In Table 2.2 we summarize some of the standard and proprietary physical-layer protocols used in RFID. Note that many additional standards exist to define higher-level issues such as the organization and meaning of the unique tag identifier, and conformance test requirements.

Table 2.2: Some RFID Air Interface Protocols

Tag Type	Frequency					
	125/134 KHz	5–7 MHz	13.56 MHz	303/433 MHz	860–960 MHz	2.45 GHz
Passive	ISO 11784/5, 14223 ISO18000-2 HiTag	ISO10536 iPico DF/ iPX	MIFARE ISO14443 Tag-IT ISO15693 ISO18000-3 TIRIS Icode		ISO18000-6A,B,C EPC class 0 EPC class 1 Intellitag Title 21 AAR S918 Ucode AAR S918	ISO18000-4 Intellitag μ-chip
Semi-passive						ISO18000-4
					Title 21 EZPass Intelleflex Maxim	Alien BAP
Active				ANSI 371.2 ISO18000-7 RFCode		ISO18000-4 ANSI 371.1

Each of these mysterious numbers, acronyms, or (in some cases) company/product names is attached to a laborious and hopefully exhaustive compilation of the necessary conventions for moving information from reader to tag and perhaps back again. For example, ISO 11784 and 11785 are designed for identification of livestock. A reader powers up nearby tags with a 50 ms unmodulated transmission at 134 KHz, and then (in the half-duplex option) listens for a frequency-modulated reply at 125/134 KHz. ISO 14443 supports smart cards at 13.56 MHz. This standard has two incompatible flavors, A and B, one using 100%-deep pulse position modulation and the other 10%-deep on-off modulation to send commands to the tags. Type A uses a binary tree walk to resolve collisions between tags, whereas B uses a slotted-Aloha scheme. The reader may infer from the discussion above how these sorts of discrepancies arise. ISO 15693 also supports smart cards at 13 MHz but uses still another set of schemes for uplink, downlink, and collision resolution. California Title 21, designed for automobile toll collection, and EPCglobal class 0, designed for supply chain applications, both can operate within the US ISM band at 902–928 MHz and both employ frequency modulation to send data from the tag back to the reader, but Title 21 tags use 0.6 and 1.2 MHz frequency offsets, whereas EPCglobal class 0 tags instead employ 2.2 and 3.3 MHz. (See chapter 8 for explanations of some of the terms used above.)

It should be apparent that most of the protocols in the table are mutually incompatible, even when they come from the same standards activity, though the degree of distinction does vary somewhat. In the majority of cases, a tag understands only one of these protocols, either because there is no need for interoperability, or to minimize tag cost and complexity. When applications require multiprotocol interoperability, the burden generally falls on the reader: readers are relatively small in number, expensive, provided with a power source, and computationally capable. Multiprotocol capability is relatively easy to provide within a fixed frequency band (a single column of the table). Crossing frequency boundaries is much more complex; as described above, changing bands may require completely differing antenna structures, have different read ranges and data rates, and may involve large ranges in transmit voltage and received signal strength.

2.7 The Internet of Things and UHF RFID

The reader who has made it this far will appreciate that the acronym RFID encompasses a broad range of distinct and often incompatible technologies, each with its own strengths and weaknesses. Every potential application has a different mix of requirements and thus a different optimal technology to support it. No particular choice of technology will satisfy all users.

That being said, it is apparent that only a few types of applications can generate a demand for very large numbers of tags. The total number of industrial gas cylinders in the world is unlikely to exceed a few tens of millions in the foreseeable future; in contrast, the total volume of corrugated cardboard manufactured for boxes and containers averages around 400 million square meters *per day*. Marking cases of goods, and eventually individual items, with RFID would involve a truly astronomical quantity of unique tags. The Internet of Things, should it come to pass, will encompass the largest number of uniquely human-labeled objects ever to exist.

As we have noted previously, the dominant barrier to implementing such a grandiose scheme is the cost of labeling, particularly of the tags. Passive tags, due to their simplicity of manufacture and zero maintenance, have a tremendous advantage in cost-sensitive applications. Printed-battery technologies do exist, but even were they to be widely implemented, the total energy thus made available is quite limited: batteries are volume storage devices, and printed films have little volume. For example, today's printed batteries provide around 10 mA-hours from a few square centimeters of area, and are several hundred microns thick. While this is enough power to run a passive tag IC for hundreds of hours, and thus provide several years of operation with good duty cycle management, it is entirely inadequate to support an active transmitter or computationally-intensive data processing for more than a few hours. Tag with printed batteries will almost certainly look much more like enhanced passive tags than active radios.

Passive tags can be operated at any RFID frequency. LF and HF passive tags are very widely deployed in animal identification, automobile immobilization, and smart card applications. However, LF tags will always be limited to very low data rates and are not appropriate for most supply chain applications. HF tags can support high data rates, can be very small, and can achieve read ranges of several meters — but not all at the same time. UHF tags are able to provide all these benefits in a single package. While not all supply chain applications will require long read ranges, there is a tremendous benefit in flexibility gained from having a single tag and reader technology, with the option of reading the tag at a distance when needed. This versatility is a powerful argument for the use of UHF in supply-chain applications.

As we mentioned briefly above and will discuss in detail in chapter 7, operation at UHF is very sensitive to the presence of conductive materials: metals, metal films, and aqueous solutions, all of which are very common in manufactured goods. HF tags are also sensitive to metals, though relatively tolerant of water. This proximity problem represents a significant challenge for ubiquitous implementation of UHF RFID. To solve it will require a mixture of approaches, including improved tag and reader designs, changes in packaging techniques, optimized procedures for suppliers and retailers, and perhaps semi-passive tags in some instances. Insufficient progress in this area could limit the broad use of UHF technology.

Chipless RFID technologies have also been explored. *Surface-acoustic wave* (SAW) devices have been implemented in some RFID applications, and electromagnetic techniques using resonant conductive fibers are under active investigation. However, these approaches suffer from the inability to write data to the tag, and must compete with the tremendous flexibility offered by digital logic designs and the stupendous global manufacturing competence in silicon CMOS integrated circuit technology. Printed organic integrated circuits have demonstrated improved capabilities, but it is important to note that the per-feature cost of printing technologies is considerably higher than that of silicon manufacturing: for the same complexity, it is cheaper to fabricate a circuit on silicon than to print it directly onto a plastic substrate. These constraints seem likely to limit chipless RFID technology to specialized applications where their potential for very low cost makes up for their relative inflexibility.

Finally, it is worth noting that at a certain point, low cost and ready availability often generate new markets of their own accord. Ethernet is everywhere in part because Ethernet is everywhere. As UHF tags and readers continue to fall in cost, they are likely to find applications where their use is driven by simplicity of access rather than any special virtue.

In summary, it seems likely that the most important RFID technology over the next decade will be the use of passive or semi-passive silicon-IC-based tags in the UHF bands. In the remainder of this book, we will focus almost exclusively on passive UHF technology, but the reader should now be equipped to place this choice within the context of the specific set of requirements and circumstances current here in the early years of the twenty-first century, rather than regarding it as eternal, foreordained, and unchangeable.

Further Reading

History

"Shrouds of Time: The History of RFID", Jeremy Landt and Barbara Kaplin, AIM (Association for Automatic Identification and Data Capture Technologies)

"Communication by Means of Reflected Power", Harry Stockman, Proc I.R.E., October, 1948, p. 1196

"Radio Transmission Systems with Modulatable Passive Responder", Harris, US Patent 2,927,321, filed 1952, granted 1960

"Folded Dipole Having a Direct Current Output", Crump, US Patent 2,943,189, filed 1956, granted 1960.

"The toll highway faces automation", William Arnold, Electronics, November 8, 1973, p. 74

"Electronic Road Pricing in Singapore: A Technical Overview", B. Benight and P. Chong Sun, IES-CTR Symposium on Advanced Technologies in Transportation, Hotel Equatorial, Singapore, 19 April 1996

"A History of the EPC", S. Sarma, chapter 3 of **RFID: Applications, Security, and Privacy**, S. Garfinkel &. B. Rosenberg, Addison-Wesley 2005

Exercises

1. Can a passive RFID tag be read from a satellite?* _____YES _____ NO
2. Can a 125 KHz tag be read from across a street? _____YES _____ NO
3. If you swallow a passive UHF RFID tag, the consequences will include:
   a. The tag will be read by a reader with antenna placed on your stomach
   b. The tag will be read by a reader up to 3 meters away
   c. The tag will be unreadable due to reflection and absorption by water
   d. You will have a seriously upset stomach (list all that apply): _____
4. Is your cellphone likely to interfere with a High-Frequency (13.56 MHz) reader mounted 3 meters from your desk? _____YES _____ NO
5. How often do you need to change the battery on an Alien 'squiggle' tag?
   a. Once a month
   b. Once a year
   c. Every 5 years
   d. What battery?

---

* OK, I am cheating. The answer based on the text is, of course, "NO," but it is actually physically possible to do this. However, reading passive tags from an orbiting platform would take millions of watts and tens of billions of dollars, and be very dangerous to other orbiting objects, and the result would not work very well. There are much cheaper ways to monitor folks and intrude on privacy if that is one's goal.

6. Can an EPCGlobal class-0-only reader read a class 1 tag? _____YES _____ NO
7. How many unique identifying codes are possible using 96 bits? [If you don't have a calculator available, a hint: $\log_2 10 \approx 3.32$] _____
8. Do all RFID readers need an antenna? _____YES _____ NO
9. If you can't read a specific UHF passive tag 2 meters from a reader antenna, will it continue to be invisible to the reader at any distance larger than 2 meters? _____YES _____NO

CHAPTER 3

Radio Basics for UHF RFID

3.1 Electromagnetic Waves

Recent estimates by cosmological folks suggest that around 95% of the mass in the universe is composed of dark matter and more-recently-minted dark energy, about which essentially nothing is known. Dark matter and dark energy don't appear to interact with our alternately glowing and dusty stuff except through gravitational means. Folks made of dark matter (if such were to exist) couldn't watch reruns of *American Idol* even if you forced them: they don't have any means of interacting with the broadcast signal, and probably don't want to pay for cable.

For those condemned to the world of baryons and leptons, electromagnetic waves are a fact of life. In most textbooks on electromagnetic theory, you'll wade through Maxwell's equations and possibly laborious arguments on mysterious exchanges between the electric and magnetic fields launching self-supporting structures with little Poynting vectors pointing out of them: all true but unnecessarily obscure. Before we go on to the mundane tasks of introducing the relevant terminology and technology of radio, let's share a little secret, implicit but not readily apparent in the standard texts, which the author has found to considerably simplify his view of electromagnetic radiation. It goes like this:

Everything radiates, but most things cancel.

To expand a bit: every object in the world that has an electric charge creates an **electrostatic potential**, which falls inversely as the distance. The potential sensed at some distance r corresponds to what the charged object was doing at an earlier time (r/c), because signals move at the speed of light $c = 3 \times 10^8$ m/s. The total electric potential in the space between your nose and the pages of the book you're reading depends on the amount of charge on the fur of a cat in Bulgaria (or Wisconsin, if you happen to be in Dobrich). However, we almost never care, because electric charge comes in two flavors, positive and negative, and the amount of energy associated with an isolated charge of only one type is enormous: a microgram of hydrogen, split into its constituent protons and electrons and separated by 1 meter, could support a mass of 8 million kilograms against the gravitational attraction of the entire earth. So in almost every case, adjacent to each electron with a negative charge is a proton with a positive charge, such that the two cancel, and

The RF in RFID
© 2013 Elsevier Inc. All rights reserved.

have no net effect on your cellphone conversation. Electric currents similarly give rise to a *magnetic vector potential* in the direction of the current flow, which again exists everywhere with amplitude decreasing with distance, at a correspondingly delayed time. Similar arguments show that most currents don't have any effect on distant objects: if a current is flowing in one direction, with no compensating countercurrent, charge must be accumulating somewhere, leading after a while to enormous energies (voltages). Most electric currents flow in a balanced loop: the potential from current flowing up cancels that from current flowing down, and again no net effect results on distant observers. These points are made pictorially in Figure 3.1, where we also introduce a bit of the mathematical terminology associated with the subject.

At first glance, we're left with no potentials and no waves, but of course this is not correct. For example, we can run an uncompensated current for a little while before charge accumulation causes too much voltage to build up, and then turn it around. This uncompensated current will lead to a detectable signal at a distance. In addition, cancellation will often fail to be exact when the charges and currents are changing in time,

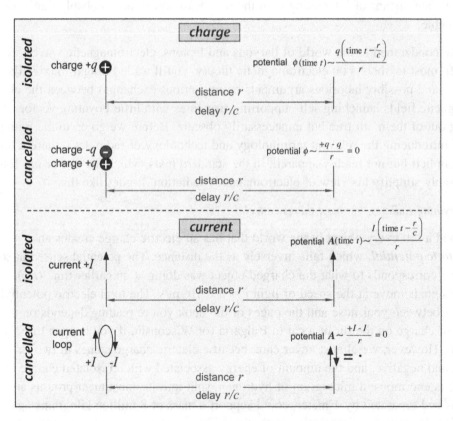

Figure 3.1
Potentials from charges and currents usually cancel.

because of the slight differences in delays due to the finite size of the region over which the currents flow. For example, if in Figure 3.2 the loop current is suddenly turned on all around the loop at some time $t = 0$, the potential from the downward-flowing current arrives at r just a bit sooner than that from the upward-flowing current. Cancellation fails, and an observer sees some resulting potential: ***radiation*** has occurred.

This leads to our second key observation:

> An ***antenna*** *is a device to produce currents and charges whose effects don't cancel for a distant observer.*

For an antenna to work, it should be apparent that something has to change: radiation is the result of the transient failure of delayed signals to cancel each other. In order to create a continuous signal, currents flowing on an antenna must continuously change, without actually getting anywhere: that is, currents and charges are usually ***periodic*** functions of time, alternately increasing and decreasing but returning to the same state again and again after the same interval. Periodic functions have a ***period*** − a time duration over which the signal is exactly repeated − and a ***frequency***, conventionally measured in ***Hertz*** (Hz) and equal to (1/period). Thus a signal that repeats itself every second has a frequency 1 Hz. The sine and cosine are archetypal periodic functions, widely used in science and electrical engineering; in electrical engineering these are often combined into a complex exponential function, which absorbs both frequency and delay (phase) into one expression: $e^{ix} = \cos(x) + i\sin(x)$, where i is the imaginary unit $\sqrt{(-1)}$. (The reader who wishes to follow the subsequent discussion in detail, but who is not familiar with these functions, may find it useful to refer to Appendix 2 for a brief introduction to the terminology and characteristics of these ***harmonic*** functions. However, the main conclusions will be presented in pictorial form, and the reader new to the field may find it more convenient to absorb the images and defer their mathematical underpinning to a future date.)

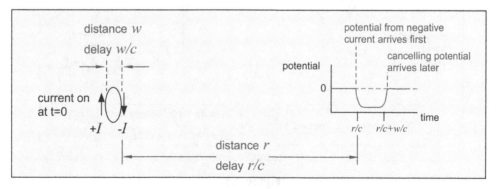

Figure 3.2
Changing currents on a structure of finite size disrupts cancellation.

We should note that instead of arranging the currents on an antenna so as to frustrate cancellation at a distance, we can place the observer (the receiving antenna) so close to the transmitting antenna that cancellation is defeated simply because some currents on the transmitting antenna are close to the receiving antenna, and have a larger effect than those more distant. This sort of interaction is known as ***near-field coupling*** or alternatively as ***inductive coupling***. We can think of inductive coupling as being fundamentally about differences in distance between differing parts of an antenna, whereas radiation is usually more closely related to differences in propagation time (phase) from one part of an antenna to another.

Armed with an antenna carrying a periodic current, we can create electromagnetic waves, propagating at the speed of light and falling in amplitude inversely with the distance (Figure 3.3). The waves induce a voltage in the receiving circuit, periodic with the same frequency as the transmitted signal, whose magnitude is inversely proportional to the distance between the transmitter and receiver. Using harmonic notation, the delay in time of Figure 3.1 becomes a phase offset by the wavenumber k multiplied by the distance r. (The absolute phase is often not readily observable or controllable in practical radio systems, so we can generally drop this term.) It is this voltage we make use of to transmit information — in the case of RFID, from a reader to a tag and back. How should we measure and describe it?

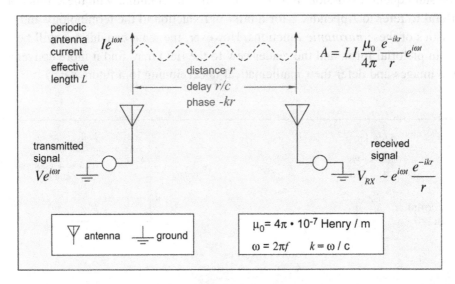

Figure 3.3
Radiated waves launched by a transmitting antenna give rise to a voltage in a receiving antenna.
See Appendix 2 for definitions of the harmonic functions used here.

3.2 Describing Signal Voltage and Power

In most radio systems, we are interested in periodic currents and voltages, since unchanging currents or voltages don't radiate as discussed above. Thus a time-dependent signal voltage is usually written as the product of a magnitude (here v_0) and a periodic function like the sine or cosine:

$$V(t) = v_0\cos(\omega t) \tag{3.1}$$

with an analogous expression for periodic currents. The instantaneous power dissipated into a load is the product of the voltage across the load and the current flowing through it. For a resistive load with a DC current flowing, we find:

$$P = I \cdot V = \underbrace{\left(\frac{V}{R}\right)}_{\text{Ohm's law}} V = \frac{V^2}{R} \tag{3.2}$$

To get the average power for a periodic signal, we add up the total power over a cycle and divide by the cycle time. This gives us a factor of (1/2), the average value of $\cos^2$:

$$P_{av} = \frac{v_0{}^2}{2R} \tag{3.3}$$

Sometimes people introduce the ***root-mean-square*** (RMS) voltage to eliminate the extra factor of (1/2) from the expression for power:

$$v_{rms} = \frac{v_0}{\sqrt{2}} \quad \rightarrow \quad P_{av} = \frac{v_{rms}{}^2}{R} \tag{3.4}$$

It isn't always obvious which definition of voltage is being used; confusion on this score leads to erroneous factors of 2 floating around. In this book, we will always use peak voltages and currents rather than RMS quantities, and thus explicitly include the factor of (1/2) in calculating average power. It is often of interest to display the amount of power associated with a sinusoidal signal of a given frequency as a ***power spectrum***; a simple example of such a display for the single-frequency signal of equation (3.1) is shown in Figure 3.4.

Signal power can vary over a huge range in typical radio practice: power dissipated into a typical 50-ohm load can range from tens of watts to 0.000000000000001 (10^{-15}) watt. Related quantities, such as voltage, current, and gains and losses, span similar ranges. It is inconvenient to write out and manipulate such quantities as decimal numbers; instead, we use logarithmic notation. Recall that the base-10 logarithm is defined as:

$$10^{\log(x)} = x \tag{3.5}$$

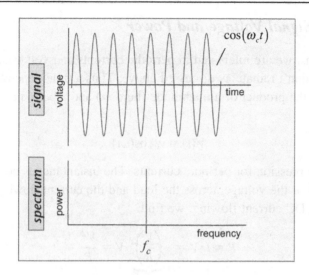

Figure 3.4
A single-frequency sinusoidal signal and corresponding power spectrum.

For example, $\log(10) = 1$, and $\log(1000) = 3$. Negative logarithms denote numbers less than 1: $\log(0.001) = -3$.

It is traditional to use not raw logarithms, but **deciBels** (dB) in communications engineering. The ratio of two powers — for example, the ratio of the output power from an amplifier to the power that went in, which is the **power gain** of the amplifier — can be written in dB as:

$$G_{dB} = 10\log\frac{P_2}{P_1} \tag{3.6}$$

Now, recall that the power is proportional to the square of the voltage. If we wanted to express the powers in equation (3.6) in terms of the corresponding voltages, we would get:

$$G_{dB} = 10\log\frac{V_2^2}{V_1^2} = \underbrace{10\log\left(\left[\frac{V_2}{V_1}\right]^2\right)}_{\log(x^2)\,=\,2\log x} = 20\log\frac{V_2}{V_1} \tag{3.7}$$

That is: deciBels are **defined differently** depending on the physical nature of the quantity being measured:

$$G_{dB} = 10\log\underbrace{\left(\frac{P_2}{P_1}\right)}_{\text{power ratio}} = 20\log\underbrace{\left(\frac{V_2}{V_1}\right)}_{\text{voltage ratio}} = 20\log\underbrace{\left(\frac{I_2}{I_1}\right)}_{\text{current ratio}} \tag{3.8}$$

Table 3.1: Power Gain in dB

Gain	Gain (dB)
1	0
10	10
100	20
1000	30

Table 3.2: Power in dBm

Power (W)	Power (dBm)	Peak Voltage in 50 Ω Load
10	40	32
1	30	10
0.1	20	3.2
0.001	0	0.32
10^{-6} (1 μwatt)	−30	0.01
10^{-12} (1 pwatt)	−90	10^{-5}

To define absolute power in dB, we need to decide on a reference level. In microwave engineering, the most common reference level is 1 milliwatt (mW), and power measured in dB relative to one mW is referred to as **dBm**:

$$dBm = 10\log\left(\frac{P}{1 \text{ mW}}\right) \tag{3.9}$$

Some practical examples of logarithmic notation are shown in Table 3.1 and Table 3.2.

Because of the fact that $\log(ab) = \log(a) + \log(b)$, deciBels add when the corresponding numbers multiply. A one-microWatt signal that is passed through an amplifier with a power gain of 1000 produces one milliWatt of output; we can equivalently say "−30 dBm + 30 dB = 0 dBm".

At the cost of memorizing a few quantities, one can almost obviate computation in converting from dB to numbers. A factor of 10 in power is 10 dB, and a factor of 2 is very nearly 3 dB. Since logarithms add, that means a factor of 4 = 6 dB, and a factor of 8 = 9 dB. Knowing these points allows quick, reasonably accurate estimates: for example, 50 μW = 1 mW/(10*2) = 0 dBm −10 dB −3 dB =−13 dBm.

3.3 Information, Modulation, and Multiplexing

A periodic signal that persists indefinitely, without changing its amplitude, frequency, or phase − a **continuous wave** (CW) signal − carries no information other than the fact that it is present. In order to convey data, a signal needs to change. We normally think of this change as a relatively slowly-changing variation – *modulation* – imposed on the periodic signal, for example:

$$V(t) = \underbrace{m(t)}_{\substack{\text{slowly-varying}\\\text{modulation}}} \cdot \underbrace{\cos(\omega_c t)}_{\text{carrier frequency}} \tag{3.10}$$

The function $m(t)$ is said to contain the **baseband** information, and the relatively high-frequency cosine function is the **carrier.** When the function $m(t)$ is another sine or cosine (presumably of much lower frequency) we can make use of trigonometric identities (see Appendix 2) to rewrite the signal in a revealing fashion:

$$V(t) = \underbrace{\cos(\omega_m t)}_{\substack{\text{slowly-varying}\\\text{modulation}}} \cdot \underbrace{\cos(\omega_c t)}_{\text{carrier frequency}} \quad \{\omega_m \ll \omega_c\}$$

$$= \frac{1}{2}\left\{ \underbrace{\cos([\omega_c + \omega_m]t)}_{\text{upper sideband}} + \underbrace{\cos([\omega_c - \omega_m]t)}_{\text{lower sideband}} \right\} \tag{3.11}$$

A sinusoidal modulation splits the carrier wave into two signals, called *sidebands*, one above and one below the carrier, each displaced by the modulating frequency (Figure 3.5). While a continuous sinusoidal modulation is hardly more interesting or useful than a CW signal, this result suggests that when a signal is modulated, the resulting frequency spectrum becomes wider.

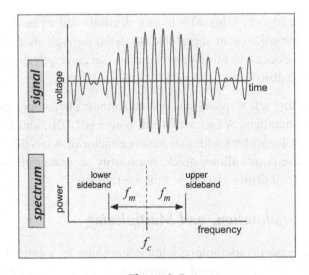

Figure 3.5
Sinusoidally-modulated carrier wave and corresponding frequency spectrum; f_c is the carrier frequency.

Signals of interest for RFID are generally *digitally modulated*. A digitally-modulated signal is a stream of distinct *symbols*. A simple example with substantial relevance for RFID is *on-off keying* (OOK). The signal power is kept large ($m = 1$) to indicate a binary '1' and small or zero ($m = 0$) to represent a binary '0'. An example is shown in Figure 3.6. In OOK each symbol is a period of fixed duration in which the signal power is either high or low. Each OOK symbol represents one binary bit, though other types of symbols can convey more than one bit each. Any circuit that can change the output power, such as a simple switch, can be used to create an OOK signal, and any circuit that can detect power levels can *demodulate* (extract the data from) the signal. For example, a *diode* – an electrical component that passes electrical current only in one direction and blocks current flow in the opposite direction – can rectify a high-frequency signal, turning it into pulses of DC. These pulses can be smoothed with a storage capacitor to produce an output signal that looks very much like the baseband signal $m(t)$ (see Figure 2.17 in chapter 2). If the diode responds rapidly, it can be used at very high frequencies. Modern diodes can operate up to over 1 GHz, allowing passive RFID tags to demodulate a reader signal using only a diode and capacitor.

Unmodified OOK is admirably simple and seems promising as a method of modulating a reader signal. However, there is a problem with OOK for passive RFID. As we noted in chapter 2, a passive RFID tag depends on power obtained from the reader to run its circuitry. If that power is interrupted the tag cannot operate. However, imagine the case of an OOK signal containing a long string of binary 0's: in this case, $m = 0$ for as long as the data remains 0. The tag will receive no power during this time. If the data remains '0' for too long, the tag will power off and need to be restarted, a situation not likely to be

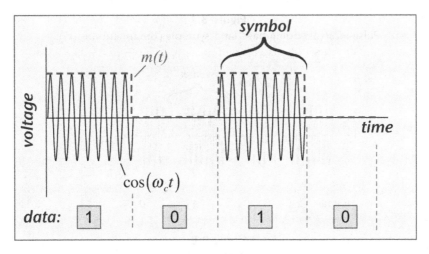

Figure 3.6
On-off-keyed signal.

conducive to reliable operation. Even when some binary 1's are present, the power level delivered to the tag is strongly data-dependent, an undesirable trait.

A common solution to the power problem is to *code* the binary data prior to modulation. One RFID coding approach is known as *pulse-interval encoding* (PIE). A binary '1' is coded as a short power-off pulse following a long full-power interval, and a binary '0' is coded as a shorter full-power interval with the same power-off pulse (Figure 3.7). The resulting coded baseband signal *m(t)* is then used to modulate the carrier (Figure 3.8). PIE using equal low and high pulses for a '0' ensures that at least 50% of the maximum power is delivered to the tag even when the data being transmitted contains long strings of zeros, and if the high is three times as long for a '1', a random stream of equally-mixed binary data will provide about 63% of peak power. Note that in this case the data rate becomes dependent on the data: a stream of binary 0's will be transmitted more rapidly than a stream of binary 1's. A single symbol has two features — the off-time and on-time — but still conveys only one binary bit. (This scheme is used in EPCglobal Class 1 Generation 2 readers. Other passive RFID standards use slightly different coding schemes, all generally characterized by the desire to have the reader power on as much as possible to power the tag.)

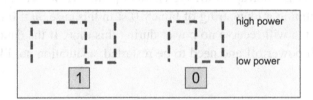

Figure 3.7
Pulse-interval coding baseband symbols (the function m(t)).

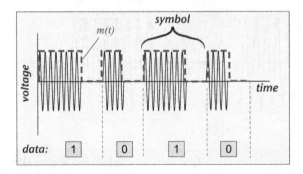

Figure 3.8
Pulse-interval coding with OOK modulation of a carrier wave.

In fixing the problem with transmitted power by replacing OOK with PIE, we've made another problem worse. Radio waves travel everywhere, so in some sense the radio medium is ***shared*** between various users. For example, I would like to be able to read tags on packages in my storeroom despite that fact that the storeroom is also illuminated by the local broadcast radio and television stations, cellular phone basestations, the radio link from the taxi across the street, and the satellite downlink to the neighborhood cable TV system. Using a single medium for many signals is known as ***multiplexing***. The most common form of multiplexing in radio, in use for almost a century, is ***frequency-division multiple access*** (FDMA): different users transmit using different carrier frequencies, and receivers are adapted to capture only the frequency of interest. (Signals can also be multiplexed in time and in coding. In RFID, time multiplexing is employed when a reader uses an anti-collision algorithm to poll tags one at a time; see chapter 8 for more details.) We will discuss the means used to filter the desired frequencies from the received signal in more detail in chapter 4; for the present, it suffices to know that this operation can be accomplished. An RFID reader transmits on a frequency within the band at 902−928 MHz (in the United States), and listens to responses only within that band, rejecting the AM radio broadcast at 1 MHz, the television transmission at 52 MHz, the cellular transmission at 874 MHz, and so on.

This scheme would seem to allow an unlimited number of users to share the electromagnetic spectrum. However, recall that a signal must be modulated in order to convey information. When we modulate the signal, we increase the ***signal bandwidth***. We saw an indication that this would be so in examining analog sinusoidal modulation of a signal (Figure 3.5). A modulated signal occupies a finite region of frequency, and neighbors must be separated by something like that amount in frequency to avoid interference.

Furthermore, choices we make in modulation affect how much bandwidth we use. For example, if we modulate the signal faster by making the individual symbols take less time − that is, if we increase the data rate − we use more bandwidth. This phenomenon is illustrated in Figure 3.9[1], where we show the power spectrum of a modulated signal, and we have made use of the deciBel notation for spectral power introduced in section 2 above. The spectrum has its largest power near the carrier frequency f_c, but a considerable amount of power is transmitted at frequencies rather far from the carrier, as we might have suspected from Figure 3.5 above. The distance from the carrier frequency to the first major 'dip' in the spectrum is inversely proportional to the symbol time τ − that is, it is

---

[1] It is worth noting that in this and the next few figures, the spectra are calculated for a series of about 80 random data bits, only a few of which are shown in the upper "signal" display, in order to keep the diagrams intelligible. If we calculated the frequency spectra over a larger number of bits, they would be smoother, but the spectra shown are reasonably representative of the kind of data actually obtained when the output of a typical frequency-hopping RFID reader is examined over short time scales.

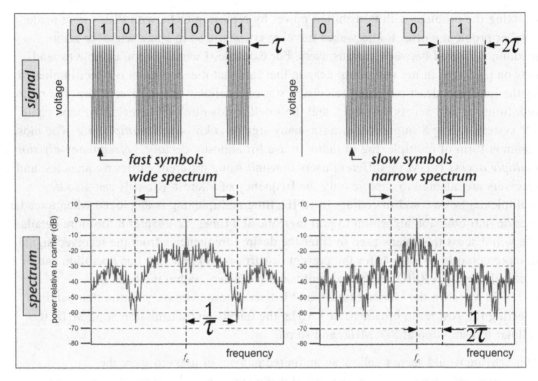

Figure 3.9
Faster modulation = wider spectrum.

the same as the data rate R = 1/τ for OOK. The shorter the symbol time, the faster we can send data, but the more bandwidth we use.

How we send symbols also matters. An abrupt step at the edge of each symbol gives more power far from the carrier than a smooth transition between low and high power states, as depicted in Figure 3.10. (Note that the residual power shown far from the carrier for the smooth symbols in this figure is affected by the specific method of smoothing the symbol and the accuracy of the numerical model.) Of course, the ability to smooth the transitions is limited by the duration of the symbols: at some point changes happen so slowly that fully-on or fully-off states are never reached, causing the transmitted power to fall (and become data-dependent). Smoothing the signals also makes the receiver's problem harder. It doesn't really matter when you test the voltage of a signal like that in left side of Figure 3.10 as long as you are within the symbol, but the smoothed signal on the right side is best sampled exactly at the center of the symbol, where the power is either at its maximum value or nearly zero. Sampling at any other times will result in more power for a nominal '0' or less power for a nominal '1': that is, the measured *modulation depth* is reduced. Thus, the receiver needs to do a better job of synchronizing with the incoming signal if that signal is smoothed.

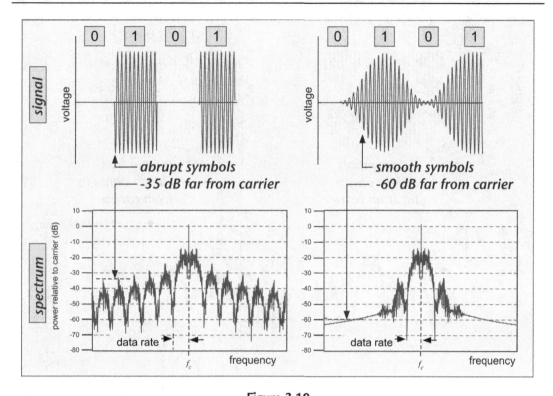

Figure 3.10
Abrupt symbols have more power at frequencies far from the carrier. (The exact levels shown here are somewhat dependent on the modeling algorithm.)

Finally, the way we code the signal also matters. By examination of Figure 3.6 and Figure 3.8 we can see that pulse interval encoding will result in shorter pulses than on-off keying for the same data rate, so from Figure 3.9 it seems likely that PIE would have a wider spectrum than OOK for the same data rate. This expectation is confirmed in Figure 3.11: substituting a stream of PIE symbols at the same average data rate for OOK symbols results in reduced power very near the carrier, but more power far from the carrier, In particular, a strong, narrow emission is seen at a frequency which turns out to correspond to (1/duration of a binary '0'); as depicted by the inset in the figure, the strong resemblance of a '0' symbol to a sine function results in a concentration of power at the corresponding frequency. The more diffuse band at half this offset results from the binary '1' symbol.

To clarify why this sort of thing matters in real applications, let's look at a practical example. In the United States, unlicensed readers randomly hop from one frequency to another within the ISM band from 902−928 MHz. Typically, RFID readers use channels that are 500 KHz wide, and separated by 500 KHz. When a reader is trying to hear a tag, it transmits a signal of constant amplitude and phase. If reader #1 on channel 10 is trying to hear a tag, while reader

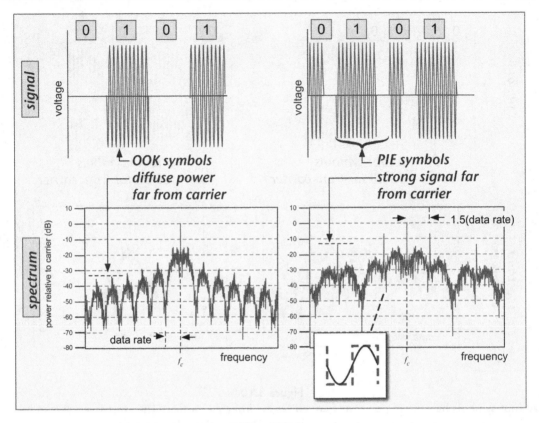

Figure 3.11

Coding data as PIE produces a strong narrow emission far from the carrier, as well as a higher average signal power far from the carrier; the inset shows how this band arises from the '0' symbol. (The exact position of these features relative to the data rate varies depending on the duration of a binary '1'.)

#2 on channel 11 is producing an emission spectra like those shown in Figure 3.11, the situation would look something like Figure 3.12, where the spectrum from reader #2 is scaled for a data rate of about 100 Kbps and a distance of about 20 meters. In section 5 below, we will find that for typical distances, a tag signal is likely to be 40–90 dB smaller than the CW signal from the reader. The leakage from reader #2 into reader #1's channel is thus comparable to or even larger than the tag signal; it will be difficult to detect the tag when reader #2 is transmitting data. Note this is happening despite the fact that the tags are only 1–3 meters from the reader, much closer than the interfering reader!

Even worse, if one of the readers happened to be near the edge of the ISM band, some of this power may be radiated outside of the allowed frequency range, potentially interfering with users of licensed frequencies, who have often paid for the privilege of exclusive use of

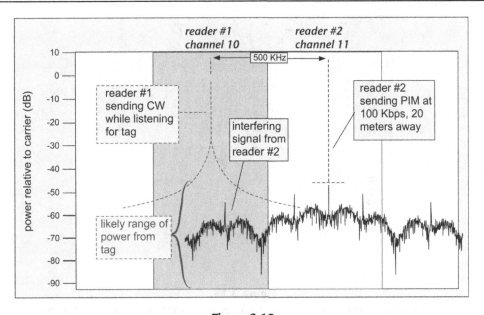

Figure 3.12

Power far from the carrier of reader #2 is in the channel of reader #1 if data rate is high and unsmoothed PIE is used.

said spectrum and get upset when they encounter freeloaders. In the US, the FCC requires that all radios be tested to ensure that such out-of-band radiation is minimized. Interference and out-of-band emissions represent important limits on how fast data can be transmitted by a reader, and on coding and modulation employed, because the speed and method of modulation determine the bandwidth of the resulting signal.

Let us pause for a bit of mathematics to clarify the frequency scales of the figures above. An ideal abrupt pulse (an OOK binary '1') of duration τ has a spectrum:

$$\tilde{f}(\omega) = \sqrt{\frac{2}{\pi}} \frac{\sin(\omega\tau/2)}{\omega} \tag{3.12}$$

This function has some useful special values:

$$\tilde{f}(0) = \sqrt{\frac{2}{\pi}} \frac{\tau}{2}; \quad \tilde{f}(\omega_n) = 0 \text{ for } \omega_n = \frac{2n\pi}{\tau} \left[f_n = \frac{n}{\tau} \right], \quad n \neq 0 \tag{3.13}$$

In particular, the first zero of this function is at a frequency of $(1/\tau)$, where τ is the duration of the pulse. When the signal is a modulated carrier wave, the spectrum is centered around the carrier frequency, and the zeros are displaced from the carrier by $(1/\tau)$ (Figure 3.13).

A stream of binary pulses — an OOK signal as in Figure 3.6 — is just the sum of a number of these pulses, each with the same spectrum, so the full data stream will also have a

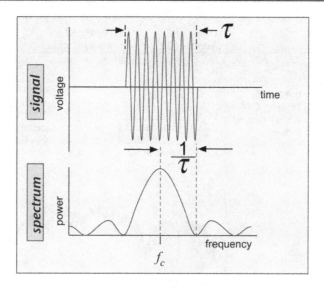

Figure 3.13
A pulse-modulated carrier and corresponding power spectrum; square-root of power is shown for clarity.

spectrum with zero value at the same frequency offset from the carrier. These first zeros determine the width of the main lobe of the signal spectrum, and are indicated by the dashed lines in Figure 3.9. Most of the power in the spectrum is contained within the region about half this wide, that is within a frequency range of $(f_c - 1/(2\tau))$ to $(f_c + 1/(2\tau))$. Thus the narrowest channel that makes sense for an OOK signal is about twice as wide as the inverse of the data rate; we need 200 KHz to fit in 100 Kbps.

PIE is much less efficient, because the shortest pulse – the high part of a binary '0', Figure 3.8 – is about 1/3 as long as an OOK pulse for the same data rate, so roughly three times as much spectrum is needed. To fit the main lobe of the spectrum within a 500 KHz channel, we can only use a data rate of around 85 Kbps – which, as we will see in chapter 8, is just about the upper limit on reader data rates in US operation, using unfiltered PIE-like modulations.

To summarize:

1. to convey information on a signal, the signal must be modulated
2. modulation causes the signal's spectrum to expand, requiring allocation of bandwidth in order to avoid interference
3. the peculiar requirements of passive RFID lead to modulation and coding of binary data that are relatively inefficient in spectral use, limiting reader data rates

It is important to note that more sophisticated radio systems, such as cellular telephony or IEEE 802.11 (WiFi), employ modulation techniques that are substantially more efficient users of spectrum than PIE or OOK. However, these methods generally depend on the ability of the receiver to detect changes in the phase of the high-frequency signal rather than simply determining the power level, which passive RFID tags generally cannot do. As we will discuss in more detail in chapters 4 and 8, *single sideband* (SSB) and *phase-reversal ASK* (PR-ASK) modulations, which use phase information at the reader but require only amplitude detection from the tag, can be used to improve the spectral efficiency.

3.4 Backscatter Radio Links

Passive and semi-passive RFID tags do not use a radio transmitter; instead, they use modulation of the reflected power from the tag antenna. Reflection of radio waves from an object have been a subject of active study since the development of radar began in the 1930's, and the use of backscattered radio for communications since Harry Stockman's work (see chapter 2) in 1949.

A very simple way to understand backscatter modulation is shown schematically in Figure 3.14: current flowing on a transmitting antenna leads to a voltage induced on a receiving antenna. If the antenna is connected to a load which presents little impediment to current flow, it seems reasonable that a current will be induced on the receiving antenna. In the figure, the smallest possible load, a short circuit, is illustrated. This induced current is no different from the current on the transmitting antenna that started things out in the first place: it leads to radiation. (A principle of electromagnetic theory almost always valid in the ordinary world, the *principle of reciprocity*, says that any structure that receives a wave can also transmit a wave. We shall make use of this principle in discussing antennas in greater detail shortly.) The radiated wave can make its way back to the transmitting antenna, induce a voltage, and therefore produce a signal that can be detected: a *backscattered* signal. On the other hand, if instead a load that permits little current to flow − that is, a load with a large *impedance* − is placed between the antenna and ground, it seems reasonable that little or no induced current will result. In Figure 3.14 we show the largest possible load, an open circuit (no connection at all). Since it is currents on the antenna that lead to radiation, there will be no backscattered signal in this case. Therefore, the signal on the transmitting antenna is sensitive to the load connected to the receiving antenna.

To construct a practical communications link using this scheme, we can attach a transistor as the antenna load (Figure 3.15). When the transistor gate contact is held at the appropriate potential to turn the transistor on, current travels readily through the channel, similar to a short circuit. When the gate is turned off, the channel becomes substantially

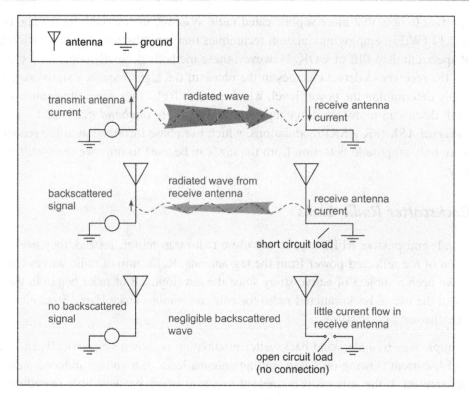

Figure 3.14
Simplified physics of backscatter signaling.

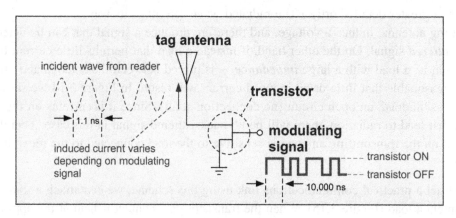

Figure 3.15
Modulated backscatter using a transistor as a switch.

non-conductive. Since the current induced on the antenna, and thus the backscattered wave received at the reader, depend on the load presented to the antenna, this scheme creates a modulated backscattered wave at the reader. Note that the modulating signal presented to the transistor is a ***baseband*** signal at a low frequency of a few hundred KHz at most, even though the reflected signal to the reader may be at 915 MHz. The use of the backscatter link means that the modulation switching circuitry in the tag only needs to operate at modest frequencies comparable to the data, not the carrier frequency, resulting in savings of cost and power. (Real RFID tag ICs are not quite this simple, and may use a small change in capacitance to modulate the antenna current instead, for reasons we will discuss in chapter 5.)

Notice that in order to implement a backscattered scheme, the reader must transmit a signal. In many radio systems, the transmitter turns off when the receiver is trying to acquire a signal; this scheme is known as ***half-duplex*** to distinguish it from the case where the transmitter and receiver may operate simultaneously (known as a ***full duplex*** radio). In a passive RFID system, the transmitter does not turn itself off, but instead transmits CW during the time the receiver is listening for the tag signal. RFID radios use specialized components known as circulators or couplers to allow only reflected signals to get to the receiver, which might otherwise be saturated by the huge transmitted signal. However, in a single-antenna system, the transmitted signal from the reader bounces off its own antenna back into the receiver, and the transmitted wave from the antenna bounces off any nearby objects such as desks, tables, people, coffee cups, metal boxes, and all the other junk that real environments are filled with, in addition to the poor little tag antenna we're trying to see (Figure 3.16). If two antennas are used (one for transmit and one for receive), there is still typically some signal power that leaks directly from one to the other, as well as the aforesaid spurious reflections from objects in the neighborhood.

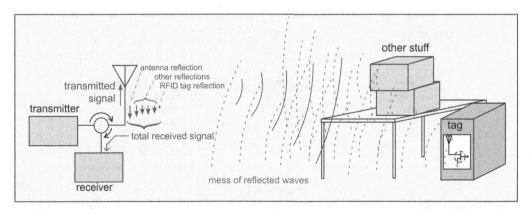

Figure 3.16
Realistic environments create many reflected waves in addition to that from the wanted tag.

The total signal at the receiver is the **vector** sum of all these contributions, most much larger than the wanted tag signal, with appropriate amplitudes and phases, most of which are unpredictable *a priori*. Thus the actual effect of a given change in the load on the tag antenna on the receiver signal is completely unpredictable and uncontrollable. For example, modulating the size of the tag antenna current (amplitude modulation) may not result in the same kind of change in the reader signal. In Figure 3.17, we show a case where changing the tag reflection from a large amplitude (HI) to a small amplitude (LO) causes the received signal to increase in magnitude without changing phase (the "AM" case). Changing the phase of the tag signal without changing the size of the reflected signal in order to symbolize a LO state may change the amplitude of the reader signal at constant phase (Figure 3.17, "PSK" case). The only thing we can say with any confidence is that when we make a change in the state of the tag antenna, something about the phase or amplitude of the reader signal will change. In order to make a backscatter link work, we need to choose a way to code the data that can be interpreted based only on these changes, and not on their direction or on whether they are changes in phase or amplitude.

As a consequence, all approaches to coding the tag signal are based on counting the number of changes in tag state in a given time interval, or equivalently on changing the frequency of the tag's state changes. Therefore all tag codes are variations of *frequency-shift keying* (FSK). It is important to note that the frequency being referred to here is not the radio carrier frequency of (say) 900 MHz, but the tag (baseband) frequency of perhaps 100 or 200 KHz. A binary '1' might be coded by having the tag flip its state 100 times per millisecond, and a binary '0' might have 50 flips per millisecond. Because the frequency being changed is the frequency at which a carrier is being amplitude-modulated, techniques like this are sometimes known as *subcarrier modulation*.

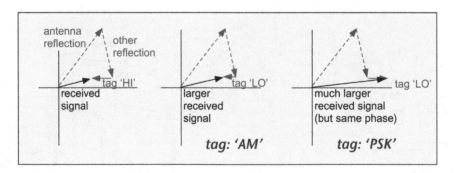

Figure 3.17
The received signal is not simply correlated to the tag signal. The AM case assumes the tag reduces its scattered magnitude without changing phase; the PSK case assumes phase inversion without amplitude change.

Let's look at one specific example of tag coding, usually known as *FM0* (Figure 3.18). In FM0, the tag state changes at the beginning and end of every symbol. In addition, a binary 0 has an additional state change in the middle of the symbol. Note that, unlike OOK, the actual tag state does not reliably correspond to the binary bit: for example, in the left-hand side of the figure, two of the binary '1' symbols have the tag in the LO state and another '1' symbol has the tag in the HI state. Remember the reader can't reliably distinguish which state is which, but can only count transitions between them. The right side of the figure shows the baseband signal corresponding to a series of identical binary bits, to clarify the correspondence of binary '0's with a frequency twice as high as that of binary '1's.

Different tag coding schemes can be used to adjust the offset from the carrier frequency at which the signal from the tags is found. As we will find in chapter 4, readers have an easier time seeing a tag signal when it is well-separated from their own carrier frequency, so higher subcarrier frequencies help improve the ability to read a tag signal. However, if the separation is large compared to the channel size, the tag signal might lie on the signal of another reader in a different channel. Just as with readers, increasing the data rate of a tag signal tends to spread the spectrum out in frequency. To have a flexible choice of tag data rates while minimizing noise, the reader needs to be able to adapt the band of frequencies it tries to receive, adding cost and complexity.

In real receivers, noise and interference may be present as well as the desired signal. A certain minimum *signal-to-noise* ratio (S/N) is necessary for each type of modulation in order that it can be reliably decoded by the receiver. The exact (S/N) threshold depends on how accurate you're trying to be, and to a lesser extent on the algorithms used for demodulation/ decoding. For RFID using FM0, (S/N) of around 10 or better (10 dB or more) is usually sufficient. (Requirements for demodulation of reader symbols, like PIE, in the tag are generally similar.) As we will see in chapter 8, modern protocols provide alternative modulations that can operate with smaller (S/N) ratios, at the cost of a reduction in the tag data rate.

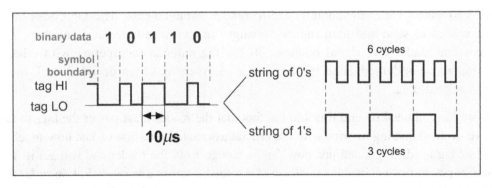

Figure 3.18
FM0 encoding of tag data.

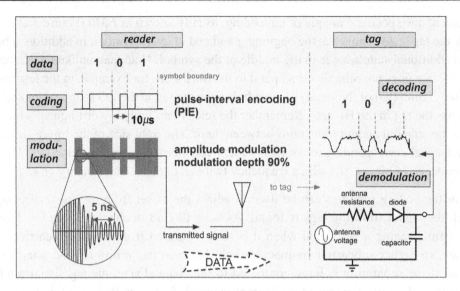

Figure 3.19
Schematic depiction of reader-to-tag data link.

3.5 Link Budgets

Let's summarize the message of the last couple of sections. To transmit to a tag, a reader uses amplitude modulation to send a series of digital symbols. The symbols are coded to ensure that sufficient power is always being transmitted regardless of the data contained within in. The received signal can be demodulated using a very simple power detection scheme to produce a baseband voltage, which is then decoded by the tag logic. The whole scheme is depicted in Figure 3.19.

Figure 3.20 shows the corresponding tag-to-reader arrangements. The tag codes the data it wishes to send and then induces changes in the impedance state of the antenna. The reader CW signal bounces off the tag antenna (competing with other reflections) and is demodulated by the reader receiver, and then decoded back into the transmitted data.

While we have alluded several times to the fact that the reader must power the tag, so far we have avoided coming to grips with the crucial associated question of just how much power the tag needs to get and just how far we can go from the reader and still get it. The amount of power that one needs to deliver to a receiver across a wireless link in order that the transmitted data be successfully received is known as the *link budget*. Since readers and tags both talk, for an RFID system there are two separate link budgets, one associated with

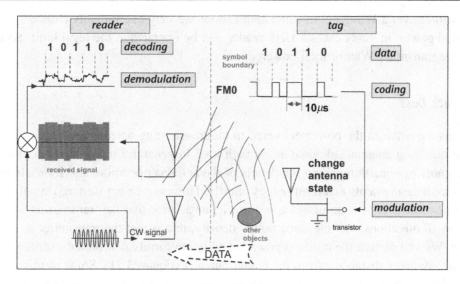

Figure 3.20

Schematic depiction of tag-to-reader data link (a separate receive antenna is shown for clarity).

the reader-to-tag communication (the *forward* link budget) and one with the tag reply to the reader (the *reverse* link budget)[2].

In order to find the forward link budget we need to know the following:

- how much power can the reader transmit?
- how much power does the tag receive as a function of distance from the reader?
- how much power does the tag need to turn on?
- how much power does the tag need to decode the reader signal?

Let's examine each question in turn.

3.5.1 Reader Transmit Power

The reader transmit power is set by a combination of practicality and regulation. Most RFID equipment operates in spectrum set aside for unlicensed use by the governmental body that regulates radio operation in a given jurisdiction. For example, in the United States, the FCC allows operation in the band 902–928 MHz without requiring that the person operating the equipment have a license to do so. However, the equipment itself must obey certain operating limitations in order to allow unlicensed use. Relevant for us at the moment is the maximum transmit power, which cannot exceed 1 Watt. While not all

---

[2] EPCglobal discourages the use of the terms forward and reverse link for readers and tags, but these terms are widely employed in other areas of wireless networking where an asymmetric link is under consideration, and seem perfectly applicable to RFID.

readers will deliver a Watt, and in some applications we may intentionally reduce transmitted power, in many cases a UHF reader will be operated at the legal limit. So let's assume we transmit 1 Watt of total power.

3.5.2 Path Loss

The difference between the power delivered to the transmitting antenna and that obtained from the receiving antenna is known as the path loss. In general, finding the path loss requires knowing something about the details of the antenna operation, and we shall discuss the relevant measurements and terminology shortly. However, to get started, we will use the simplest possible (not very accurate) approach: let us assume that the transmitting antenna radiates in all directions with the same power density, that is that the transmitter is *isotropic*. We can picture the radiated power as being uniformly distributed over a spherical surface at any given distance r from the reader antenna (Figure 3.21). Some of this power can be collected by a tag antenna. It is reasonable to guess that the amount of power collected should be proportional to the density of power impinging on the tag, and dimensionally necessary that the constant of proportionality be an area, often known as the ***effective aperture*** A_e of the tag antenna.

Since in the isotropic case the power density at a distance r is the ratio of the transmitted power P_{TX} to the sphere area, we can find the power received by the tag P_{RX}:

$$P_{RX} = P_{TX} \frac{A_e}{4\pi r^2} \tag{3.14}$$

In order to get numbers out we need a value for the effective aperture. It is not trivial to derive what this area should be, but it is plausible (and correct) to guess that the effective

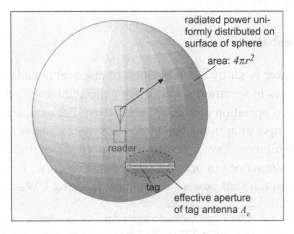

Figure 3.21
An isotropic antenna radiates power uniformly over the surface of a sphere.

aperture of an antenna around a half-wavelength long might correspond to a square around a half-wavelength on a side. (The interested reader is referred to Balanis or Kraus and Marhevka in Further Reading, section 9 of this chapter, for more information on how these areas are obtained.) The actual answer for an isotropic antenna (which a tag isn't quite) is

$$A_e = \frac{\lambda^2}{4\pi} \approx 86 \text{ cm}^2 @ 915 \text{ MHz} \tag{3.15}$$

With a value for the aperture, we can now obtain an estimate of the path loss for our proposed isotropic link. At a distance of 1 meter, the spherical surface has an area of 12.6 m^2, so for 1 watt of transmit power we get about $1(86)/(126,000) = 7 \times 10^{-4} = 0.7$ mW (-1.6 dBm). Since we started with a watt or 30 dBm, the path loss is about 32 dB.

Since the area scales with the square of the radius, we can very easily scale path loss, especially in dB: a factor of 10 in area adds 20 dB to the path loss ("20 dB/decade"). A factor of 3 is worth just a bit less than half of this (about 9.5 dB). So at 3 meters, the path loss is about $(32 + 9.5) \approx 41$ dB, and at 10 meters it is about 52 dB.

3.5.3 Tag Power Requirement

The tag antenna needs to deliver enough power to turn the tag IC on. We will consider this problem in some detail in chapter 5; for the present, it suffices to give the results. Modern tag ICs actually consume around $10-30 \mu$W to operate when being read (much more power is required to write new data to the tag memory). This power must be supplied by a rectifying circuit, which is about 30% efficient, due primarily to the substantial turn-on voltage required to make current flow through the diodes (see chapter 5). As a consequence, tags require about $30-100 \mu$W of power to be delivered from the antenna to provide the required $10-30 \mu$W of power to the chip. For simplicity, let us for the moment use a rather conservative 100μW (-10 dBm) as the required threshold power. If we started at the transmitter with 1 watt (30 dBm), and we need to end up with -10 dBm, we have room for a path loss of $(30 - (-10)) = 40$ dB. By reference to the previous paragraph, this corresponds to a distance of just less than 3 meters. Thus, we expect the *forward-link-limited range* of a 1-watt reader connected to an isotropic antenna to be no more than about 3 meters, for a tag that requires 100μW to power up. Most RFID readers employ modulation depths (the extent to which the power is reduced in the low-power state of e.g. Figure 3.6 or Figure 3.8) of nearly 100%, so it is reasonable to guess that any time the tag has enough power to turn the IC on, it also receives more than enough signal power to interpret the data being sent by the reader.

The calculation is depicted graphically in Figure 3.22. We construct a line of slope -20 dB/decade (-6 dB/octave) and adjust the height of the line to give -1.5 dBm at

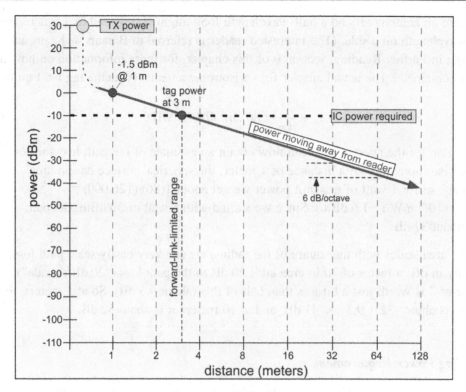

Figure 3.22

Forward link budget calculation for passive tag, US operation. (Note simple scaling is not valid when the tag is within a wavelength of the antenna, here shown as a dotted line.).

1 meter. We can then immediately obtain the range as the intersection of this line with the required power for the IC, here taken as −10 dBm.

To perform the analogous calculation for the reverse link, we need to give thought to two additional issues:

- how much power does the tag send?
- how much power must the reader receive to demodulate and decode the tag data?

As we noted in section 4, a passive tag does not generate its own carrier but simply modifies the amount of the incident radiation it backscatters. It is in principle possible for the tag to backscatter up to four times as much power as it could absorb − but if it does so, the IC will receive no power at all. It is in principle possible to simultaneously deliver slightly less than the maximum absorbed power (e.g. −10 dBm in Figure 3.22) to the IC and scatter about the same amount of power back to the reader. In practice this is challenging to accomplish. Actual modulation efficiency varies from one design to another;

a reasonable estimate for our purposes is to assume a modulated backscatter power around 1/3 of the absorbed power (that's −5 dB).

The amount of power the reader needs to receive is also complex and depends on a number of details of implementation we shall consider somewhat more thoroughly in chapter 4. For the present purposes we shall suggest a plausible and convenient lower limit of around −75 dBm (0.03 nW), deferring justification of this value until later. With the reader's indulgence, we shall proceed to use these unjustified assumptions to construct a diagram of the reverse link power in the same fashion as that previously constructed for the forward link; the result is depicted in Figure 3.23. We construct a second line, like the first but starting at 5 dB less than the tag received power. Note that in this case, as the line descends, we are physically moving back towards the reader. If we move back 3 meters (to intercept the dotted vertical line labeled 'forward-link-limited range') we find the reader receives about −55 dBm, about 20 dB in excess of the power required by the reader's

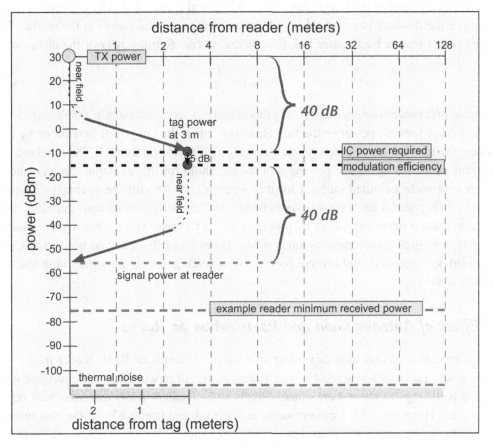

Figure 3.23
Forward- and reverse link budget calculation for passive tag, US operation.

receiver. In fact, a receiver could be an additional 29 meters away before the signal would fall so low as to fail to be received for this threshold value.

While the details of our simplified calculations are hardly authoritative, the observation that passive tags are forward-link-limited has historically been generally correct. The reason is that tag IC power requirements of tens or hundred of microwatts are actually monstrously large compared to the tiny signal powers that can be detected by a good-quality radio receiver. However, as the required power delivered to the IC is decreased with continued progress in IC technology, this may change.

To understand why, we need to understand how the power returned to the reader scales with tag-reader distance. Note in Figure 3.23 that the starting power for the tag scales with the received power. If we double the distance to the tag, the power the tag receives falls by a factor of 4, and thus the transmit power associated with the tag (the reverse link power) also falls by a factor of 4. But this power has to travel twice as far to get back to the reader, so the received power at the reader falls by an additional factor of 4. The net result for a doubling of the distance is a 16-fold (2^4) decrease in the received power at the reader. The received power from a backscatter link falls as the inverse fourth power of the distance:

$$P_{RX,back} \sim \frac{1}{r^4} \qquad (3.16)$$

In the case of a power-hungry passive tag this scaling is rendered moot by the need to provide a fixed forward power to the tag. However, when the tag power is reduced by (say) 10 times, the forward-link-limited range increases by a factor of about 3. The received signal thus decreases by 20 dB, placing it at the threshold for this example receiver: the tag becomes reverse-link-limited (at least for this receiver). As we will see in chapter 5, reader sensitivity is dependent on several design choices, particularly antenna configuration, and will become more important as tag IC power is scaled to lower values. For a semi-passive tag, the forward-link requirement is much more lenient since the received power must only be decoded not exploited, and inverse-fourth-power scaling is very important in determining the range of the tag.

3.6 Effect of Antenna Gain and Polarization on Range

We have been able to conclude that using an isotropic antenna, an RFID reader might achieve a read range of a few meters with 1 watt of output power. This configuration might be fine if RFID tags of interest are equally likely to be located in any direction with respect to the reader. However, such a circumstance is itself rather improbable. In the vast majority of cases, the reader antenna is placed at the edge of some region of interest, and the tags are to be located more or less centrally within this region, at some fairly well-defined angular relationship with the reader antenna. The power that is then being radiated in other directions is wasted (or worse, is reading tags outside the region of interest and confusing rather than enlightening the user). We could make better use of the transmitted power if we could cause the antenna to radiate preferentially along the directions in which tags are most likely to be found.

Fortunately this is entirely possible to achieve. An antenna that performs this trick is known as a ***directional*** antenna. The operation of such an antenna is often depicted by showing an ***antenna radiation pattern***; an example of such a pattern is depicted in Figure 3.24. For any direction *d* relative to the center of the antenna, the distance to the pattern surface represents the relative power density radiated by the antenna in that direction. The radiation pattern is an intuitively appealing method to represent the way a directional antenna concentrates its radiated power in a beam propagating in a particular direction.

The ratio of the radiation intensity in any direction *d* to the intensity averaged over all directions is the ***directive gain*** of the antenna in that direction. The directive gain along the direction in which that quantity is maximized is known as the ***directivity*** of the antenna, and the directivity multiplied by the radiation efficiency is the ***power gain*** of the antenna (very often just referred to as the gain, *G*). In the direction of maximum radiated power density, we get *G* times more power than we would have obtained from an isotropic antenna of the type discussed in connection with Figures 3.21−3.23.

A note of caution is appropriate in considering the terminology we have just introduced. Antennas are passive devices and have no gain, in the sense that they can only radiate the power that is put into them, no more. The term antenna gain refers to the fact that, for a receiving antenna fortunate enough to be located along the direction of maximum power density, the received power is increased relative to that of an isotropic antenna just as if the output power of the directive antenna had been increased (isotropically) by a factor of *G*. Of course, this has not actually happened; the radiated power has just been rearranged, and receiving antennas located in less fortunate directions receive much less power than would have been the case with an isotropic radiator.

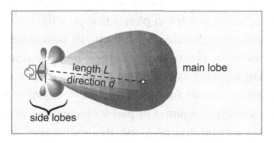

Figure 3.24
Pseudo-3D radiation pattern for directional antenna.

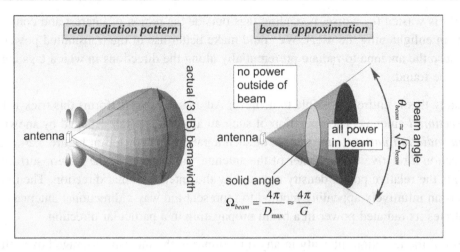

Figure 3.25
Beam approximation for the radiation pattern of a directional antenna.

The higher the gain of a directional antenna, the more narrowly-focused is the energy radiated from it. We can express the relationship mathematically by making the approximation that all the energy radiated by the antenna is uniformly distributed across a beam with some solid angle Ω_{beam}, and no energy is radiated elsewhere. In this case, the directivity of the antenna must be equal to the ratio of the beam solid angle to the total area of the unit sphere (4π), so we find that the solid angle is inversely proportional to the directivity (Figure 3.25). If the antenna radiates most of the energy it receives (which is usually the case for antennas with high directivity) the gain and directivity are about the same, so the size of the beam is inversely proportional to the gain. The beam angle is roughly the square root of the beam solid angle when the beam is reasonably symmetric.

Pseudo-3D depictions of the radiation pattern are helpful to visualize complex geometries, but are difficult to obtain quantitative information from when printed. It is traditional to extract slices of the true radiation pattern in planes that pass through symmetry axes of the antenna. These may be labeled as altitude and azimuth, or sometimes E-plane and H-plane patterns (the notation refers to the planes in which the electric and magnetic fields are located and needn't concern us here). An example of such a pattern diagram for a real commercial directional antenna usable for RFID readers is shown in Figure 3.26. This particular antenna used is known as a *panel* or *patch* antenna, because it is constructed of a metal patch suspended over a metal ground plane, though the user cannot see these details unless they have the courage to slice up the nice-looking plastic casing. This particular pattern is plotted on a logarithmic radial scale, but linear scales are also used. By simply

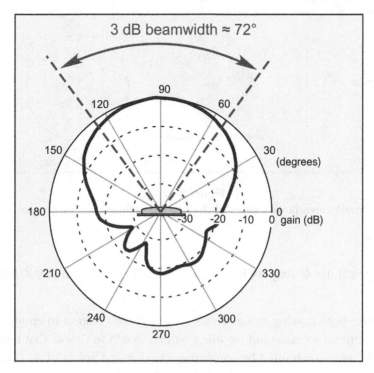

Figure 3.26
Example azimuth pattern for a commercial directional panel antenna (Maxrad MP9026CPR).

finding the locations at which the gain is reduced by 3 dB from the maximum value in the center of the beam, we can extract the 3 dB beamwidth, as has been done in this figure. Since 72° ≈ 1.25 radians, we can estimate the beam solid angle to be about $(1.25)^2 = 1.6$ steradians, so the antenna gain must be roughly G ≈ 4π/1.6 ≈ 8, or 9 dB. (The actual gain of this antenna as reported on the data sheet is about 8.5 dB, so our simple calculation has produced a quite acceptably accurate result. However, the gain is also influenced by the power in the sidelobes and deviations of a couple of dB from this simple formula are not uncommon.) Practical, usable commercial antennas can provide us with quite substantial gains relative to an isotropic antenna.

Not all antennas are highly directional. Though it turns out to be impossible to fabricate a truly isotropic antenna, one can come fairly close to this ideal. A very common example of a not-very-directional antenna is the *dipole* antenna (Figure 3.27). A dipole is constructed of two pieces of collinear wire driven by opposed voltages. Many RFID tag antennas are variants of a simple dipole. Dipole antennas do not radiate along their axes, but radiate equally well in every direction perpendicular to the axis. Thus the radiation pattern looks rather like a donut (or a bagel, depending on your nutritional inclinations).

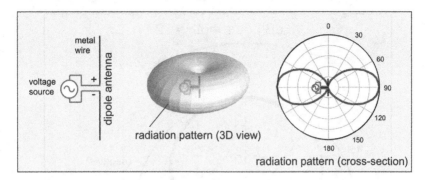

Figure 3.27
Dipole antenna with views of the corresponding radiation pattern.

The gain of a typical dipole roughly half a wavelength long − 16 cm at 900 MHz − is about 2.2 dB.

The gains we have been quoting so far are all measured with respect to an ideal (non-existent) isotropic antenna, and are often written as *dBi* to denote that reference state. In practice, gain is measured by comparing the received power of an antenna under test to a reference antenna, the latter often being a standard dipole antenna. Thus it is easy to measure and report the gain of an antenna relative to a dipole, and this is sometimes done; such gains are usually written as *dBd*. Since a dipole has 2.2 dBi of gain, gain referenced to a dipole is 2.2 dB less than gain referenced to an isotropic antenna: dBd = dBi − 2.2.

Given the gain and transmit power of an antenna, we can calculate how much power we would need to put into an isotropic antenna to get the same peak power as we get in the main beam of a directional antenna (Figure 3.28). This power is called the *effective isotropic radiated power* (EIRP). The EIRP is larger than the actual power by the antenna gain, or in dBm:

$$EIRP = P_{TX}(dBm) + G_{TX}(dBi) \qquad (3.17)$$

EIRP is often either explicitly or implicitly used as a regulatory limitation on radio operations, because it is the EIRP rather than the transmitted power which determines the peak power density transmitted by a reader, and thus the likelihood that it will interfere with other users of the same frequency bands. For example, FCC regulations in the United States allow an unlicensed transmitter to employ up to 1 watt of power with an antenna with 6 dBi of gain; for each dB of additional antenna gain, the transmit power must be reduced by 1 dB. In effect, the FCC is requiring that the EIRP not exceed 36 dBm (30 dBm + 6 dBi).

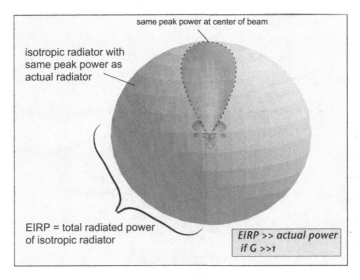

Figure 3.28
Definition of effective isotropic radiated power.

A closely-related quantity, the ***effective radiated power*** (ERP) is also used in similar contexts. However, this term is used rather more loosely: web references can be found in which it is defined in an identical fashion to EIRP, though the US FCC defines ERP as being referenced to a half-wave dipole antenna. In this book we will define ERP following the FCC definition:

$$ERP = P_{TX}(dBm) + G_{TX}(dBd) \tag{3.18}$$

where as the reader will recall, gain in dBd is defined relative to a standard dipole antenna rather than relative to an isotropic antenna. However, we shall generally encourage the use of EIRP rather than ERP since the former is unambiguously defined.

Recall that the purpose of this digression into antenna behavior was to see if we could improve the performance of our theoretical RFID reader by using a directional antenna. If we use a directional antenna to transmit the one watt of allowed power, and the RFID tag of interest is located within the main beam of that antenna, we would expect the transmitted power density to be increased by the gain of the antenna. The result ought to be an increase in the read range. The argument is depicted graphically in Figure 3.29, for an antenna with 6 dBi of gain.

The forward-link-limited range has doubled, from 3 to 6 meters, relative to that obtained in Figure 3.22. This is what we'd expect: we increased the signal power by a factor of 4, but power falls as the square of the distance, so this only provides us with a factor of 2 in range. At the same time, we've reduced our ability to see tags outside the main beam,

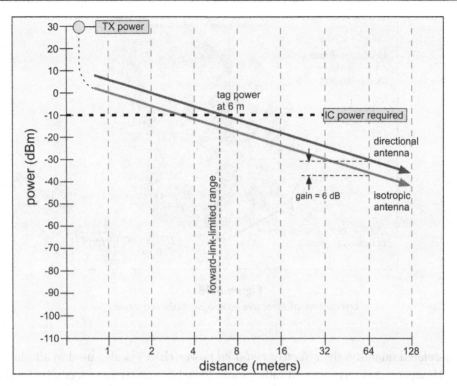

Figure 3.29
Forward link budget using a directional antenna with 6 dBi gain.

presumably around 80–100° wide here, which is usually desirable: by employing a directional antenna we are able to (mostly) select the region in which tags can be read, and thus exclude tags that are not of interest.

What about the reverse link? In considering the action of an antenna as a receiver, we have heretofore asserted without detailed proof that the antenna collects energy from some effective aperture, and given a typical size. In fact, the size of the receiving aperture of any antenna is directly proportional to the gain of the antenna when used as a transmitter. This is a consequence of the ***principle of reciprocity***, briefly alluded to previously, which for our purposes we can state as: transmitting from antenna 1 and receiving with antenna 2 ought to give the same result as transmitting from antenna 2 and receiving with antenna 1. Since we have already cited the effective aperture for an isotropic antenna (equation (3.15)), we can write:

$$A_e = G\left(\frac{\lambda^2}{4\pi}\right) \tag{3.19}$$

where the gain G is measured relative to an isotropic antenna, that is in dBi. Using this relationship, we can write a very general equation for the power received from a transmitting antenna *TX* by a receiving antenna *RX* if both gains and the distance between them are known:

$$P_{RX} = P_{TX}G_{TX}\frac{A_{e,RX}}{4\pi r^2} = P_{TX}G_{TX}G_{RX}\frac{\lambda^2/4\pi}{4\pi r^2}$$

$$= \underbrace{P_{TX}G_{TX}G_{RX}\left(\frac{\lambda}{4\pi r}\right)^2}_{\text{Friis equation}} \tag{3.20}$$

The last form of the relationship is known as the **Friis equation**, a very convenient way to state the expected received power. Note that this equation does not imply, as is sometimes erroneously asserted, that waves fail to propagate as wavelength decreases; from the derivation it should be apparent that the factor of λ^2 arises from the effective aperture of the receiving antenna and is not related at all to propagation in the intervening space.

With the Friis equation in hand, we can immediately draw the reverse-link diagram for a directional antenna: the received power is simply increased by the antenna gain, just as the transmit power was. The result is given graphically in Figure 3.30. The received power is the same as in the isotropic case, even though the tag is twice as far away, because the

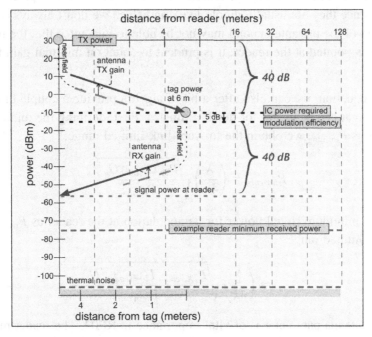

Figure 3.30
Forward and reverse link budgets for directional antenna.

power at the tag is the same in both cases, and the received power is decreased by 6 dB due to the larger distance, but increased by 6 dB due to the receiver antenna gain.

We can also construct a mathematical statement of the same relationships using the Friis equation. We define the gain of the tag antenna G_{tag}, and a backscatter transmission loss T_b ($=1/3$ or -5 dB here). We then have:

$$P_{TX,tag} = P_{TX,reader} G_{reader} G_{tag} \left(\frac{\lambda}{4\pi r}\right)^2 T_b$$

$$P_{RX,reader} = P_{TX,tag} G_{tag} G_{reader} \left(\frac{\lambda}{4\pi r}\right)^2 \rightarrow \quad\quad (3.21)$$

$$P_{RX,reader} = P_{TX,reader} T_b G_{reader}{}^2 G_{tag}{}^2 \left(\frac{\lambda}{4\pi r}\right)^4$$

As promised, in the most general case the power received at the reader goes as the inverse fourth power of the (symmetric) distance. It is also proportional to the square of the antenna gains, so when reverse link power is important (e.g. when a semi-passive tag, or an unpowered device like a surface-acoustic-wave (SAW) tag, is used) the antenna gain plays a very large role in achievable read range. We have previously treated the tag antennas as having a gain of 1 (0 dBi). Real tag antennas have some gain, but it is typically modest (around 2 dBi, since they are usually dipole-like), and since we don't always control the exact orientation of the tag antenna and may not be able to guarantee that the main beam of the tag antenna is pointed at the reader, it is prudent to count on minimal gain from the tag antenna.

Using the Friis equation, we can also after a bit of algebra provide a couple of convenient range equations that can be useful for quick estimates. First, defining the minimum power the tag requires as $P_{min,tag}$ we obtain the forward-link-limited range:

$$R_{forward} = \left(\frac{\lambda}{4\pi}\right) \sqrt{\frac{P_{TX} G_{reader} G_{tag}}{P_{min,tag}}} \quad\quad (3.22)$$

and defining the minimum signal power for demodulation at the reader as $P_{min,rdr}$ we obtain the reverse-link-limited range:

$$R_{reverse} = \left(\frac{\lambda}{4\pi}\right) \sqrt[4]{\frac{P_{TX,reader} T_b G_{reader}{}^2 G_{tag}{}^2}{P_{min,rdr}}} \quad\quad (3.23)$$

One additional antenna parameter is of vital importance in RFID. The reader may perhaps recall that, back in section 1 of this chapter, we described the radiated magnetic vector potential as being in the direction of the current from which it radiates. The vector potential has a direction at each point in space. The *electric field*, which is derived from the vector and scalar potentials, describes the effect these potentials have on electrons in a wire. It is always pointed along that part of the vector potential that is perpendicular to the direction of propagation. This isn't as scary as it sounds: it just means that electromagnetic waves are normally *transverse* waves. Like a wave on water, the effect associated with the wave is perpendicular to the direction in which the wave is propagating. When the wake from a boat strikes a buoy, the buoy (mostly) moves up and down, and only slightly towards or away from the passing boat. An electromagnetic wave moves electrons in the plane perpendicular to the direction of propagation, not along the direction of propagation. The direction in which the field points determines the *polarization* of the radiated wave. When this direction is constant in time, the wave is said to be *linearly polarized*.

Unlike water waves, electromagnetic waves are not influenced by gravity and the electric field can point in any direction in the plane perpendicular to the direction of propagation. Because human beings are gravitationally challenged, it is most common to orient linearly-polarized antennas either vertically or horizontally (Figure 3.31). However, any intermediate angle is also possible.

It is also possible for the direction of polarization to be time-dependent. For example, the electric field can rotate around the axis of propagation as a function of time, without

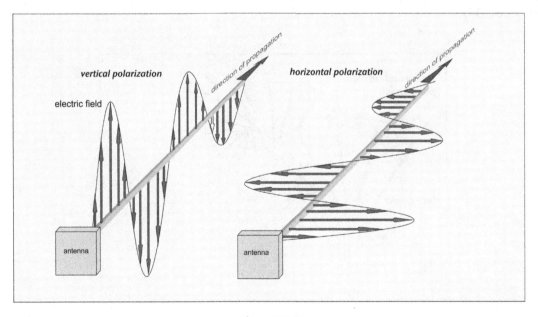

Figure 3.31
Linearly polarized radiation.

changing its magnitude, producing ***circularly polarized*** radiation (Figure 3.32). Depending on the sense of rotation, we obtain either right-handed or left-handed polarization.

Note that the electric field of a circularly-polarized wave still points in a specific direction at each moment in time, or at each location along the wave. Circular polarization does not refer to circulating fields or potentials, but merely to the time dependent orientation of the field. Circularly polarized radiation can be regarded as the sum of vertical and horizontal polarized waves that are out of phase by 90°. By adjusting the ratio of horizontal and vertical components, and their phase relationship, we can produce ***elliptically polarized*** waves of arbitrary orientation, extending from pure circular to pure linear polarization.

The importance of polarization in RFID is simple to grasp: many RFID tag antennas consist primarily of narrow wire-like metal lines in one direction. If the electric field is directed along the wire, it can act to push electrons back and forth from one end of the wire to the other, inducing a voltage that is used to power the IC and allow the tag to reply. If the electric field is directed perpendicular to the wire axis, it merely moves electrons back and forth across the diameter of the wire, producing negligible current, no detectable voltage at the IC, and thus no power (Figure 3.33).

When a circularly-polarized wave impinges on a linear antenna, only the component of the wave along the antenna axis has any effect. Thus a circularly polarized wave will interact with a linear antenna tilted at any angle within the plane perpendicular to the axis of propagation, but in every case only half the transmitted power can be received (Figure 3.34).

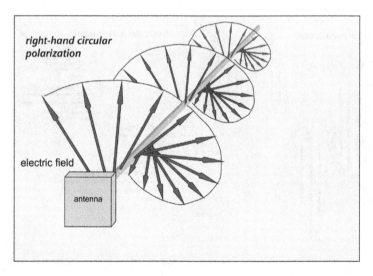

Figure 3.32
Right-hand-circular polarization.

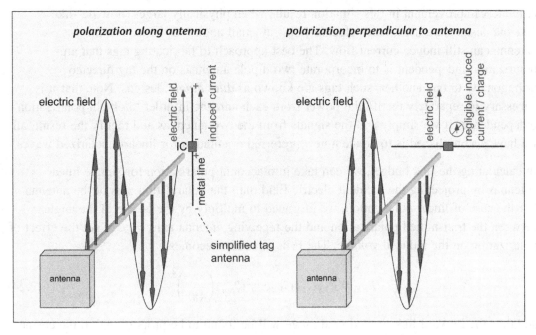

Figure 3.33
Linearly polarized wave interacting with linear antenna.

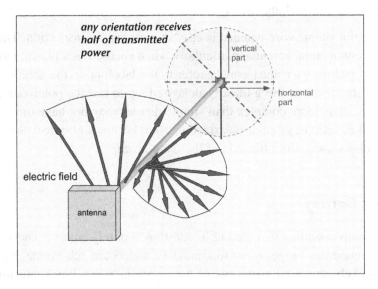

Figure 3.34
Circular polarization interacting with linear antenna.

A modest improvement in this situation results when physically larger 'bow-tie'-like antenna designs are used, since electric fields at small angles to the axis of the antenna can still induce current flow. The best approach to fabricating tags that are polarization-independent is to incorporate two dipole antennas on the tag directed orthogonally to one another; such tags are known as *dual dipole* designs. Note that it is necessary to separately rectify the power from each antenna in order to obtain polarization independence; if we simply add the signals from the two antennas and rectify the result, all we have accomplished is to create a new preferred orientation for linearly polarized waves.

In calculating the link budget, we can take into account polarization for simple linear antennas by projecting the incident electric field onto the polarization axis of the antenna. For the case of linear polarization, we just need to multiply by the cosine of the angle between the transmitted polarization and the receiving antenna axis, θ_{pol}, to get the effect of polarization on the induced voltage. The Friis equation becomes:

$$P_{RX} = P_{TX} G_{TX} G_{RX} \cos^2(\theta_{pol}) \left(\frac{\lambda}{4\pi r}\right)^2 \tag{3.24}$$

and thus the forward-link-limited read range will be found to be proportional to the cosine of the misalignment angle. Note, finally, that because electromagnetic waves are transverse, there is no electric field along the direction of propagation. A simple linear tag antenna oriented along the direction of propagation — that is, pointing towards the reader antenna — sees no electric field along the wire axis and therefore receives no power. We have alluded to this fact previously (Figure 3.27) in connection with transmitted power: a dipole antenna does not transmit or receive along its axis.

The polarization of a simple wire antenna is easy to establish by inspection. The polarization of a commercial antenna, particularly when encased in a plastic radome, is not so obvious, and the user must usually refer to the labeling on the antenna or the manufacturer's data sheets, or use a linearly-polarized tag to test the polarization of the radiated field. Antennas more complex than simple dipoles may not have the same polarization in all directions; circular polarization often becomes elliptical as the direction of observation moves away from the axis of the main beam.

3.7 Adding a Battery

So far, we have been assuming that the tag in question is purely passive: cheap, simple, and dumb. We have found that ranges of up to around 10 meters are achievable, but a hundred meters seems unlikely, and a kilometer out of the question. What if an application really requires much longer range than a passive tag can provide, and the customer is willing to pay for improved performance?

The first step is to graduate to a semi-passive, or battery-assisted, tag, described schematically in Figures 2.16 and 2.19. The simplest battery-assisted tags work rather like passive tags, in that they can receive a signal from the reader and backscatter their reply. Therefore, there is a forward- and reverse-link budget just as for the passive tag. However, a battery-assisted tag has (guess what?) a battery to power its internal logic, so it does not need to receive nearly as much power as a pure passive tag does. The received power only needs to be large enough to allow data to be decoded.

To figure out how large that is, we can make some simplifying assumptions. Even with a battery this is still an RFID tag. It has to live a long time in the field. It probably does not have a low-noise amplifier (and we will see in a moment that it would not help much if it did). It probably relies on some diodes to rectify the RF signal into a DC amplitude, which would be directed into a comparator to see if it represents a high or low power state. For small signals, a diode acts as a *square-law detector*: that is, the output DC-like voltage is proportional to the square of the input voltage. An example for a realistic diode detector (in this case, the Skyworks APN1014) is shown in Figure 3.35. An incoming signal power of −30 dBm (1 microwatt) produces an output signal of about 3 mV from a single diode. If two diodes are combined to form a voltage doubler (about which we will have more to say in chapter 5), an input power of −40 dBm (0.1 microwatt) produces about the same DC voltage out. Practical comparators can achieve about 1 mW of offset over process and temperature variations, so it is certainly necessary to have 2−3 mW of signal to reliably

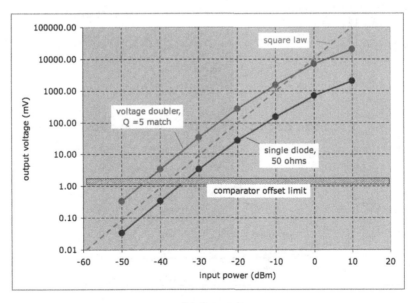

Figure 3.35
Output signal for a commercial diode detector.

distinguish between high and low amplitudes. Thus, we would expect the receive sensitivity of a battery-assisted tag to be in the neighborhood of −30 to −40 dBm. That is a heck of a lot better than a passive tag, but still much worse than a full-fledged radio, with its own amplifiers, mixers, and local oscillator.

Since the battery-assisted tag still uses backscatter for the reverse link, an improvement in receive sensitivity represents a decrease in the amount of power available to talk back to the reader. At the outer edge of its range, a battery-assisted tag only has about − 30 to −40 dBm to scatter back to the reader, 10−100 times less than a conventional passive tag. The decreased power and increased path loss implies very small signals at the reader. An example of the link budget in both directions is shown in Figure 3.36. The improved sensitivity (here taken to − 35 dBm) results in a forward-link-limited range of around 100 meters. However, the consequent return signal, assuming plausible modulation efficiency, is around − 100 dBm. Such a small signal will not be readily detected at the reader unless separate antennas are used for transmitting and receiving (the bistatic configuration, to be discussed in chapter 4). With good antenna and source design, the sensitivity will be limited by thermal noise, which will be around − 110 dBm for a bandwidth of 200 kHz and receiver noise figure of around 10 dB. The receiver signal-to-noise ratio, around 10 dB, is just enough to decode a simple shift-keyed signal. (A more detailed discussion of noise and sensitivity may be found in chapter 4.) We can do better, at

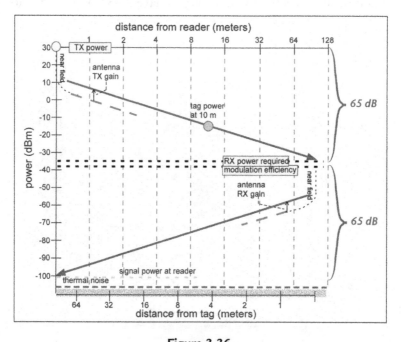

Figure 3.36

Typical forward- and reverse-link budget for a battery-assisted tag, assuming reader antenna directive gain of 6 dB.

the cost of slower communications, by using a lower data rate: less bandwidth is required, and therefore less noise is received.

We can conclude that a simple battery-assisted tag can achieve a range on the order of 100 meters, but that an expensive, high-quality reader receiver will be needed, and both the forward and reverse links must be optimized for good performance. You can see that adding a low-noise amplifier in front of the diode detector would not really help that much: if we increase the sensitivity of the tag receiver by (say) 15 dB, we need to provide an additional 15 dB at the reader receiver, which in turn requires reduced bandwidth in order to avoid getting lost in the thermal noise. As always, we could make the tag antenna more directional, but then we would not see it if it is oriented the wrong way.

If a battery-assisted tag is not good enough, we can go on to a fully-active tag, with its own radio transmitter and receiver. The tag receive sensitivity is now that of a radio (albeit typically a cheap one!). The sensitivity depends on the protocol and data rate in use, but for low-data-rate applications a reasonable value is -90 dBm. The forward-link range is greatly expanded by this improvement in the receiver, as shown in Figure 3.37. It is now possible to receive the reader signal from a substantial distance, here 50 km.

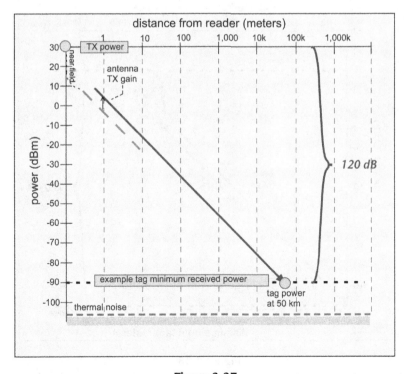

Figure 3.37
Forward-link budget for an active tag, assuming 1 W transmit power, 6 dB reader TX/RX gain, and operation around 900 MHz.

Before you get too excited about this, there are a lot of practical issues to consider. The computation here applies only when a clear line of sight exists between the reader and the tag. That is pretty sensible when the two are 3 meters away, but as you can imagine, it becomes a lot less likely at 50 km. Unless you are floating in space or pointing straight up, you need to account for buildings, mountains, people, cars, the curvature of the earth, scattering from rain and fog, and so on. It is indeed possible to construct radio links at kilometer ranges using ordinary battery-powered devices, but to do so you need antennas on tall masts or mountains. This is not a normal usage model for RFID.

Furthermore, if you could reach the tag at 10 km, you could not hear it. This is because the transmit power of the battery-powered active tag is likely to be much less than that of the plugged-in reader. An example is depicted in Figure 3.38. A battery-powered device might transmit at 10 dBm (10 mW), though power of 1 mW or less is also common. This is 100 times less than the reader transmit power. The received power at the reader is thus 100 times lower than that at the tag, around −110 dBm. The reader can have more expensive receiver components that add less noise, and perform more elaborate signal processing; these advantages generally account for around 10−12 dB improvements, but not

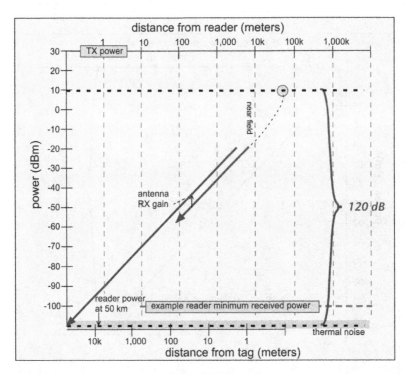

Figure 3.38

Reverse-link budget for an active tag, assuming 10 mW transmit power, with other conditions as in Figure 3.37.

20 dB. The reader is going to have a hard time hearing the tag. This is an example of an *unbalanced* link. The range is limited by the reverse rather than the forward link, as was the case for passive tags.

It is clear that a battery-assisted tag, or an active tag, will go much farther than a passive tag under the same conditions, though 50 or even 5 km is more than one would expect in realistic environments. The challenges of propagation in the real world are discussed in more detail in the next section. The problem with both of these alternatives is, of course, the battery. Either we need a battery that will last for the expected useful lifetime of the device, or we need to replace or recharge the battery, which is often costly and inconvenient. To make the battery last as long as possible, we would like to turn the tag off when we do not need it, but how is the tag to know when it is needed? We will return to the problem of letting tags get their rest in chapter 5.

Note that the reader sensitivity shown here is substantially better than the values used for the passive tags discussed previously. That is because there is no need for the reader to transmit a 1-watt signal at the same time it is trying to receive the tiny tag signal, so the sensitivity is limited mainly by thermal noise rather than by a nearby very-high-power transmitted signal.

3.8 Propagation in the Real World

All the calculations we've performed so far assume that a wave leaves the antenna and strikes the tag, interacting with no other objects. This kind of calculation is very sensible if the tag and reader are placed in a specially-designed *anechoic chamber*, or perhaps suspended high in the air from (non-metallized) balloons. In the actual circumstances in which most readers and tags are used, the wave emitted from a reader antenna is likely to interact with many other objects besides the tag.

The interaction between waves that travel along the *direct* path between the reader and the tag, and those that are *scattered* or *reflected*, is of counter-intuitively large importance because it is voltages not powers that add. Let us consider, for example, the addition of a direct *beam* and two reflected beams, perhaps from the floor and a distant wall (Figure 3.39), each of which contains only 1/10 of the power of the direct signal. We can write the resulting voltage as:

$$V_{total}\cos(\omega t) = \underbrace{v_{dir}\cos(\omega t)}_{direct} + \underbrace{v_{r1}\cos(\omega t + \delta_1) + v_{r2}\cos(\omega t + \delta_2)}_{reflected} \qquad (3.25)$$

Here the δ's are the phase differences between the reflected waves and the direct wave. The phase difference depends on the relative length of the path traveled by each wave; a change in that path of 8 cm (a quarter of a wave) corresponds to a 90-degree phase shift (from a

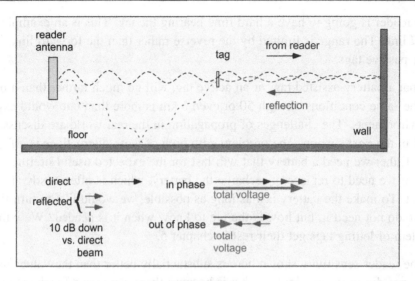

Figure 3.39
Direct and reflected beams can interfere.

maximum value to zero or vice versa) for the beam traveling that path. It is unlikely (!)
that we can measure or control the position of every object in the room to within a couple
of centimeters, so we must consider these phase delays as being generally unpredictable
and uncontrollable. Thus, the best we can do is to examine the extreme cases. First of
all, what if the reflected beams happen to both be in phase with the direct beam ($\delta = 0°$)?
We get:

$$V_{total} = v_{dir} + \frac{v_{dir}}{3.2} + \frac{v_{dir}}{3.2} \approx \left(1 + \frac{2}{3.2}\right)v_{dir} \approx 1.63v_{dir};$$

$$\frac{P_{total}}{P_{dir}} = 1.63^2 \approx 2.7 \ (4.2 \ dB)$$

(3.26)

The received power is about 4 dB **higher** than in the absence of reflections. On the other
hand, if the reflected beams are exactly out of phase (that is, $\delta = 180°$), we find:

$$V_{total} = v_{dir} - \frac{v_{dir}}{3.2} - \frac{v_{dir}}{3.2} \approx \left(1 - \frac{2}{3.2}\right)v_{dir} \approx 0.375v_{dir};$$

$$\frac{P_{total}}{P_{dir}} = 0.375^2 \approx 0.14 \ (-8.5 \ dB)$$

(3.27)

That is, the total received power can change by $(4.2 + 8.5) = 12.7$ dB — a factor of 20! —
even though the reflected beams' **combined power** is only 20% of that of the direct beam.

This sort of wild variation in received signal strength with small displacements in position or frequency is known as *fading*, and is an ubiquitous problem in all radio systems. It is exacerbated in RFID because during the CW portion of an exchange, the reader transmits a very narrow spectrum with essentially only one frequency. Thus nearly perfect cancellation is possible if directed and reflected beams happen to be of the right magnitudes and interfere destructively.

In US operation, the reader will soon hop to a different frequency within the 902–928 MHz ISM band. The change in phase due to a hop from f_1 to some other frequency f_2 is proportional to the difference in frequency and the difference in the length of the various paths. For example, in Figure 3.39, the path length of the direct path might be (say) 1.5 meters, and the path length to the wall and back to the tag would be 4.5 meters; the path length difference is 3 meters. Imagine that a tag is in a deep fade at some frequency f_1. If we shift the frequency by 10 MHz, the phase difference will change by $2\pi(10 \text{ MHz})$ (3 meters)$/c = (6.28)(10^7)(3)/(3 \times 10^8) = 0.63$ radian or about 36°. This is more than enough to ensure that the signals no longer cancel at the new frequency, though the received power may still be below that of the direct beam on its own. However, in other jurisdictions much less bandwidth is available, and hopping may be impossible or ineffectual in defeating fades. For this reason, it is usually necessary to attempt multiple reads of a tag population in different physical configurations, for example by moving the tags or rotating the objects to which they are attached, in order to ensure that all tags are read.

In RFID operation, the most important single reflector is the floor: RFID reader antennas, unlike many other communications systems, are typically oriented to transmit horizontally and are located within a meter or two of the floor. In many facilities floors are constructed of concrete, which has a refractive index of around 2.5 at microwave frequencies, and can act as an effective reflector of incident radiation.

Since the concrete acts as a dielectric, the angle of incidence and the polarization of the incident beam are both of importance in determining the reflection coefficient. Vertically-polarized radiation incident on a horizontal floor will experience no reflection at all at a particular angle of incidence, *Brewster's angle*, which is around 65° (measured with respect to the vertical) for microwave reflection from concrete (Figure 3.40). Horizontally polarized radiation benefits from no such effect; the reflection coefficient increases monotonically (and monotonously) with increasing angle from the normal. For a reader antenna placed 0.75 meter above the floor, the Brewster's angle reflection point is about 1.5 meters away, so the specularly reflected location is 3 meters away, generally within the range of a typical UHF passive tag. A vertically polarized antenna will experience no floor reflection at this distance, and thus produce little local fading (at least due to the floor reflection), whereas a horizontally polarized antenna will produce strong fading in this distance range (Figure 3.41). Thus, a reader employing a linearly polarized antenna will produce more

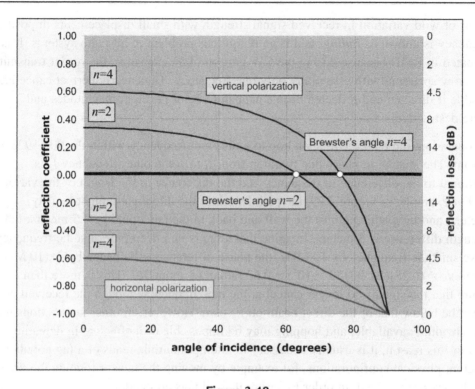

Figure 3.40
Reflection coefficient from a horizontal plane as a function of incident angle and polarization, for refractive indices of 2 and 4.

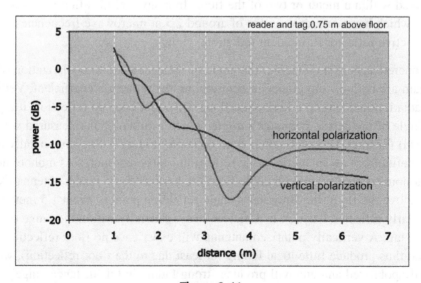

Figure 3.41
Relative received power considering only direct beam and floor reflection, vs. reader antenna polarization; n(floor) = 2.5, beam width = 75 degrees.

reproducible read results if the antenna is vertically polarized; on the other hand, a horizontally polarized antenna will display more prominent fades but also (sporadically) read more distant tags.

In the general case, with many irregular scatterers and reflectors present in uncontrolled positions and orientations, a propagation environment can be very complex and quite unpredictable. A simplified example is depicted in Figure 3.42 in which only reflection from walls and floor of a cubical room is incorporated. It is apparent that the signal strength varies in a complex fashion over size scales comparable to 1/2-wavelength even for a fairly simple environment with no people or furniture. A tag moving within this environment will be easily read in certain regions, and very difficult to read when displaced in an arbitrary direction by 10 or 12 cm. In realistic environments it is often necessary to attempt to read tags multiple times in differing physical configurations to ensure that all the tags are read.

We have so far considered only unimpeded straight-line propagation, and specular reflection (where the angle of incidence and the angle of reflection are the same). In UHF RFID, the typical wavelengths of around 32 cm are comparable to the size of many obstacles present in the environment, so to fully treat the propagation environment we must account not only for propagation and reflection, but *diffraction*: the ability of obstacles of

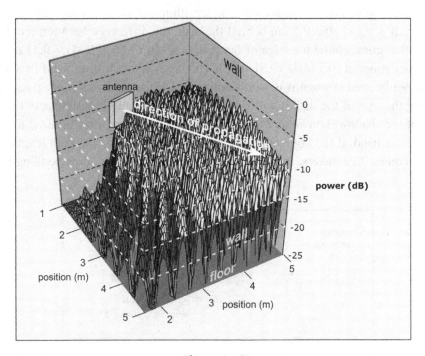

Figure 3.42

Simple simulation of received power distribution for 5 × 5 × 5 meter cubical room with reflecting walls and floor (n = 2.5); vertical polarization, transmitter at x = 1, y = 2.5, z = 1.

finite size to scatter the incident radiation in directions other than specular. The full treatment of diffraction is rather complex, and not as important for passive RFID as for other communications fields since the forward link budget is so small. The importance of diffraction may be roughly estimated by calculating the effective size of an obstacle in terms of the phase difference between the shortest and longest paths through the obstacle (Figure 3.43). In the figure, the shortest distance is the direct path (which passes right through the obstacle) and the longest distance goes around the edge of the obstacle. The phase difference is the difference in these path lengths multiplied by the wavenumber $k = \omega/c$:

$$\delta\phi = k(L_1 - L) \tag{3.28}$$

(A typical value of k for UHF RFID is about $19-20$ radians/meter.) This difference, measured in half-wavelengths (i.e. $\delta\phi/\pi$, since there are 2π radians of phase in one wavelength), is the number of **Fresnel zones** subtended by the object. When this number is small (on the order of $1-2$ or less), diffraction is important, and the received intensity is a complex function of position, with no well-defined shadow region. When the obstacle subtends many Fresnel zones ($>3-5$), it is able to form a fairly well-defined shadow, and tags in that region are unlikely to be visible to the reader.

Consider, for example, a disk of diameter 1 meter, illuminated by a reader antenna 2 meters from the disk. If a tag is placed 5 cm behind the disk, the difference between the direct path and the path that goes around the edge of the disk is about $(2.56 - 2.05) = 0.51$ m, which is about 3 Fresnel zones at 915 MHz (0.51/0.16). The tag is likely to find itself in a deep shadow and not be read (though it is worth noting that if the tag is carefully positioned exactly along the axis of the disk it will find itself in a relatively high-intensity region in the middle of the shadow, known as **Poisson's bright spot**, which may allow it to power up). On the other hand, if the tag is placed 2 meters from the disk, the path length difference becomes 0.12 meters, rather less than one Fresnel zone, corresponding to weak

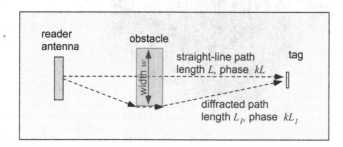

Figure 3.43
Rough estimate of diffraction behavior of an obstacle is based on phase difference of shortest and longest paths.

and complex shadowing. In this position the tag will move in and out of faded regions as its position relative to the disk and reader changes. In practice such weak shadowing often simply adds to the complex fading behavior resulting from walls, floors, and other obstacles that can be treated as specular reflectors. Thus, obstacles that are small relative to a wavelength, or distant from both the transmitter and the receiver (reader antenna and tag) have modest though non-negligible effects, and tags may be read even though the straight-line path from reader to tag is obstructed. Obstacles that are large compared to a wavelength, and close to either the reader antenna or the tag, are likely to prevent passive tags from being read. It is in this sense that RFID is a non-line-of-sight technology even for metallic obstacles.

3.9 Capsule Summary: Chapter 3

Electrical currents and charges radiate, but the net effects usually cancel; an antenna is a special structure arranged to avoid such cancellation and create electromagnetic waves from electrical currents and voltages. These waves are usually periodic in nature and characterized in terms of sines and cosines, or complex exponential functions. They are converted into voltages and currents in electrical circuits. The size of the voltages and related power varies over a large range, so power and gain are usually measured logarithmically, using dBm and dB.

Wave must be modulated in order to transmit information. When a signal is modulated, the width of its frequency spectrum increases. Modulations used for RFID readers are constrained by the need to provide power to passive tags, and are thus profligate users of spectrum relative to the amount of information transmitted.

The currents induced in tag antennas, like other uncompensated currents, radiate, leading to backscattered waves. The load connected to the antenna can be varied to change the amount of induced current and thus the backscattered wave, enabling a tag to communicate with a reader even though it has no transmitter. This reflected signal adds to other, larger reflections from the system and ambient, so there is no simple relationship between the tag state and reader signal; thus tag modulations are all variants of frequency-shift keying.

The amount of power needed to turn on a tag IC is the main limit on the range of passive tags. Directional antennas increase the power that reaches tags in the main beam of the antenna, but regulations limit the transmitter power and antenna gain that can be used, so the range in air is typically only a few meters. Radiation from antennas is polarized, and if the polarization from the reader antenna does not agree with the polarization the tag can receive, the power received is reduced, and the tag may not be read.

In realistic environments, propagation is greatly complicated by reflections from surfaces as well as diffraction around obstacles, leading to local fading and requiring that tags and reader be moved in some fashion to ensure that all the tags have a chance to be read.

Further Reading

Signal and Signal Processing

Digital Modulation and Coding, S. Wilson, Prentice-Hall 1996. *For the serious student; the fundamentals of signal modulation and detection, developed with considerably more rigor than we have employed here.*

Backscatter Links

"Communication by Means of Reflected Power", Harry Stockman, Proc I.R.E., October, 1948, p. 1196

Antennas

Antenna Theory (3rd Edition), C. Balanis, Wiley 2005
Antenna Theory and Design (2nd Edition), W. Stutzman and G. Thiele, Wiley 1997
Antennas (3rd Edition), J. Kraus and R. Marhefka, McGraw-Hill 2001
RF Engineering for Wireless Networks, D. Dobkin, Elsevier 2004, chapter 5; see also p. 350 for references covering the microwave properties of common construction materials.

Reflection from Dielectric Surfaces

Physics of Waves, W. Elmore and M. Heald, Dover 1985, chapter 8
Classical Electricity and Magnetism (2nd Edition), W. Panofsky and M. Philips, Addison Wesley 1962, chapter 11

Exercises

Frequency and wavelength:

1. RFID operation worldwide extends from about 860 MHz to 960 MHzFind the corresponding wavelength in meters. Use c (speed of light) $= 3 \times 10^8$ meters/second.

    _____ at 860   _____ at 960

2, A typical commercial tag might have an antenna that is 9.5 cm long. How big a fraction of the wavelength is this at 860 MHz? 960 MHz?

    _____ at 860   _____ at 960

Voltage and power, path loss:

3. FCC regulations allow a reader to transmit 1 watt. What is the power in dBm?

    _____ dBm

4. A tag receives 20 microwatts of power from a 1-watt reader. What is the path loss in dB?

    _____ dB

5. Assume the reader in problem (2) above uses a perfect isotropic antenna and the tag antenna has an effective area of 50 square centimeters. What is the tag-reader distance?

    _____ meters

6. Let's give the reader a directional antenna. Assume the gain of the reader antenna is 6 dBi, and that of the tag antenna is 1.5 dBi. How much power does the tag receive at the distance you found in problem (3) above, assuming a frequency of 924 MHz? What is the EIRP of the reader?

    _____ microwatts _____ dBm EIRP: _____ dBm

CHAPTER 4

UHF RFID Readers

4.1 A Radio's Days (And Nights)

An RFID reader is, at heart, a radio transceiver: a transmitter and receiver that work together to communicate with the tag. As such, it faces the same challenges all radios encounter, plus a few specialized problems unusual in wireless communications but well-known to practitioners of that other passive communications technology, radar.

Every radio transmitter must deliver:

- *Accuracy*: the transmitter must accurately modulate the carrier frequency with the desired baseband signal, and maintain the carrier at the desired frequency.
- *Efficiency*: the transmitter must deliver this undistorted signal at the desired absolute output power without wasting too much DC power. The final amplifier of the transmitter is often the single largest consumer of DC power in a radio.
- *Low spurious radiation*: distortion of the transmitted signal can lead to radiation at frequencies outside the authorized bands, which potentially can interfere with licensed users and is frowned upon by most regulatory authorities. (We'll discuss in more detail how this *spurious* output arises in section 3 of this chapter.) Production of clean, *spur*-free signals is often a tradeoff between the amount of RF power to be transmitted and the amount of DC power available for the purpose.
- *Flexibility*: the transmitter should turn off when not in use to save power and avoid creating a large interfering signal, and turn back on again quickly, so as to be responsive when there are tags to be read.

Any radio receiver needs to provide:

- *Sensitivity*: a good radio must successfully receive and interpret very small signals. The ultimate limit on radio sensitivity is thermal noise. In a 1 MHz bandwidth, the thermal noise at room temperature is about -114 dBm or about 4 femtoWatts (4×10^{-15} W). This is much smaller than the received signal power from a passive tag at typical ranges (see e.g. figure 3.30). As we will learn in sections 4 and 5 below, other factors can degrade the sensitivity of an RFID receiver, but in many cases passive tags are forward-link limited and radio sensitivity is less important than in many other communications systems (though continuing improvement in tag IC's will make the reverse link more important as time goes on.) On the other hand, if we use semi-passive tags, the reverse link becomes the limiting factor, and good receive sensitivity is of paramount importance.

- *Selectivity*: an RFID radio needs to detect the tag signal in the presence of often vastly more powerful *interferers*. In a facility with many RFID readers operating simultaneously, the signals from other readers are likely to be much larger than the signals from the tags the reader is trying to communicate with. In addition, other sources of RF radiation such as cellphones and cellphone basestations, cordless phones, and older local area network devices operate in the same or nearby frequency bands. The receiver needs to reject signals outside the channel it is trying to receive, even if they are large compared to the wanted signal.

- *Dynamic range*: the same reader must receive and interpret signals from a tag 3 meters from the antenna and a tag 30 cm away — approximately a factor of 10,000 difference in received power. Much greater demands are placed on the receiver if a semi-passive tag is to be read at distances of tens of meters.

- *Flexibility*: In passive RFID procotols, the transmitter sends an amplitude-modulated signal and then transmits CW while it awaits a tag response. The receiver must recover quickly from any disturbance resulting from the portion of the modulated signal that leaks into it, in order to hear the small tag response.

An RFID reader radio also has to deal with special challenges. Most RFID readers operate **unlicensed**: that is, the user of the RFID radio does not require a license from a regulatory body to operate the reader, nor does the seller of the reader own a specific dedicated frequency band solely for their readers. Instead, the reader radio must obey certain restrictions in addition to those imposed on any device. In the United States, radio transmitters and receivers are regulated by the **Federal Communications Commission** (FCC). A radio operating in the unlicensed ISM band from 902–928 MHz is required to either employ wideband "digital modulation", or to hop from one frequency channel to another within the band. US RFID readers are for the most part frequency-hopping devices. Such a reader must execute a frequency hop to a different channel within the ISM band no less often than once every 0.4 seconds. In a high-speed reading application, it is important to make this hop as rapidly as possible so as to minimize disruption to tag reading.

In Europe, national regulatory bodies generally follow the recommendations of the European Telecommunications Standards Institute (ETSI). ETSI's recommendation EN302 208 provides a specific set of four high-power operating channels for unlicensed RFID operation, in the range from 865–868 MHz, and specifically requires that such readers listen to a given frequency channel before transmitting on it. ETSI also places stringent requirements on spurious radiation from RFID transmitters. (See Appendix 1 for a more detailed discussion of regulatory requirements, or if you are experiencing an attack of insomnia.)

The EPCglobal Generation 2 standard imposes special requirements on the spectral width of radiation from an RFID reader that is to be certified for operation in *multiple interrogator* or *dense interrogator* environments. These requirements are generally more stringent than those required by the regulatory bodies cited above, though a vendor is under no legal obligation to seek such certification to sell its products.

RFID radios are also required to operate in an unusual manner that puts some special demands on their design. Radios are generally described as either *half-duplex* or *full-duplex*, depending on whether they transmit and receive in sequence, or are able to do both simultaneously. Full-duplex radios almost always use separate frequency bands for transmitting and receiving; for example, cellular phones in the United States transmit between 825 and 849 MHz, and receive signals in the band from 869–894 MHz. Half-duplex phones (using the *TDMA* or *GSM* protocols) switch their transmitters off when it is time to receive a signal from the basestation. Those phones that operate in full-duplex mode (typically those that employ *code-division multiple access*, CDMA) use a special filter called a *duplexer* to allow the received signal to pass into the receiver while directing the transmitted signal to the antenna.

An RFID reader communicating with a passive or semi-passive tag must operate in full-duplex mode, in the sense that it must transmit CW for the tag to backscatter while listening for the tag response, but since the tag signal is at essentially the same frequency as the reader's own signal, a duplexer cannot be used to remove the transmitted signal. Leakage from the transmitter to the receiver can be an important limit on receiver sensitivity. This leakage can be minimized external to the radio itself by using separate antennas to transmit the reader signal and receive the tag signal; borrowing from radar terminology, such an arrangement is often known as a *bistatic* configuration (Figure 4.1).

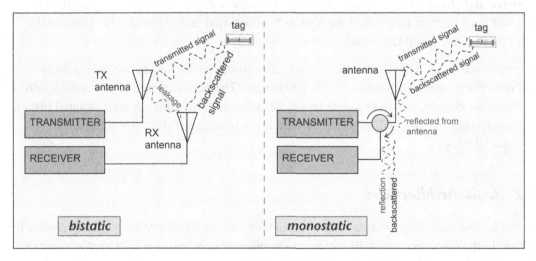

Figure 4.1
Bistatic and monostatic antenna configurations.

Such a configuration can ensure that very little of the transmitted signal enters the receiver, if the antennas are properly positioned and no near-antenna obstacles are present to scatter transmitted radiation into the receiving antenna. However, the use of two antennas involves additional size, complexity, and expense, and is obviously impractical in some applications, such as a handheld reader. Alternatively, a single antenna can be used for both transmission and reception: a *monostatic* configuration. In this case, it is likely that the receiver input will be exposed to a substantial signal from the transmitter, due at least to reflection from the antenna, so the receiver must be designed to detect the tag signal despite the incursion of unwanted transmitted leakage.

Let us briefly examine the magnitude of the quantities involved for the monostatic case. The transmitted signal, as we have seen, is typically limited by regulatory considerations; in the US, it might be 1 watt or 30 dBm. The backscattered signal from a passive tag is essentially determined by the power required to turn the tag IC on, and for a tag a few meters away ends up being about -50 to -70 dBm at the reader (figure 3.30). A good-quality antenna will reflect about 3% of the power incident on it: such an antenna is said to have a *return loss* of 15 dB. If the transmitted power is 30 dBm, the reflected power is then $(30 - 15) = 15$ dBm. This is about $(15 - (-50)) = 65$ dB higher than the tag signal for tag ranges of a few meters. The receiver must be able to deal with an unwanted reflected power over a million times larger than the tag signal in order to read the tag!

These numbers are generally improved for the bistatic case, as $40-50$ dB of isolation between antennas is not unreasonable to achieve. The leakage signal from the transmitter might be $(30 - 40) = -10$ dBm: still 40 dB ($10,000\times$) larger than the tag signal. Note, however, that if any conductive objects are close to the antennas, or they are not properly mounted and aligned, the isolation may be substantially degraded. In general, an RFID receiver must tolerate large *blocking* signals from internal and external reflections while still detecting the small tag signal.

A semi-passive tag, which uses a battery or other power source to operate the circuitry, can potentially operate at ranges of $50-100$ meters. The backscattered signal, which falls at 40 dB per decade, will in this case be on the order of -100 dBm. A well-isolated bistatic antenna configuration is indispensable to take full advantage of the capabilities of semi-passive tags.

4.2 Radio Architectures

In this section we will provide a brief overview of the two basic alternative designs used for radios. In the discussion, we will allude to a number of concepts that will be discussed in more detail later in the chapter; the reader who is at times puzzled is encouraged to charge bravely on in the hopeful expectation that later sections will provide clarification.

As we saw in chapter 3, data is usually sent over a radio link by imposing a slow modulation on a fast carrier signal. In the case of a typical RFID reader, the transmitted data rate is typically less than 100 Kbps, limited largely by the amount of frequency spectrum allocated to the reader. Somewhat faster modulations are often used by the tags: up to 640 Kbps for EPCglobal Generation 2 tags. All these signals are much lower in frequency than the carrier signal, typically around 900 MHz. These low frequency signals that describe the way the carrier is to be modulated are generally known as *baseband* signals. One of the key tasks for a radio is to impose this desired baseband modulation on the transmitted signal, and to extract if from the received signal. These operations are collectively referred to as *frequency conversion*.

There are two basic frequency-conversion architectures: *direct conversion*, also known as *homodyne*, and the multiple conversion or *heterodyne* configuration. Direct conversion schemes proceed directly, as suggested by their name, from the baseband signal to the radio frequency (RF) signal and back again. Heterodyne methods use an *intermediate frequency* (IF) in between the carrier and baseband frequencies. In Figure 4.2 we depict schematically these alternatives for the receiver, in which case these are known as *downconversion* operations. The analogous transmitter case is obtained by flipping the arrows around: *upconversion*.

The heterodyne architecture was invented by the remarkable American engineer Edwin Armstrong around 1917. It was the dominant approach to the design of radio transmitters and receivers throughout the 20th century, and continues to be important today in many applications. Its popularity stems from the tremendous flexibility that is obtained when multiple conversion steps are used. For example, it is always necessary to filter out signals at frequencies other than the wanted channel (selectivity). Filtering of a channel of a given

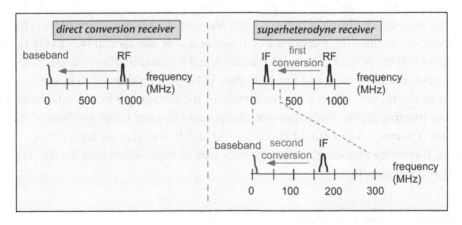

Figure 4.2
Frequency conversion approaches.

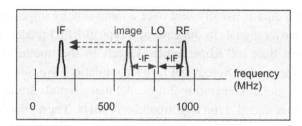

Figure 4.3
Both the wanted RF signal and the image are converted to the same IF.

width gets harder as we go up in frequency. To select a channel 500 KHz wide at 180 MHz (a plausible value for the IF of a reader receiver), we need a filter with a bandwidth of about 3% of the center frequency. To perform the same trick at the original carrier frequency of (say) 915 MHz, the filter must have a window only 0.5% of the center frequency: a very difficult object to fabricate. Furthermore, by adjusting the frequency of the *local oscillator*[1] (LO), which is fairly easy to do as we'll see in section 3, we can arrange for the IF signal to be at the same frequency no matter what RF channel we're trying to receive. This precaution allows us to use a fixed-frequency IF filter (easy) instead of a variable-frequency RF filter (hard) to reject unwanted signals.

On the other hand, the superheterodyne architecture introduces some special problems. As we will see when we cover mixers and frequency conversion in section 3, the conversion process depends on the difference between the frequency of the local oscillator and the input signal (the RF signal when receiving, or the IF signal when transmitting). For example, we can use a local oscillator frequency of 735 MHz to convert an RF frequency of 915 MHz into an IF at 180 MHz: $(915 - 735) = 180$. However, an input signal at 555 MHz will also be converted to the same IF: $(735 - 555) = 180$. The unwanted but converted frequency of 555 MHz is an *image* frequency (Figure 4.3). A similar problem exists when transmitting. Images are always a potential problem in heterodyne architectures. The designer must either filter out the offending frequencies, or use special (relatively complex and expensive) mixer designs that reject the unwanted bands. Filtering is easier if the separation between the wanted and image signals is large, tempting the designer to use a large value of the IF, but a high IF vitiates many of the advantages of the superheterodyne architecture: filtering and amplification are cheaper and simpler for low values of the intermediate frequency. The optimal choice of LO and IF frequencies for a given application, *frequency planning*, is an important part of superheterodyne design, involving a complex set of tradeoffs.

---

[1] The local oscillator is what its name implies: a signal source within the radio that generates a high frequency used to perform conversions. The frequency of the local oscillator is usually adjustable, enabling the radio to transmit or receive across a band of frequencies.

Direct conversion dispenses with messy intermediate steps and goes directly from baseband to carrier and back again. Since all the IF components are eliminated, direct conversion radios are generally more compact and often less expensive than their superheterodyne counterparts. Direct conversion transmission is relatively simple to implement and is popular in many radio systems. The complexity of implementing direct conversion for the transmitter depends on the modulation requirements. If an OOK signal (chapter 3 section 3) is all that is desired, one may simply interpose a switch between a continuously-operating local oscillator and the output amplifier to switch the output on or off. As we noted in chapter 3, symbols with abrupt edges tend to use more bandwidth than is strictly necessary, so it is usually desirable to employ an analog means of modulation, such as a variable attenuator (a device whose loss is continuously adjustable between a small and large value) instead of a switch. Alternatively, one can adjust the voltage or current available to the output amplifier to modulate the signal; this trick has been used for many years in AM radio transmission. More sophisticated modulation approaches must be used if the phase of the transmitted signal is also to be modulated.

A basic disadvantage of direct conversion relative to superheterodyne configurations is the problem of filtering: noise and spurious signals generated in the upconversion process can be more readily filtered at an intermediate frequency than at the final carrier frequency. When phase is also to be modulated, a direct conversion transmitter must control delays through the system with a precision comparable to the RF cycle time. For example, if we are transmitting a phase-modulated signal at 915 MHz we need to ensure that the output of the modulator is accurate to within a small fraction of the RF period of 1.1 nsec, whereas if we use a 180 MHz IF, the IF period of 5.5 nsec is much more forgiving of errors or misalignments.

The simplest sort of direct receiver uses a diode and capacitor to produce an output voltage proportional to the peak value of the input RF signal. This approach is known as an ***envelope detector***, since the output voltage follows the envelope of the radio signal. Envelope detection is cheap since no local oscillator is required; ***crystal radios***, available as children's toys in the author's youth, were essentially envelope detectors with an antenna and earphone. Envelope detection is rarely employed in reader radio designs because it has poor sensitivity and selectivity. However, this technique is widely used in passive tag ICs and will be discussed in more detail in that context in chapter 5.

Most direct conversion receivers use a local oscillator signal tuned to the same frequency as the incoming RF signal to convert the received signal to baseband. In consequence, direct conversion receivers avoid the problem of images: the wanted and image frequencies are identical. However, they encounter other challenges. Since there is no IF gain and RF gain is expensive, a very large amount of amplification is used on the baseband signal. Small DC offsets, which result from fixed-amplitude signals such as the reflected signal from the antenna in a monostatic system, may when amplified reach the limits of allowed input voltage to one of the amplifiers and cause it to turn fully on or fully off (in this case the signal is said to be *clipped*, undesirable except for audio amplifiers at concerts frequented by teenagers). An amplifier that is clipped, being e.g. fully off, is unaffected by the small additional signal from the tag, which is thus lost. Even when the amplifier is not completely saturated, its gain for small signals may be reduced – the radio becomes *desensitized* to the wanted signal. Offsets must be eliminated either by filtering or compensation to allow the wanted small signals to be amplified and recovered. In addition, at low frequencies many electronic components are much noisier than at typical IF and RF frequencies. Low-frequency noise sources in the ambient may also find their way into the radio.

Passive RFID readers encounter a special set of conditions that favor the use of direct conversion approaches for both transmission and reception. Readers usually employ variants of on-off keying for modulation of signals sent to the tag, so it is easy to implement direct conversion transmitters. Furthermore, since the reader must transmit a CW signal to provide something for the tag to backscatter, and the backscattered signal is always at essentially the same frequency as the transmitted signal, it makes sense to use part of the CW signal as a local oscillator to directly convert the backscattered signal from a tag into the baseband data signal. UHF RFID readers usually use a homodyne receiver. Like other direct conversion radios, RFID readers must deal with filtering challenges for the transmitter and offset problems for the receiver.

4.3 Radio Components

All radios must perform certain generic functions. Small signals must be made larger (*amplification*), high-frequency signals must be generated (*oscillation*), signals at different frequencies must be combined to create new frequencies (*mixing*), and signals at wanted frequencies must be accepted while other frequencies are rejected (*filtering*). The components that perform these functions are unsurprisingly known as amplifiers, oscillators, mixers, and filters. In modern radios, signals must always be converted to and from digital form, though in RFID radios it can a bit grandiose to assign the terminology *analog-to-digital conversion* to the simple comparators and switches that are often sufficient for these purposes.

In Figure 4.4 we introduce common symbols for these components, and depict how each component modifies the signals it encounters. Depending on the behavior of the reader antenna, the signal entering the reader may be a mixture of a wide variety of frequencies due to many sources of RF energy at differing frequencies and power levels (1). The band select filter ideally removes most of the signals outside of the band of interest (e.g. 902–928 MHz for RFID in the US), leaving signals from one or more transmitters at RF frequencies in the wanted band (2). These signals are optionally amplified (3) and then

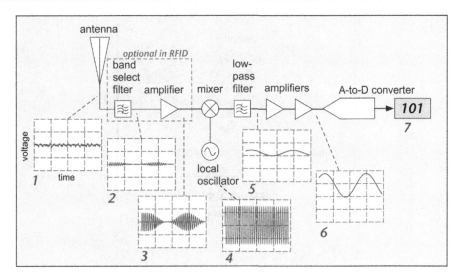

Figure 4.4
A generic direct conversion receiver, with the signal schematically depicted at each stage.

presented to a mixer, where they are mixed with a CW signal from the local oscillator at a constant frequency and amplitude (4). The result, after low-pass filtering (which removes frequencies above some cutoff, chosen well below the carrier frequency so that the remaining RF and harmonics are eliminated), is a low-frequency signal whose amplitude reflects the average signal strength of the high-frequency signal: the signal envelope (5). This low-frequency signal is amplified (6) and then converted by an analog-to-digital converter (*ADC*) into a bitstream describing the baseband signal (7).

As we discussed in chapter 3, passive RFID tags are usually limited by the amount of power received from the reader: as long as the tag is operating, it is relatively close to the reader and able to provide a fairly large return signal. Thus the RF amplifier may not be needed, and may be undesirable in a monostatic system where it amplifies the antenna reflection and thus increases the likelihood of saturating the mixer. We will also find that practical filters can't select the narrow RFID bands without including some nearby frequencies, so it may be better to use very linear RF components and filter the signal after conversion to the baseband. In order to understand how these choices of architecture are made, we need to examine how the components in a radio work: the task of the next several sections.

4.3.1 Amplifiers

Amplifiers are described by five key parameters: gain, power, bandwidth, distortion, and noise. Let us examine each of these properties.

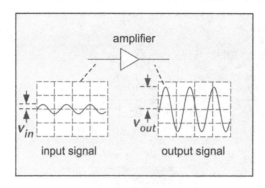

Figure 4.5
Amplifiers amplify an input signal.

4.3.1.1 Gain

The basic purpose of an amplifier is to make a signal bigger, so the most important and fundamental parameter describing an amplifier is the **gain** (Figure 4.5). Gain can be measured and reported in several different ways. The **voltage gain** is the ratio of the magnitude of the output voltage to the input voltage:

$$G_v = \frac{v_{out}}{v_{in}} \tag{4.1}$$

The **power gain** is the ratio of the output to the input power. By reference to chapter 3, section 2, we find that the power gain is the square of the voltage gain (assuming the input and output have the same load resistance R):

$$\left.\begin{array}{l} P_{in} = \dfrac{v_{in}^2}{2R} \\[2.5ex] P_{out} = \dfrac{v_{out}^2}{2R} \end{array}\right\} \rightarrow G = \frac{P_{out}}{P_{in}} = \frac{\frac{v_{out}^2}{2R}}{\frac{v_{in}^2}{2R}} = \frac{v_{out}^2}{v_{in}^2} = G_v^2 \tag{4.2}$$

Naturally, gain can be reported in dB. Because of the way deciBels are defined, this number is the same whether we are reporting the power or voltage gain:

$$G_{dB} = 10\log\frac{P_{out}}{P_{in}} = 20\log\frac{v_{out}}{v_{in}} \tag{4.3}$$

Amplifiers that operate at microwave frequencies typically provide 10−20 dB of gain (a factor of 10−1000 increase in the input power). Once the signal has been converted to the baseband frequency range of less than 1 MHz for most RFID systems, gain of 30−50 dB is readily available from a single amplifier stage.

Gain is important because the signals of interest may be small. For example, in an RFID radio receiver, recall from chapter 3 that the received signal from a tag may be as small as -60 dBm when a tag is a few meters from the transmitter (figure 3.30). This amount of power corresponds to a peak voltage of around 0.3 mV. To digitize this signal easily we might like it to have a magnitude of around (say) 0.3 V, so we need a system voltage gain of 1000 or 60 dB to get the input signal big enough to conveniently convert. We will need more gain than that to make up for the losses encountered in the filters and the mixer, so a total of 90–100 dB of gain might be present in the radio chain of an RFID receiver.

4.3.1.2 Power

The second important parameter for an amplifier is the ***maximum output power***, often denoted P_{sat} indicating that the output of the amplifier is saturated and can increase no further. Clearly, when the amplifier is saturated, it has no gain: a small change in the input voltage or power has no effect on the output power, which is as high as it can get. A closely related parameter frequently employed is the 1-dB-***compressed*** power P_{1dB}: this is the input power at which the gain is 1 dB less than measured at very small input powers. These quantities are depicted in Figure 4.6.

In many types of radios, the designer need give little thought to power handling in the receiver. However, in RFID readers, particularly when using a monostatic antenna, a large reflected signal is always present, and the designer must ensure that this signal does not saturate the amplifiers in the receiver and prevent the tag signal from being read. We will have more to say on this issue in section 5 below.

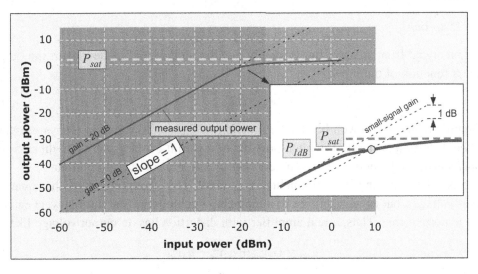

Figure 4.6
Output power vs. input power (logarithmic scales) illustrating saturated and 1-dB-compressed output power.

In designing an RFID transmitter, the output power rating of the final output amplifier determines the maximum possible power the reader could provide. As we've seen in chapter 3, the reader power is a key factor in determining the read range for passive tags, so the output *power amplifier* (PA) must be rated for sufficient output power to provide as much signal as the local government allows. Due to these high-power requirements the PA is often the single largest user of DC power in a reader.

4.3.1.3 Bandwidth

The **bandwidth** of an amplifier is the range of frequencies it can amplify. Silicon integrated circuit amplifiers with substantial gain up to frequencies over 1 GHz are commercially available, and bandwidths of tens of MHz are provided by any modern transistor technology. As we will see, in an RFID receiver, most or all of the gain will be placed in the baseband section and requires quite modest bandwidths. The transmitter amplifiers must operate at the full UHF frequency. Appropriate components are no longer a major challenge, particularly due to the huge market for components used in cellular telephony, which operates in similar frequency bands.

It is important to note that, while sufficient bandwidth is good, excess bandwidth may not be. An amplifier with gain up to frequencies much higher than the intended frequency of operation may be susceptible to unintended parasitic oscillations, due to accidental feedback paths within the circuitry. Such oscillations lead to mysterious and frustrating problems such as sporadic discontinuous changes in output power or gain, and if the resulting signals escape from the radio, they may cause the system to fail to meet regulatory standards for out-of-band emissions.

4.3.1.4 Distortion

The output signal from an amplifier is ideally just the input signal made bigger: the output is a linear function of the input:

$$v_{out} = G_v v_{in} \qquad (4.4)$$

Real amplifiers are not so well-behaved: **distortion** of the output signal is always present. When the distortion is small, it is often possible to regard it as the result of slight curvature or **non-linearity** in the output. The simplest way to add a bit of curvature to the characteristic of equation (4.4) is to include a quadratic term (proportional to the square of the input voltage), but as we shall see in a moment, a cubic term (the third power) can also play an important role. Thus, a real amplifier with distortion has an output voltage like:

$$v_{out} = G_v v_{in} + a_2 v_{in}^2 - a_3 v_{in}^3 \qquad (4.5)$$

where the a's are small for a good-quality linear amplifier. The resulting curvature of the relationship between output and input voltage, the **transfer characteristic**, is shown in greatly exaggerated form in Figure 4.7. (The a's can be of either sign, depending on the amplifier; the signs above are chosen to be consistent with the figure.)

The effect of polynomial distortions becomes apparent when we examine the effect of such a distorted amplifier on the frequencies present in a signal. Recall from chapter 3 (Figure 3.5) that a simple modulated carrier can be regarded as two frequencies, one above the carrier and one below:

$$v_{\text{mod}} = v_s \cos(\omega_{hi}t) + v_s \cos(\omega_{lo}t) \tag{4.6}$$

If such a signal is amplified by an ideal linear amplifier, all that happens is that the value of v_s increases. However, something more subtle and often more troubling occurs when distortion is present. Let us first examine the case of a second-order (quadratic) distortion ($a_3 = 0$ in equation (4.5)). The output signal is:

$$v_{out} = \underbrace{G_v v_s[\cos(\omega_{hi}t) + \cos(\omega_{lo}t)]}_{\text{linear part}} + \underbrace{a_2 v_s^2[\cos(\omega_{hi}t) + v_s\cos(\omega_{lo}t)]^2}_{\text{distortion}} \tag{4.7}$$

If we multiply out the 'distortion' part we obtain:

$$a_2 v_s{}^2[\cos(\omega_{hi}t) + \cos(\omega_{lo}t)]^2 = a_2\{\cos^2(\omega_{hi}t) + \cos^2(\omega_{lo}t) + 2\cos(\omega_{hi}t)\cos(\omega_{lo}t)\} \tag{4.8}$$

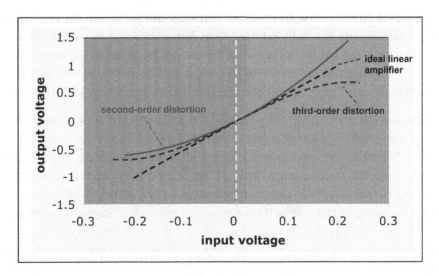

Figure 4.7

Comparison of transfer characteristics (output vs. input) for ideal amplifier and simple polynomial distortions.

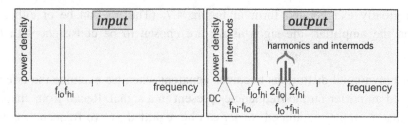

Figure 4.8
Spectrum of a two-tone signal before and after second-order distortion.

Now we employ some trigonometric identities (see Appendix 2):

$$\cos(a)\cos(b) = \frac{1}{2}\cos(a+b) + \frac{1}{2}\cos(a-b)$$

$$\cos^2(a) = \frac{1}{2}\cos(2a) + \frac{1}{2}$$

(4.9)

to obtain:

$$a_2 v_s{}^2[\cos(\omega_{hi}t) + \cos(\omega_{lo}t)]^2$$

$$= a_2 v_s{}^2 \left\{ 1 + \frac{1}{2}[\cos(2\omega_{hi}t) + \cos(2\omega_{lo}t)] + \cos([\omega_{hi} + \omega_{lo}]t) + \cos([\omega_{hi} - \omega_{lo}]t) \right\}$$

(4.10)

The distorted signal contains **new frequencies** that were not present in the original signal. These new frequencies are generally referred to as **harmonics** and **intermodulation products**. In this case, the harmonics are at twice each of the input frequencies, and there are **intermods** at the sum and difference of the original frequencies, as well as a constant term ($=$ zero frequency). These new frequencies are shown pictorially in Figure 4.8.

To clarify how this happens, the transformation of equation (4.10) is illustrated in Figure 4.9 for the simplified case of a single input frequency. When the input signal is 0, the output is 0, but when the input swings either high or low, the output voltage is always positive. This means the average output voltage must be positive, even though the average input voltage was 0: the quadratic distortion introduces a DC offset. A half-cycle of the input signal with voltage greater than 0 (towards the right in the diagram) produces exactly the same result as a half-cycle with voltage less than 0 (towards the left), so each half-cycle of the input must result in a full cycle of the output: that is, the frequency of the input has been doubled.

The good news about second-order distortion is that it is often possible to use filters to remove the distortion products, because their frequencies are very different from the input

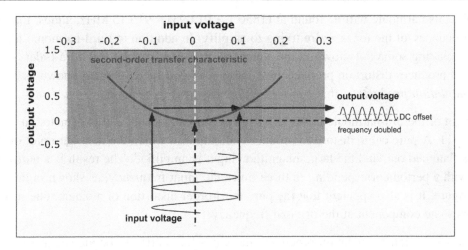

Figure 4.9
A sinusoidal input to a pure second-order transfer characteristic.

frequency. For example, if we imagine the input signal to be a 915-MHz carrier modulated at 50 KHz (so that the frequencies of the tones are 915.05 and 914.95 MHz), the harmonics will be at 1830 MHz, 1830.1 MHz, 1829.9 MHz, and 100 KHz, in addition to a DC component. It is easy to create a filter that allows the amplified signal at 915 MHz to pass through while rejecting signals at twice that frequency, and at very low frequencies.

What about third-order distortion? Here the situation is both more complex and less amenable to correction. The mathematics is similar to that of second-order distortion but noticeably more laborious, so let's focus on the most important part. When we multiply out the cube of the input voltage, we will get terms like:

$$a_3 v_s{}^3 [\cos(\omega_{hi}t) + \cos(\omega_{lo}t)]^3 = a_3 v_s{}^3 \{\ldots + \cos^2(\omega_{hi}t)\cos(\omega_{lo}t) + \ldots\} \qquad (4.11)$$

We can expand the squared part and use the product identity (equation (4.9)) twice to find:

$$\cos^2(\omega_{hi}t)\cos(\omega_{lo}t)$$

$$= \frac{1}{2}\cos(\omega_{hi}t)\{\cos([\omega_{hi}+\omega_{lo}]t) + \cos([\omega_{hi}-\omega_{lo}]t)\}$$

$$= \frac{1}{4}\left\{ \underbrace{\cos([2\omega_{hi}+\omega_{lo}]t)}_{2f_{hi}+f_{lo}} + 2\cos(\omega_{lo}t) + \underbrace{\cos([2\omega_{hi}-2\omega_{lo}]t)}_{2f_{hi}-f_{lo}} \right\} \qquad (4.12)$$

The harmonic at $(2f_{hi}+f_{lo})$ is again much higher than the fundamental frequency; using the same tones as before, this frequency would be $(1830.1 + 914.95) = 2745.05$ MHz, easily removed by a filter. However, the output at $(2f_{hi}-f_{lo})$ is a much tougher problem. In our example, this harmonic will be found at $(1830.1 - 914.95) = 915.15$ MHz, almost identical to the frequency of the tones we're trying to amplify. In addition to third-harmonic-like terms (including some not shown in the equations above for simplicity), third-order distortion produces distortion products near the input frequencies – often known as *intermodulation products* – that *cannot be filtered out* (Figure 4.10).

The origin of the 3rd-harmonic terms is fairly easy to understand, and is depicted in Figure 4.11. A pure cubic distortion produces an output in which the near-zero part of the input is flattened out, and the large-magnitude input is amplified. The result is a narrowed bump with a period corresponding to three times the input frequency, as shown in the inset of the figure. It is also apparent that the pure third-order distortion of a single tone still contains some component at the original frequency.

The near-carrier intermodulation products (intermods for short) can be viewed as the result of third-order distortion of the modulation signal. Remember that the sum of two tones is

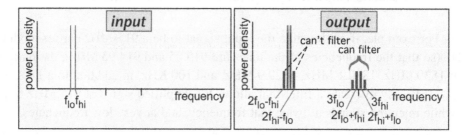

Figure 4.10
Spectrum of a two-tone signal before and after third-order distortion.

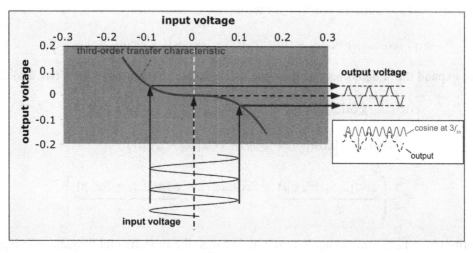

Figure 4.11
A sinusoidal input to a pure third-order transfer characteristic.

the same as the product of a low-frequency modulation and a high-frequency carrier (equation (3.11), repeated below for convenience as equation (4.13)). The modulation frequency is equal to half the difference between the frequencies of the two tones.

$$V(t) = \underbrace{\cos(\omega_m t)}_{\substack{\text{slowly-varying} \\ \text{modulation}}} \cdot \underbrace{\cos(\omega_c t)}_{\text{carrier frequency}} = \frac{1}{2} \left\{ \underbrace{\cos([\omega_{hi}]t)}_{\omega_{hi} = \omega_c + \omega_m} + \underbrace{\cos([\omega_{lo}]t)}_{\omega_{hi} = \omega_c - \omega_m} \right\} \qquad (4.13)$$

If we imagine that the modulating signal is passed through a third-order distortion, it will acquire a component at three times the original modulating frequency, but this is exactly what the near-carrier distortion tones do:

$$(2f_{hi} - f_{lo}) - (2f_{lo} - f_{hi}) = 3(f_{hi} - f_{lo}) \qquad (4.14)$$

Amplifier distortion can be characterized by simply reporting the magnitude of one of the distortion tones, or the total distortion, at a given input power. However, it is somewhat more convenient to provide a parameter that allows the designer to estimate the distortion at any input power level. Thus amplifier linearity is often described in terms of distortion *intercepts*. Recall from equations (4.10) and (4.11) that the terms resulting from second- and third-order distortion scale as the square and cube of the input voltages, respectively. If this scaling were to be extrapolated to large input voltages, at some input power or voltage the distortion would become equal in magnitude to the linearly amplified signal. This concept is depicted for second-order distortion in Figure 4.12. In the figure we have plotted input and output power on a logarithmic scale using dBm, so that second-order scaling of

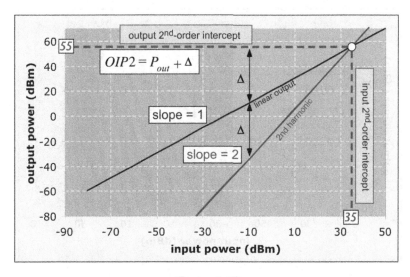

Figure 4.12
Definition of the second-order intercept.

the output power corresponds to a line of slope = 2, making it easy to locate the intercept point. The ***output second-order intercept*** (OIP2) is the output power at which the linear and harmonic components are of equal magnitude. We can also define an ***input*** second-order intercept (IIP2); the two differ by the gain of the amplifier.

The ***output third-order intercept*** (OIP3) is defined in an analogous fashion (Figure 4.13). Here the 3rd-harmonic contribution has a slope of 3 on a log-log (dB) scale. Once the intercept is given, the distortion power can be found at any input power by scaling from the intercept power. This operation is conveniently described in terms of backing off (away from) the intercept: at a ***backoff*** of 30 dB from the intercept point, the second-order distortion is reduced by 60 dB, and the third-order distortion is reduced by 90 dB. Thus for example, referring to Figure 4.12, at an output power of 15 dBm, we are backed off from the second-order intercept by $(55 - 15) = 40$ dB. The second-order distortion at this output power is $(55 - 80) = -25$ dBm. The distortion power is 40 dB below the linear output power. At the same output power, from Figure 4.13, the third-order backoff is 15 dB, and the third-order distortion power is $(30 - 45) = -5$ dBm, 30 dB below the linear output.

It is important to note that these intercepts are found by extrapolation of measurements at very low power. The intercept points are often at power levels exceeding the saturated output power of the amplifier, which means that other distortion effects would dominate the output signal long before the input power could be increased to the intercept point. An intercept is merely a convenient means for summarizing distortion data at varying power levels.

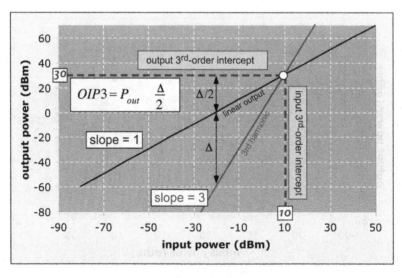

Figure 4.13
Definition of the third-order intercept.

Reported intercept data can also be misleading. As the output power nears the saturated power, higher-order distortion effects become important, leading to a complex response with the possibility that in some range of power, different distortions will cancel. If we happen to measure an amplifier at this power level, we will get a wonderfully optimistic idea of its linearity, which will not provide accurate guidance when the power level is changed. Intercepts should be measured by extrapolating distortion power measured at two or more differing input power levels, to verify that the distortion scales in the expected fashion, but this is not always done. Various possible definitions of the intercepts are also possible. As we have noted, one can quote either input or output intercept power, the values being different by the amplifier gain. The second-order intercept can be defined in terms of the second-harmonic power due to a single input tone, or due to a two-tone input: these approaches are conceptually similar but produce values that differ by 6 dB. Similarly, the third-order intercept can be defined in terms of the third harmonic of a single input tone, or the intermodulation products due two input tones, the latter being 9 dB lower. (In practice the intermodulation products are usually used to define OIP3.) Furthermore, one can define the intercept point as the power at which the power in an intermodulation tone equals the total input power, or at which the power in an intermod is the same as one of the tones in the input, a more modest 1-dB-distinction in this case.

In many applications, the most important distortion issue is the effect of third-order distortion on the near-carrier output spectrum. Recall from chapter 3 that the data rate of a reader is limited by the width of the transmitted spectrum; if we try to transmit too fast, we use so much bandwidth that we interfere with neighboring channels. Let us imagine that we have gone to the trouble of smoothing the signal to minimize the width of the transmitted spectrum (as in figure 3.10). If we amplify the resulting signal using an amplifier with substantial third-order distortion, the output spectrum will be wider than the input spectrum, because of the creation of additional modulation tones as in Figure 4.10. This effect, often known as *spectral regrowth*, is depicted qualitatively in Figure 4.14. Excessive distortion in a transmitter power amplifier will cause the transmitted spectrum to spill over into neighboring channels, and possibly also into neighboring bands, leading to interference with both other unlicensed and

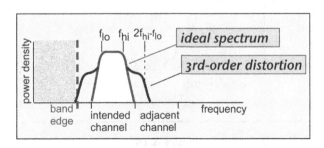

Figure 4.14
Third-order distortion results in increased bandwidth.

licensed users. The amount of power radiated into an adjacent channel, relative to that in the intended channel, is the ***adjacent channel power ratio*** (ACPR).

Distortion can also play a role in the receiver. Recall that readers in the US often divide the allowed band (902−928 MHz) into 500 KHz channels. Let us imagine that our reader happens to be transmitted and receiving at (say) 915 MHz, listening for the small signal backscattered from a tag. What if two other readers are transmitting at equally-spaced frequencies, for example 917 MHz and 919 MHz? Remember that it is difficult to filter individual channels at the carrier frequency, so these signals will be passed into the receiver and amplified (if an amplifier is present) prior to conversion to the baseband. If the receiver is perfectly linear, the interfering signals will be easily filtered out after conversion: the tag reflected signal will be at e.g. around 100 KHz, whereas the interfering readers are at 2 and 4 MHz relative to our carrier. However, what if the receiving amplifiers have third-order distortion? When we put in two tones at 917 and 919 MHz, the output will contain a new tone at $(2(917) - 919) = 915$ MHz − right on top of our wanted signal (Figure 4.15)! This interfering signal can no longer be filtered out and may prevent us from demodulating the tag signal. Good linearity (low distortion) in the receiver is particularly important in a dense-reader environment, where many interfering signals are likely to be present, or when nearby signals from other bands (e.g. in the US, cellular telephony basestations at 869−894 MHz) impinge on the reader.

4.3.1.5 Noise

Any dissipative electrical system creates a certain amount of electrical noise due to the finite thermal excitation of the electrons. This noise is closely related to the blackbody emissions produced by objects at a finite temperature, and has a similar frequency spectrum. At typical

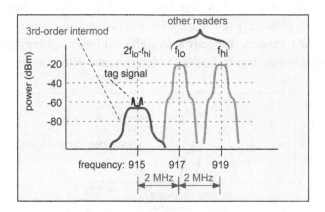

Figure 4.15
Receiver third-order distortion creates interfering signal from readers on distant but equally-spaced channels.

operating temperatures, the peak emission is in the mid-infrared, much higher than any frequency we are interested in, and so for microwave purposes the noise can be regarded as frequency-independent. For any source resistance R, the maximum noise power is delivered to a matched load of the same resistance, and in this case is independent of the value of the resistance:

$$N = kT[BW] \tag{4.15}$$

where k is Boltzmann's constant, 1.38×10^{-23} J/K, T is the temperature in Kelvin ($=°C + 273$), and [BW] is the bandwidth in Hz. At 300 K (a cozy spot by the fire at the ski lodge for those who can still afford such recreations), this is 4×10^{-21} W/Hz, or *−174 dBm/Hz*: a quantity worth memorizing if you will be thinking about radio operation a lot. To send 100 Kbps of data, a tag will use around 200 KHz of bandwidth. In logarithmic terms, the amount of noise competing with the tag signal in an ideal radio, limited only by thermal noise, will be $(-174 + 10\log(200,000)) = -121$ dBm. A quick check of e.g. figure 3.30 from chapter 3 shows that this amount of noise is much less than the signal we expect from a passive tag that is close enough to get turned on: thermal noise is of little consequence for a receiver designed for today's passive tags. However, a semi-passive tag may not be limited by the forward link, and the ability to decipher much smaller signals may be important.

Real amplifiers emit more noise than the ideal thermal limit. The noise from an amplifier is usually characterized by the ratio of the signal-to-noise ratio on the output of the amplifier to that on the input, known as the *noise factor F*:

$$F = \frac{\left(\frac{S_{in}}{N_{in}}\right)}{\left(\frac{S_{out}}{N_{out}}\right)} \tag{4.16}$$

Since the noise factor is always greater than 1 for a real device (the noise on the output can't be any less than the noise on the input for the same bandwidth), we can also report the noise factor in dB; this latter quantity is usually known as the *noise figure NF*:

$$NF = 10\log(F) \tag{4.17}$$

The noise figure is a convenient quantity, as it can be simply added to the noise floor from equation (4.15) in dB to find the equivalent input noise of the amplifier. The noise figure is always greater than 0 for real devices. Good-quality amplifiers designed for low noise performance have noise figures from 1 to 3 dB at around 1 GHz. Amplifiers designed for high power output may have noise figures of $10-20$ dB. Thus, for a *low-noise amplifier* (LNA) with a noise figure of 3 dB, the equivalent thermal noise level in the 200-KHz channel we are using to listen for a tag will be about $(-121 + 3) = -118$ dBm. Recall from chapter 3 that FM0 requires a (S/N) ratio of about 10 dB for reliable demodulation.

If thermal noise is the only noise source that is important, the tag signal needs to be about −108 dBm or larger to be decoded. Using equation (3.23) with a transmitted power of 1 watt at 915 MHz, a 6 dBi reader antenna, −5 dB tag modulation efficiency, and ignoring any tag antenna gain, we obtain a reverse-link-limited range of about 110 meters, greatly in excess of the range achievable with a forward-link-limited passive tag. Note that because of the very strong dependence of signal strength on distance $(1/r^4)$, this result is not very sensitive to the exact value of the noise figure or transmit power.

We will find that other noise sources typically limit the sensitivity of a monostatic RFID receiver to a greater extent than thermal noise. A bistatic receiver configuration with good transmit-receive isolation can produce thermal-noise-limited sensitivity, and is more appropriate for use with semi-passive tags.

4.3.2 Mixers

The operation of a mixer is rather more subtle than that of an amplifier, and deserves some discussion. The purpose of a mixer is to convert a signal from one frequency to another while preserving the modulation information contained in the signal. This operation is accomplished by mixing the wanted signal with a *local oscillator* (LO) signal that is not modulated − that is, the amplitude and phase of the LO are constant, so it contains no information of its own − as shown schematically in Figure 4.16.

In order to effect this transformation, some sort of nonlinear relationship between the input and output voltages is needed. While elementary treatments of mixer operation often begin with the same sort of polynomial approximations we used in treating amplifier distortion in section 3.2, practical mixers are more realistically approximated as switches, as shown in Figure 4.17. When the switch is on, the input signal is passed directly to the output.

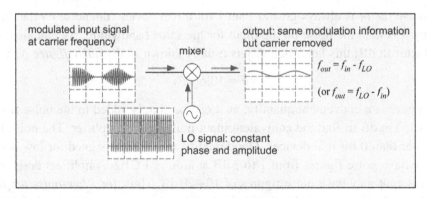

Figure 4.16
Schematic depiction of mixer functional operation. Note that filtering would normally be needed to remove the high-frequency component of the output (not shown here).

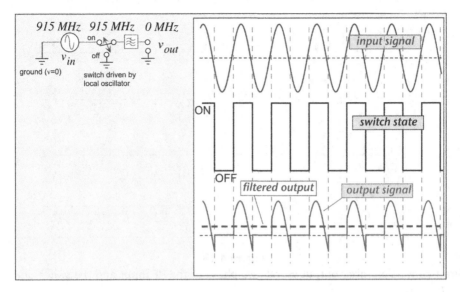

Figure 4.17
Operation of a simple switch mixer used in a direct conversion receiver.

When the switch is off, the output signal is 0. If the switch is turned on and off at the same frequency as the incoming signal, the output voltage will be a series of pulses of exactly the same shape. When we filter the output to remove the high-frequency (rapidly-changing) parts, or equivalently average the output voltage over a cycle or two, we obtain a constant output voltage. That is, we have converted an RF signal at (in this example) 915 MHz into a DC signal at 0 MHz using a switch (mixer) driven at 915 MHz.

It is important to note that the output voltage depends on the relative *phase* of the switch state and the input signal. If the RF input and the switch are exactly in phase, a maximum value of output voltage is obtained; if they are 90° out of phase, the output voltage is 0, and if they are out 180° out of phase a negative voltage is obtained (Figure 4.18). It is easy to demonstrate that the output voltage is proportional to the cosine of the difference in phase. If we use a direct-conversion mixer to receive a tag signal, and the tag signal happens to be 90° out of phase with the local oscillator, we get zero output signal even if the tag signal is large. We will discuss some solutions to this problem in section 5 below.

Since the output signal depends on the relative phase of the RF signal and the switch (which is the same as the local oscillator phase), if these relative phases change with time, so will the output signal. If the LO frequency is different from that of the RF signal, this is exactly what will happen: the relative phase will change with time, and so will the value of the filtered output voltage (Figure 4.19). In fact, it is easy to see that the rate of change of the relative phase corresponds to the difference between the two frequencies, and thus a

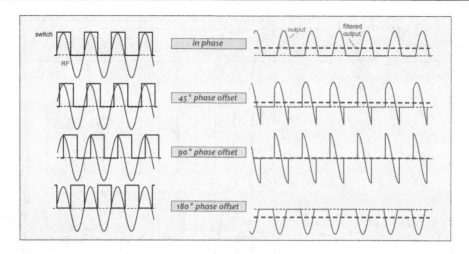

Figure 4.18
Direct-conversion mixer output vs. relative phase of the RF input and the switch states.

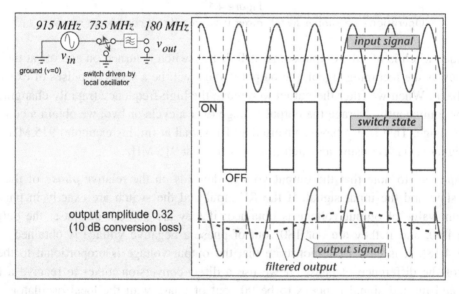

Figure 4.19
Operation of a simple switch mixer with differing RF and LO frequencies (heterodyne configuration).

mixer converts an RF input to an IF that is the ***difference between the RF and LO frequencies***, as we suggested in Figure 4.3.

$$\Delta(phase) = \omega_{RF}t - \omega_{LO}t = (\omega_{RF} - \omega_{LO})t$$
$$v_{out,filtered} \sim \cos(\Delta(phase)) = \cos((\omega_{RF} - \omega_{LO})t) = \cos(\omega_{IF}t) \tag{4.18}$$

Since the cosine is an even function its argument, $\cos(-x) = \cos(x)$, it doesn't matter whether the RF frequency is above or below the LO frequency. Therefore if we use a mixer to downconvert a received signal to a finite IF, both frequencies $f_{LO} + f_{IF}$ and $f_{LO}\text{-}f_{IF}$ will be converted to the IF. Either one of these signals could be the one we want to receive, and the other becomes the unwanted image frequency discussed in section 2. The case where we want the sum frequency is known as **low-side injection** (since the LO frequency is below that of the wanted signal); the configuration that accepts the difference is known as **high-side injection**.

A mixer can also be used to upconvert an information-containing signal (baseband or IF) from a low to a high frequency (Figure 4.20). In this case the mixer is being used as a **modulator**.

Real mixers are implemented using transistors or diodes as switching elements. The LO voltage is connected to the gate or base of the transistor, and switches it between the full-on and full-off states. Transistors have finite capacitance and switching speed, and the input voltage from the local oscillator is typically sinusoidal, so the transition takes a finite time, instead of being instantaneous as we have shown above. Furthermore, real transistors have some finite loss when on, and some leakage when off. Transistors also contribute distortion to the signals, since the conducting regions are not perfectly linear.

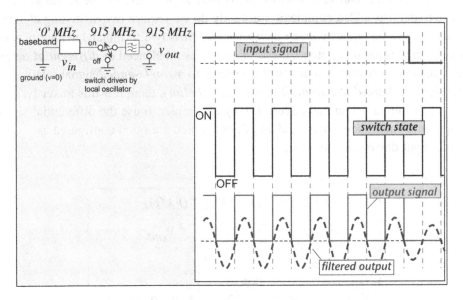

Figure 4.20

Simple switch mixer used to upconvert a low-frequency (baseband) signal to the carrier (RF) frequency. (The baseband signal is shown with a sharp transition for clarity, but real baseband signals would be filtered and change state slowly on the scale of the carrier.)

4.3.2.1 Mixer Parameters: Conversion Loss and Noise

It is easy to see that the filtered output voltage in Figure 4.18 is quite a bit smaller than the peak voltage, even when the signal and the switch are in phase. In fact, the largest DC output obtained is about 0.32 of the input peak voltage. The output signal of a mixer at the desired frequency is in general smaller than the input signal, even when (as is the case here) the mixer has no internal losses. Mixers have *conversion loss*. (An active mixer — a mixer that incorporates an amplifier — can also have *conversion gain*, though it is always less than that of the amplifier without the mixing function.) An output voltage of 0.32 corresponds to an output power (presuming the RF signal is modulated, so that the output is not quite at DC) of about 10% of the input power; the conversion loss of this mixer is about 10 dB.

The conversion loss depends on a number of factors. For example, the switch in the simple switched mixer used in the examples above is off half of the time, so the average voltage is half of what it would otherwise be. We could make a more efficient mixer by using two switches and a crossover network, so that the output could be connected to the input in either polarity (Figure 4.21). This configuration is known as a *balanced* mixer. Now the output is always connected to the input, but the polarity of the connection is reversed each cycle, so the average voltage is doubled and the average power increased by a factor of 4. This mixer has an ideal conversion loss of 4 dB instead of 10 dB for the single-switched mixer. The tradeoff is added complexity: not only do we require two switching elements instead of one, but the output voltage now appears across two terminals, neither of which is always connected to ground potential: this is known as a balanced or *differential* output voltage. It may be necessary to convert this voltage to a *single-ended* signal relative to ground, using a *balanced-unbalanced* transformer (*balun*), though if this mixer is incorporated within an integrated circuit, it may be simpler to use the differential signal as the input into subsequent filters and amplifiers, which are often configured as differential-input devices in any case.

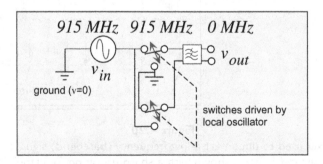

Figure 4.21
Two cross-connected switches can be used to make a balanced mixer with lower conversion loss.

It is often advantageous to also provide balanced connections to the local oscillator, in which case the mixer is said to be *double-balanced*, or even to both the input signal and local oscillator (a *triple-balanced* mixer).

Real mixer conversion loss is larger than that for ideal mixers due to the finite resistance of the switching elements in their ON state. The conversion loss is also dependent on the voltage of the LO signal, since this determines whether the switching elements are fully ON and fully OFF or only partially switched. Since higher LO voltages require more DC power and more expensive components, there is usually a tradeoff between LO voltage or power and performance. Real (passive) microwave mixers usually have conversion loss varying from about 6 dB to 11 dB.

Passive mixers generally don't add much noise of their own to the signal, fundamentally because the switching elements have little resistance in the ON state, and allow very little current in the OFF state. However, the thermal noise present in the output signal must be at least as large as that in the input if the mixer is well-matched to both the source and the load, while the output signal is reduced by the conversion loss. Therefore the output signal-to-noise ratio must be reduced by at least the conversion loss. The noise figure of a mixer is at least as large as the conversion loss — in practice a dB or two larger. Active mixers display conversion loss instead of conversion gain, but their noise figures are generally similar to those of passive mixers, because the underlying conversion loss is still present. Mixer noise figures of 8−13 dB are reasonable at microwave frequencies.

4.3.2.2 Distortion and Isolation

One of the reasons we have pictured a mixer as a switch is to simplify the discussion of mixer distortion, which is otherwise a bit puzzling: why should one worry about the non-linearity of a non-linear device? When we think of the mixer as a switch, the answer is clear: the switch ought to be completely on in the ON state and completely off in the OFF state, independent of the amplitude of the signals presented to it. If the output in the ON state is not just proportional to the input, but instead contains quadratic or cubic distortion terms (like the amplifier case discussed in connection with equation (4.5)), we can expect the same results obtained for amplifiers: harmonics and intermodulation products will appear in the output. We can define second- and third-order intercepts for mixers just as in the case of an amplifier, and use them to provide guidance about acceptable input amplitudes and distortion levels. Mixer distortion is of considerable importance for RFID readers, because of the inevitable presence of a large blocking signal from the transmitter, and the possibility of interfering signals that can't be readily filtered at RF frequencies. If the mixer has large third-order distortion, pairs of signals may mix to produce interferers at the baseband, just as in the case of an amplifier with third-order distortion (Figure 4.15).

If the OFF state isn't quite off, some of the input signal (e.g. the RF signal) will leak into the output. In addition, though we haven't shown this explicitly, a real switch such as a transistor has some coupling between the local oscillator input and the signal output, mostly due to stray capacitances present in the device and package. Leakage of the RF or LO signals into the IF port, and of the IF into the RF port when the mixer is used as an upconverter, are characterized by specifying the *isolation* of the mixer. Since there are three ports, we need to specify three isolation values (LO-IF, LO-RF, IF-RF). The local oscillator signal is normally much larger than the received RF signal, so LO-IF isolation is important in downconversion applications. Isolation is also improved by using balanced configurations. Isolation is not as critical in RFID radios as in many other applications, since direct conversion is usually used and the LO and RF signals are at the same frequency.

4.3.2.3 Spurious Output Frequencies

In our discussion of distortion, we saw that nonlinear amplification leads to output signals containing higher harmonics of the input signal, and intermodulation products involving not just harmonics but sums and differences of the input frequencies and integer multiples of the inputs. The whole purpose of a mixer is to produce new frequencies, so it should be unsurprising to find not just the wanted output, but other undesired – *spurious* – output frequencies, often known just as *spurs*. In fact, the output of a mixer can in principle contain every possible integer combination of the input frequencies:

$$f_{out} = nf_{RF} \pm mf_{LO}; \quad n, m \text{ integers} \tag{4.19}$$

The action of the switch can be regarded as multiplying the input signal by a square wave, and it can be shown that a square wave is made of the odd harmonics of the fundamental frequency, so the odd values of m are likely to be of particular significance. Like third-order distortion in amplifiers, mixer spurs can end up close to or on top of the wanted output signals, so filtering alone cannot be relied on to remove distortion products.

In addition to lowering conversion loss, balanced mixers can suppress some potential spurious frequencies by symmetry. For example, let us imagine that the switch mixer of Figure 4.17 introduces some quadratic (second-order) distortion, as shown in Figure 4.7. In the single-ended mixer, when the signal goes positive, second-order distortion will cause the output to be larger than it ought to be; when the signal goes negative the switch is off. Therefore there is an undesired offset. If instead the balanced configuration of Figure 4.21 is used, the negative half of the signal is transferred to the output through a second switch, and is a bit smaller in magnitude than it ought to be. The two distortions cancel when the output is averaged over a few cycles.

It is not trivial to calculate the amplitude of the numerous possible spurious outputs for any given mixer design and configuration. Mixer data sheets will often provide the measured power at each possible frequency at a specific set of conditions in a *spur table*, which can be compared against the specifications for a given design to set limits on allowed power levels.

4.3.3 Oscillators and Synthesizers

It should perhaps be apparent to the reader who has persisted this far that radios need oscillators: components that create a sinusoidal signal at a particular frequency. If the radio is to operate at more than one frequency, the oscillator also needs to be tunable. Finally, the absolute frequency of oscillation needs to be very precisely controlled to satisfy regulatory requirements and avoid unnecessary interference between collocated readers. How are these tasks accomplished?

To introduce the subject of oscillation let us first examine what happens if we connect the input of one amplifier to the output of another (Figure 4.22). In this figure, the little circle signifies that these are *inverting* amplifiers: that is, the output is the negative of the input voltage. As long as the input voltage to the first amplifier is 0, the output is 0, so the second amplifier also has 0 volts on input and output. However, let us imagine that some small voltage, e.g. from a bit of noise, appears on the input of the first amplifier; in the figure this is shown as 0.01 V, though the magnitude doesn't matter. This small input is amplified (by a factor of 10 in our example) and inverted, and then applied to the second amplifier, where it is again amplified by a factor of 10, and inverted again. The output of the second amplifier returns to the input of the first one as a quite substantial +1-volt signal, which will in turn be amplified and returned. It is easy to see that in some short time the output of this circuit will grow arbitrarily large and positive, until it is limited by something else, such as the power supply voltage. It is also easy to see that the same thing would happen if the initial voltage happened to be negative, save that the amplifiers would be driven to the most negative voltage available.

We can see that this simple circuit generates a signal from nothing, as it were, but the signal isn't very interesting, being fixed at either the maximum or minimum voltage once it

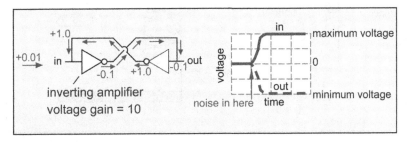

Figure 4.22
Cross-coupled amplifiers.

drifts away from 0. To make the signal do something more useful, we need to add a frequency-selective element: a ***resonant circuit***. An example of a resonant circuit is shown in Figure 4.23, along with a roughly analogous mechanical contrivance. The electrical circuit is composed of an ***inductor*** – a component in which a magnetic field is used to store energy – and a ***capacitor***, in which an electric field stores energy. The symbols for these components represent a simplified view of their construction: an inductor is often a coil of wire, and a capacitor is fabricated using pairs of closely-spaced metal plates. The capacitor is roughly analogous to a mechanical spring, in that the spring stores energy in displacement (charge ⇒ electric field), and the mass is like an inductor in that it stores energy in velocity (current ⇒ magnetic field).

The operation of this resonant circuit may be roughly understood by examining its mechanical analog. When the input force alternates very slowly (input frequency $f \ll$ resonant frequency f_{res}), the mass can readily follow the input and the spring feels no force and sits still (Figure 4.24(a)). The pivot point, which responds to the sum of the spring and mass displacements, moves with the mass. At very high frequencies, the inertia of the mass keeps it from moving much, but the spring is readily displaced, and the pivot point moves with it (Figure 4.24 (c)). However, at the resonant frequency, the spring and mass move with the same velocity, but in opposite directions. The pivot point does not move no matter how much force is applied. (Note that this discussion only applies for small displacements of the crossbar, where lateral motion can be neglected.)

The electrical resonator works in the same fashion (Figure 4.25). At frequencies well below resonance, current flows in the inductor when a voltage is applied. At high frequencies, current

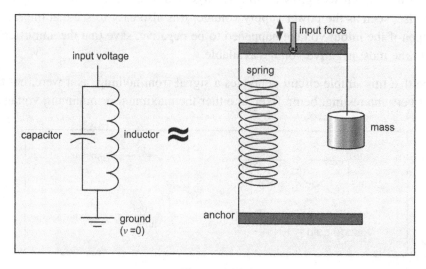

Figure 4.23
Parallel-resonant electrical circuit and mechanical analog.

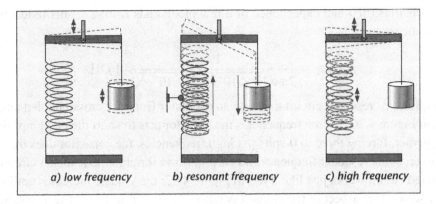

a) low frequency   b) resonant frequency   c) high frequency

Figure 4.24
Mechanical resonator response vs. frequency.

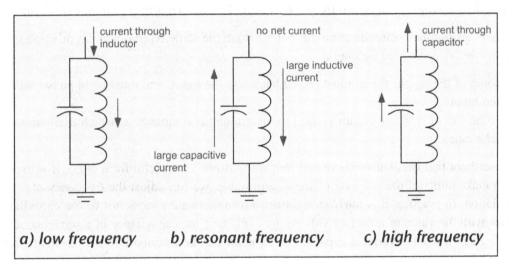

a) low frequency   b) resonant frequency   c) high frequency

Figure 4.25
Parallel resonant circuit current vs. frequency.

flows through the capacitor. However, at the resonant frequency, equal and opposite currents flow in the two components, and no net current flows into the circuit no matter how much voltage is applied. (In both mechanical and electrical cases, the actual structures have losses and other non-idealities, which allow some small residual motion / current even at resonance.) For reference, the resonant frequency of an oscillator of this type can be written as:

$$f_{res} = \frac{1}{2\pi\sqrt{LC}}$$ (4.20)

where f is the frequency in Hz, L is the value of the inductance in **Henries**, and C is the capacitance in **Farads**. In microwave applications, one more often deals with inductance of a few nanoHenries (nH) and capacitance of a few picoFarads (pF): a 12-nH inductor and a 2-pF capacitor resonate at

$$f_{res,12-2} = \frac{1}{2\pi\sqrt{12 \times 10^{-9} \cdot 2 \times 10^{-12}}} \approx 1 \text{ GHz} \qquad (4.21)$$

We can exploit this resonant circuit to create an oscillator from our cross-coupled amplifiers as shown in Figure 4.26. At low frequencies the inductor acts to short the two amplifier outputs together, forcing them to 0 volts. At high frequencies, the capacitor does the same job. However, at the resonant frequency, no current flows through the resonant circuit, so the cross-coupled circuit works just like it did in Figure 4.22: the voltage increases until it is limited by some other aspect of the circuit operation. However, now it is the alternating voltage at the resonant frequency that is large in magnitude: the circuit is acting as an oscillator, producing a CW signal at a particular frequency even when there is no input signal. This configuration is also frequently referred to as a ***negative resistance*** oscillator,

Many other oscillator circuits are used, but all share the same basic operating principles as the simple cross-coupled oscillator:

- some of the output signal must be fed back into the input, and must be in phase with the input;
- some sort of resonant circuit is used to determine the frequency at which oscillations take place.

An oscillator that oscillates only at one frequency is not very useful for a radio. However, if we make either of the resonant elements adjustable, we can adjust the frequency of oscillation. In practice, it is fairly straightforward to fabricate a capacitor whose capacitance varies with the value of a control voltage, by exploiting the capacitance of a semiconductor diode. Such a voltage-variable capacitor or ***varactor*** can be inserted into the resonant circuit

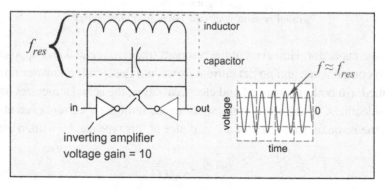

Figure 4.26
Cross-coupled oscillator.

of an oscillator to obtain a ***voltage-controlled oscillator*** (VCO). Virtually every modern radio contains one or more VCO's to create the transmitted signal and convert the received signal. In practical VCO's, the varactor may be combined with a bank of switched but fixed capacitors, to simultaneously provide a wide tuning range (coarse tune using the switched capacitors) and good tuning resolution (fine tune using the varactor) as depicted in Figure 4.27.

4.3.3.1 Phase Noise

Real circuits have losses, and as we noted in section 4.1 above, all lossy elements at a finite temperature emit thermal noise. What happens when a source of noise is present in an oscillator circuit?

A greatly simplified view of the situation is shown in Figure 4.28. We can regard the loss in the resonator as a resistor (although physically it may be due to resistance in the inductor, capacitor, or wiring). The real noisy resistor can be looked at as an ideal noiseless resistor and a noise current i_n. The interesting thing to realize is that if the oscillator is actually oscillating – producing a signal of constant amplitude at the resonant frequency – the action of the cross-coupled amplifiers must be just sufficient to cancel the effect of the resistor loss. (This is why this circuit is known as a negative-resistance oscillator.) In the figure, we can view the resistance R and the amplifier effective negative resistance $-R$ as exactly compensating for each other: thus no current flows into the bottom part of the circuit, and it can be ignored in the remainder of the analysis.

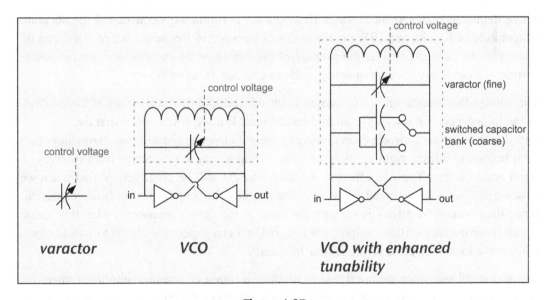

Figure 4.27
Constructing a voltage-controlled oscillator using a varactor and optional switched capacitors.

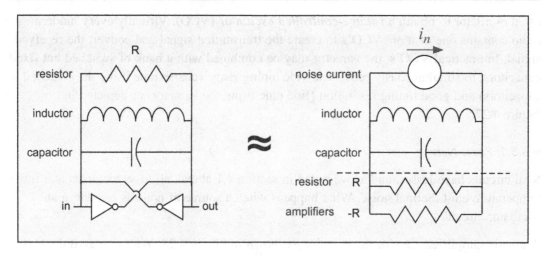

Figure 4.28
An oscillator with a lossy resonant load, and simplified equivalent circuit.

What we have left is a noise current source trying to make current flow through a parallel resonator. Recall that at microwave frequencies, thermal noise is broadband: about the same for all frequencies. So the noise current has the same average magnitude in each chunk of bandwidth; to be specific, it is:

$$\langle i_n^2 \rangle = \frac{4kT}{R}[BW] \tag{4.22}$$

where the brackets denote an average. Here again k is Boltmann's constant, T the absolute temperature in Kelvin, and [BW] is the width of the slice of frequency we're interested in. (Note that we have to consider the average of the square of the current, because the noise current is a randomly-varying quantity, with an average value of 0.)

The voltage that results when this current is introduced into the tank circuit is the product of the impedance of the tank circuit and the current. But we noted above that the impedance of the tank circuit grows larger as we get closer to the resonant frequency, in the limit becoming infinite right at resonance, producing a very large voltage from the very small noise current. Thus the effect of the noise current will be greatly amplified when we look very close to the resonant frequency, and diminish at frequencies far from resonance. Since the resonant frequency is roughly the same as the carrier frequency, what this means is that if we use an oscillator to produce a signal, we can expect the signal to contain lots of noise if we look very close to the carrier frequency.

In a real oscillator, since we are trying to produce a signal of constant amplitude, there is no reason to preserve any variations in amplitude produced by the oscillator, so the output of the oscillator is usually passed through some sort of limiting device to hold it constant.

Therefore the part of the noise that would lead to a variation in the amplitude of the output signal is suppressed, but the part of the noise that changes the frequency, or equivalently the phase, of the signal is not. Oscillators produce **phase noise**, particularly close to the carrier frequency. In an RFID radio, phase noise is of considerable importance, because the tag signal is at a frequency very close to that of the carrier. In real RFID readers, phase noise from the local oscillator can be the limit on the sensitivity of the receiver. We will examine how this arises in more detail in section 4.5 below.

It is straightforward to show that in our greatly-simplified approximation, the average value of the noise voltage at an angular frequency ω is:

$$\langle v_n^2(\omega) \rangle = 4kTR \left(\frac{\omega_c}{2Q[\omega_c - \omega]} \right)^2 [BW]; \quad Q = \frac{R}{\omega_c L} \qquad (4.23)$$

where ω_c is the carrier (resonant) frequency, and we've introduced a quantity Q, the **quality factor** of the resonator, which is equal to the ratio of the resistance and the impedance of the inductor L. When Q is large, a resonator is very nearly lossless, and responds in a narrow band of frequencies; when Q is small, the resonator loss is comparable to the energy stored in the resonator, and the circuit is hardly resonant at all. Typical values of Q for resonators built from inductors and capacitors at microwave frequencies are around $10-50$.

Since it is not the absolute noise but the amount of noise in proportion to the amount of carrier signal that matters, people generally measure this ratio. Thus phase noise is usually described in terms of the noise power in some bandwidth in deciBels relative to the carrier power, often written as **dBc**/Hz (although technically the bandwidth goes inside the logarithm). We usually plot the log of the relative phase (dBc/Hz) vs. the log of the offset of the measurement frequency from the carrier in Hz. If we divide the squared voltage in equation (4.23) by 2R we get the power; dividing by the signal power and taking the log, we find:

$$N_{phase}(dBc/Hz) = 10\log \left(\frac{2kT}{P_{sig}} \left(\frac{\omega_c}{2Q[\omega_c - \omega]} \right)^2 \right) \qquad (4.24)$$

We can see that when Q is large, the phase noise is small. If this phase noise is plotted versus the logarithm of the frequency, we would get a straight line with a slope of -2.

An example of reported phase noise for a commercial VCO is shown in Figure 4.29, plotted over the range of offset frequencies of interest for RFID. (Recall that a tag reply will typically be modulated at a few 10's to a few hundred KHz from the carrier.) The actual value of the noise is about -80 to -120 dBc/Hz, and the slope is close to the value predicted from our simple theory.

The phase noise characteristics of real oscillators are, however, often more complex than this, with a region close to the carrier that is noisier than expected from the simplistic

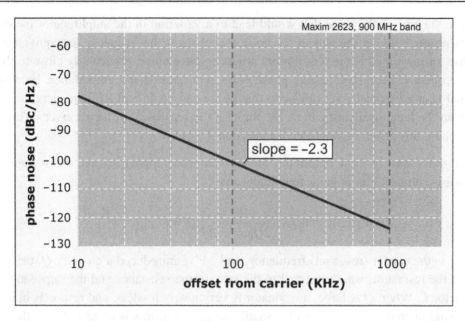

Figure 4.29
Log-log plot of reported phase noise for a commercial VCO operating in 900-MHz frequency bands.

model we have presented. The interested reader is referred to chapter 17 of Lee's book in Further Reading, at the end of this chapter.

4.3.3.2 Synthesizers

Having established that we can construct a tunable oscillator, how do we tune it to the right frequency?

If we're just trying to operate somewhere in the US ISM band (902−928 MHz) that might not be too much of a challenge. The precision required is (26/915) = 3%, so one could imagine that using 1% precision capacitors, inductors, and resistors, it would be straightforward to build an oscillator with the requisite accuracy − at least at room temperature. However, for example, the EPCglobal standard for second-generation passive tags requires the reader to operate in any of 50 distinct channels in the ISM band, so to satisfy the standard we must be able to hit channels no wider than about 500 KHz. Clearly, it doesn't do any good to define a channel and then have the frequency wander around by a substantial fraction of the channel; a variation of the center frequency of less than (say) 2% of the channel width seems likely to be fine. Now we're asking for an accuracy of 10 KHz in 915 MHz, which is 0.001% or 10 parts per million. This kind of accuracy is not achievable using fixed-value components or calibrations; we need to have some sort of feedback control of the output frequency.

In modern radios, accurate frequencies are generated by using a *phase-locked loop* (PLL) to compare the output frequency of the VCO to a very accurate reference frequency, the latter being provided by a tiny electromechanical resonator made of a bit of quartz crystal. The combination of a PLL, VCO, and appropriate control circuitry forms a *frequency synthesizer* (synthesizer for short). A simplified block diagram of a synthesizer is shown in Figure 4.30.

The unique properties of quartz allow the construction of inexpensive resonators with quality factors Q on the order of 10^6, and very little dependence of the resonant frequency on ambient temperature. Accuracy of a few parts per million is thus achievable. However, typical resonators use a bulk vibrational mode in which the frequency is controlled by the thickness of the crystal (essentially, we need to fit a half-wavelength into the crystal thickness), so higher frequencies require thinner crystals. It is straightforward to produce self-supporting crystals that resonate at 10–20 MHz, but much more difficult to produce resonators at 100 MHz, and 900 MHz resonators are not readily available. As a consequence, a quartz resonator is typically employed not directly as the source of the carrier, but as the source of a very stable reference frequency.

The output of the VCO is split, and part of the output is fed into a series of frequency dividers. Dividers are essentially flip-flops, electrical circuits with two stable states and the ability to switch between them when a pulse is applied. A single flip-flop constitutes a divide-by-two circuit: after two input pulses, it has returned to the original state. By combining flip-flops and simple logic, any necessary divider can be realized. The divided output is then sent to a phase detector, which essentially measures the time difference between (for example) the rising edge of the reference signal and of the divider output. When the time difference between the reference and the divided VCO edges is constant, the VCO output is precisely an integer multiple of the reference frequency. The output of the phase detector is filtered by a loop filter, which is essentially a low-pass filter that allows

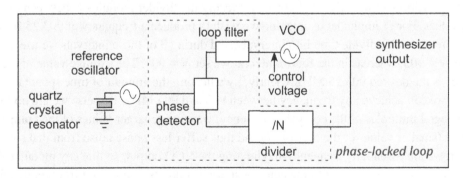

Figure 4.30
A frequency synthesizer uses a phase-locked loop to control the output of a high-frequency tunable oscillator with a low-frequency reference signal.

only slow changes in the tuning voltage so as to keep the loop stable, and used as the tuning input of the VCO.

The synthesizer in Figure 4.30 is not very flexible: for a fixed N, the output frequency will always be $N \cdot f_{ref}$. A more versatile synthesizer results if the divisor can be easily varied to allow differing output frequencies. One option is the ***integer-N*** synthesizer, in which two divisors can be employed, differing by 1: for example, 16 and 17. The first modulus N is used for (say) S cycles, and then $(N + 1)$ for $(S - F)$ cycles, for a total of F cycles. After all the cycles are done, the divider outputs one rising or falling edge. The net effect is to divide by $N_{eff} = (N + 1)S + N(F - S) = NS + S + NF - NS = NF + S$. Thus by adjusting S (which just involves setting a counter) the overall divisor and thus the output frequency can be adjusted over a wide range, with a resolution of f_{ref}.

If higher resolution is desired the reference frequency can also be divided some other integer M; in this case, we may speak separately of the reference frequency produced by the crystal and the ***compare frequency*** resulting from division of the reference. For example, a 10 MHz reference oscillator signal can be divided by 20 to produce a 500 KHz compare frequency; an integer-N synthesizer will then be able to produce channels spaced by 500 KHz, as required for US ISM band operation. However, operation under European regulations requires 200 KHz channels. We need to (at least) use a different divider for the reference oscillator to operate in both jurisdictions. We could instead divide the reference oscillator by 100 (to produce a 100 KHz reference usable for both US and European operation), but using larger divisors has a penalty: since the divider outputs an edge only after every N_{eff} cycles of the VCO output, information about the phase of the VCO signal becomes increasingly sparse as the divisor grows. The result is that the variation of phase that can occur without being suppressed by the feedback loop — the phase noise — increases with increasing N or M.

More sophisticated *fractional-N* synthesizers provide the ability to set the output frequency arbitrarily to within a fraction of a Hertz, by ***dithering*** the divider modulus N between two integer values. For example, let us imagine we wish to produce a frequency of 903.25 from a 1-MHz reference. We divide time into intervals, and during 3 of those intervals we use a modulus $N = 903$, whereas in the fourth interval we set $N = 904$. The average value of the frequency is the desired 903.25 MHz. Clearly, by adjusting the amount of time spent at each modulus, we can achieve any frequency between 903 and 904 MHz. Because we are no longer tied to integral multiples of the compare frequency, we can use larger values of the compare frequency (smaller values of the divisor M) and thus suffer less phase noise from that source. However, we are essentially frequency-modulating the VCO output, so like any modulated signal this one will contain new frequencies — spurious outputs — resulting from modulation. These spurs are in effect additional phase noise; it is important to choose the dithering frequency so that most of the noise will be out of the frequency bands of interest if possible.

4.3.3.3 Synthesizers and Phase Noise

The addition of the feedback loop and loop filter has important effects on phase noise. For frequencies near the carrier (closer than the cutoff frequency of the loop filter), the loop suppresses phase noise in the oscillator by compensating the tuning voltage to hold the output frequency at an integer multiple of the reference. Thus very close to the carrier, the phase noise of a synthesizer may fall, until the lower but still significant phase noise of the crystal reference oscillator becomes important. (The loop can't remove this noise because it locks everything to the crystal; when the reference frequency wanders, so must the output.)

At higher frequencies, the loop filter prevents any correction signal from reaching the VCO, in order to prevent instabilities. Thus as we move farther from the carrier, the phase noise becomes simply that of the VCO by itself. Moving the corner frequency of the loop filter higher allows the loop to lock to a new frequency more rapidly, and suppresses more phase noise from the VCO, at the possible cost of instability. Corner frequencies typically vary from a few 10's to a few hundred KHz. In a fractional-N synthesizer, the spurious power resulting from dithering may dominate the total output noise in the frequency range corresponding to the rate at which the modulus is changed.

4.3.4 Filters

Filters are circuits that reject some frequencies and transmit others. They are generally classified into three types: low pass, high pass, and bandpass (Figure 4.31). Filters work with mixers and the LO signal to select the wanted signals and reject both other signals and excess noise.

In a direct conversion radio, filtering is needed at the carrier frequency and at baseband (Figure 4.32). (Since there is no IF stage, no IF filters are used.) In the receiver, an RF band filter may be used to select for signals in (for example) the ISM band. Unfortunately, as we'll see below, available filters are not able to reject all frequencies near the reader band,

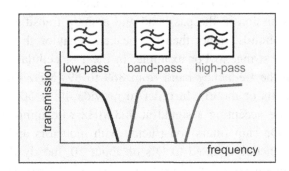

Figure 4.31
Three types of filters with common schematic symbols.

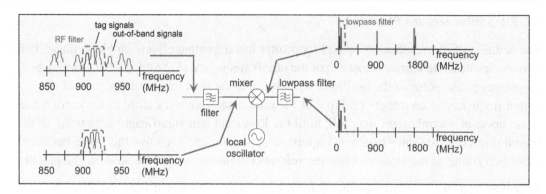

Figure 4.32
Simplified receiver block diagram with band-select and channel-select filters.

so the heavy lifting in rejecting out-of-band interferers as well as selecting the desired radio channel within the band must all be done in the baseband filters.

4.3.4.1 RF Filters

A bandpass filter made of discrete components is essentially a resonant circuit, like that of Figure 4.23. The ideal resonator shown there, with no losses, would only pass the resonant frequency, but real circuits have finite losses, characterized as we noted previously by the quality factor Q. Losses cause a resonant circuit to allow a finite band of frequencies to pass through it; the bandwidth is inversely proportional to Q (Figure 4.33). (Note that the output frequency characteristic shown in the figure is normalized to 1 ohm for convenience; a different value would change the position of the peak but not the bandwidth.) As shown in the right half of the figure, the quality factor also determines the width of the **passband** of the filter. Narrow passband filters must have very high Q: for example, an RF band filter for the 902−928 MHz ISM band must have a bandwidth on the order of 3% of the center frequency, requiring a Q of at least 30. In practice considerably higher Q is needed to make a good filter: the filter ought to have a fairly flat transmission in the passband and a sharp transition to very low transmission in the **stopbands**, rather than the peaked behavior shown in the figure. The challenges are very stringent. For example, in the US, cellular telephone basestations operate in the frequency band from 869 to 894 MHz (and may transmit at power levels of 100 watts or more!). In order to provide 40 or 50 dB of rejection of a signal at 893 MHz while accepting a signal at 903 MHz with minimal loss, we need a Q of several hundred. On-chip filters constructed with inductors and capacitors in Si CMOS processes are generally limited to Q's of about 10, mostly due to loss in the inductors. Discrete components offer Q's of up to 20−30 at these frequencies, but complex filters with many elements constructed using discrete components will become physically large and are inappropriate for GHz frequencies.

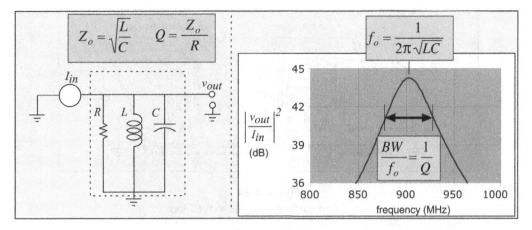

Figure 4.33
Simple filter using parallel resonant circuit, and bandpass characteristics.

On the other hand, if all that is needed is to reject second- and third-harmonic radiation from the transmitter (i.e. to pass transmissions at 915 MHz but reject 1830 MHz), discrete filters with Q around 10 are quite sufficient.

Better technologies with high Q, small physical size, and low cost are needed to provide band selection filtering. There are several methods of providing filters with high quality factors and small size at microwave frequencies. Many of these approaches depend on the fact that acoustic (mechanical) vibrations travel much more slowly than electromagnetic waves, so a resonant structure containing one or more wavelengths of sound can be much smaller than an analogous device employing electromagnetic resonances.

An important example of an electromechanical filter technology is the *surface acoustic wave* (SAW) device. (The reader may recall that we mentioned SAW-based RFID tags in chapter 2; these employ a similar technology to that of SAW filters to produce delayed encoded reflections.) SAW filters achieve Q's in the hundreds, are available in surface-mount packages, and can pass hundreds of milliwatts without damage. SAW filters are relatively expensive ($0.50 to $10). Packaged filters are on the order of 1 cm square, large enough that the number of filters must be minimized both to conserve board space and minimize cost. **Chip scale packages** with an area of only a few square millimeters have recently become available, allowing filter insertion with little penalty in area. The resonant frequency of a SAW filter is somewhat temperature-dependent; quartz filters are better in this respect than most other piezoelectric materials, but are more expensive to fabricate.

A simplified SAW filter structure is shown in Figure 4.34. The device is constructed on a piezoelectric substrate such as quartz, $LiNbO_4$, or ZnO. Electrical tranducers are constructed of a layer of a conductive metal such as aluminum, deposited and patterned on

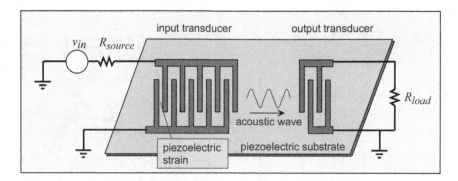

Figure 4.34
Simplified surface-acoustic-wave filter.

the surface of the piezoelectric using techniques similar to those employed in integrated circuit fabrication.

The input transducer consists of on the order of 100 interdigitated fingers, driven at alternating polarity from an RF source. Between each pair of fingers an electric field is formed within the piezoelectric material, inducing a time-dependent strain, which creates an acoustic wave. For an input frequency such that the spacing between fingers is half of the acoustic wavelength, a resonant enhancement of the wave will occur as it propagates along the transducer, as each alternating region of strain will be in phase with the wave and add to the displacement. The resulting strong acoustic wave propagates to the smaller output transducer, where the acoustic strain induces an electric field between the electrodes, resulting in an output voltage. The slice of piezoelectric is often cut at an angle to the propagation axis, so that the acoustic energy which is not converted back to electrical energy is reflected off the edges of the substrate at an odd angle and dissipated before it can interfere with the desired operation of the filter. Since the acoustic wave propagates about 10,000 times more slowly than electromagnetic radiation, wavelengths for microwave frequencies are on the order of 1 micron, making it possible to create compact, high-Q filter designs.

The performance of a fairly typical RF band (or image-reject) filter is shown in Figure 4.35, as the transmission in dB through the filter vs. frequency. Within the ISM band, the loss through the filter is only about 2.3 dB ± 0.3 dB: this transmission is known as ***insertion loss***, since it is the loss in band due to insertion of the filter in the circuit. Low insertion loss is important on both transmit and receive, though since transmit filters generally only need to provide harmonic rejection, a SAW filter may not be needed. The insertion loss of the transmit filter comes directly out of the signal power, so lossy filters mean bigger transmit power amplifiers that cost more and consume more DC power. On the receive side, the filter loss is essentially equal to its noise figure, and since the filter is typically placed

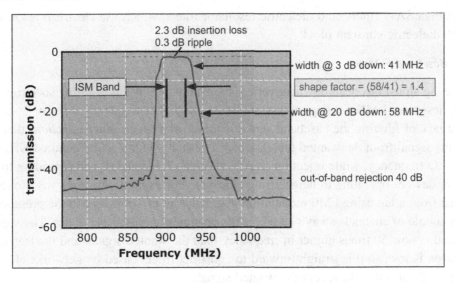

Figure 4.35
Transmission vs. frequency for a typical SAW filter centered on the US ISM band, showing definitions of insertion loss, shape factor, out of band rejection.

prior to the low-noise amplifier or mixer, the filter noise figure must be added directly to the noise figure of the receiver.

Other important properties of a filter are the sharpness with which transmission cuts off once the frequency is beyond the edges of the band, and the transmission (hopefully small, thus rejection) of out-of-band signals. The bandwidth in this case at a 3 dB decrease in transmission vs. the center frequency is about 41 MHz, rather noticeably wider than the 28-MHz ISM band: signals about 6−7 MHz outside of the band will have little rejection. Transmission falls quite rapidly thereafter: the **shape factor**, the ratio of bandwidth at 20 dB rejection to 3 dB rejection, is about 1.4. The rejection of signals far from the band edges − such as the 824−849 MHz cellphone uplink frequency band − is a substantial 40 dB or better. However, this filter only provides about 3 dB of rejection at the high end of the cellular downlink band. A nearby cellular basestation operating at the high edge of the band will not be rejected by a band select filter, but instead must be removed after baseband conversion. Such an interfering signal may combine with other interfering signals (such as other RFID readers) due to third-order distortion in the receiver front-end, to produce interference at the reader frequency that cannot be filtered out. As a consequence, RF filtering may not be sufficient to protect the reader from interferers, and instead it is necessary to ensure good linearity in the mixer and (if used) RF amplifier.

Other commercially-available filter technologies include bulk-acoustic wave (BAW) devices, which use a thin layer of piezoelectric material and can handle higher power densities than SAW filters, and dielectric resonator filters, which use electrical resonances of a high-dielectric-constant block.

4.3.4.2 Baseband Filters

Once the tag signal has been downconverted, the bandwidth of the resulting spectrum is typically less than 1 MHz, though a few protocols use higher-frequency tones of 2−3 MHz. The first task of filtering the baseband signal is to accomplish *channel selection*: that is, to pass the signal from the wanted tag(s), which is usually within a few hundred KHz of the reader's LO frequency, while rejecting signals at higher frequencies, corresponding to other readers or devices operating in neighboring channels. For example, we may wish to receive the signal from a tag using FM0 modulation at a data rate of 100 Kbps, in the presence of a reader a couple of channels away (1 MHz if the channels are 500 KHz wide). The signal to be rejected is now 50 times higher in frequency than the wanted signal, and the frequency of operation is low, so it is straightforward to construct filters based on networks of inductors and capacitors to accept the wanted signals.

It is often desirable to combine amplification and filtering; such a dual function can be readily accomplished at baseband frequencies using an inverting *differential* amplifier whose output is determined by the difference in the voltage applied to its two inputs. A simplified baseband amplifier circuit using such an amplifier is shown in Figure 4.36. When a resistor is connected between the output of such an amplifier and the input − that is, when *feedback* is introduced − the gain of the circuit becomes dependent only on the resistor values, as long as the amplifier gain is large. Such a configuration is often known as an *operational amplifier* or op amp.

If the voltage gain of the amplifier, K, is large, it is reasonable to assume that the terminal voltage of the amplifier $(v_- - v_+)$ must be small for reasonable values of the output voltage. In the case shown, since the positive terminal is grounded (fixed at 0 volts), the negative input terminal voltage must also be very close to 0. This fact enables us to analyze the circuit very simply using only Ohm's law and the further assumption that the amplifier

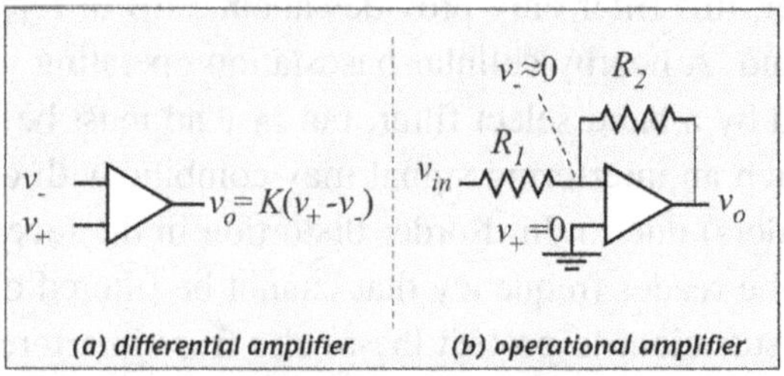

(a) differential amplifier   (b) operational amplifier

Figure 4.36
a) The output of a differential amplifier is proportional to the difference of the inputs; b) adding feedback produces an operational amplifier.

input current is very small (typically true for practical amplifiers). The current flowing through R_1 must equal the current flowing through R_2 if no current flows into the amplifier (Figure 4.37(a)). We find:

$$\frac{v_{in}}{R_1} = -\frac{v_o}{R_2} \rightarrow \frac{v_o}{v_{in}} = -\frac{R_2}{R_1} \tag{4.25}$$

This expression is not dependent on frequency (though a real operational amplifier circuit does have a frequency dependence due to the finite bandwidth of the differential amplifier). However, it is apparent that if we could reduce the feedback resistance R_2 with increasing frequency, the gain of the circuit would decrease: we would obtain a low-pass filter circuit. We can approximately accomplish this goal by placing a capacitor in parallel with the resistor (Figure 4.37(b)). Capacitors permit little current to flow at low frequencies, but pass current readily at high frequencies. Mathematically, the impedance of a capacitor falls inversely with frequency (see Appendix 3):

$$Z_c = \frac{1}{j\omega C} \tag{4.26}$$

When this impedance is large compared to the resistor, little current flows in the capacitor and the circuit acts as if the resistor were not present, providing a constant gain. When the capacitor impedance is much less than that of the resistor, current flows only through the capacitor and the gain falls inversely with frequency. The boundary between these two regions, the cutoff frequency ω_c, is the frequency at which the resistor and capacitor impedances are of equal magnitude:

$$R_2 = \frac{1}{\omega_c C_2} \rightarrow \omega_c = \frac{1}{R_2 C_2} \tag{4.27}$$

To produce a cutoff frequency of 200 KHz, we need a resistance of 5000 ohms and a capacitance of 160 pFd, both values being readily available. (In an integrated implementation it might be helpful to use a higher resistance value and less capacitance.) An op amp with these component values, and an input resistance of 1000 ohms, will

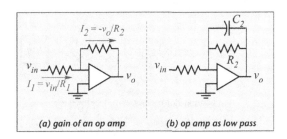

(a) gain of an op amp   (b) op amp as low pass

Figure 4.37
a) Simple analysis of circuit gain for op amp; b) frequency-dependent gain results from introducing a capacitor in the feedback circuit.

provide a voltage gain of 5 at low frequencies, and a rejection of about 13 dB for a 1 MHz signal vs. a 100 KHz signal. More elaborate feedback networks can be used to create more abrupt filter characteristics, and use of switchable resistors or capacitors can produce filters with adjustable frequency characteristics. Combining multiple stages of such active filters can be used to accurately select the bands containing most of the tag information while rejecting noise and interfering signals. Filtering can also be used in the transmitter to smooth the transmitted symbols, resulting in a narrower output spectrum.

It is also possible to dispense with most of the analog filtering of the baseband data by converting the baseband signal to a digital data stream at a sufficiently high sampling rate, and performing filtering and other signal processing digitally, bringing us to the topic of the next section.

4.3.5 Digital-Analog Conversion

All RFID readers, like most other modern radios, use a digital data stream to create the transmitted signals, and convert the received signal into digital data for decoding and interpretation. Because of the limitations of passive tags, the corresponding modulations used in reading them are simple variations of on-off keying for the downlink (reader-to-tag) and frequency-shift keying for the uplink (tag-to-reader). It is therefore possible to use very simple means to perform the requisite conversions, as shown in Figure 4.38. On the transmit side, a digital output (perhaps through a buffer amplifier) can be used to control a switch, turning the transmit power on and off as needed. On the receive side, a *comparator* − a device whose output is some fixed voltage $+V$ when the input is greater than a threshold, and 0 when the input is less than that same threshold − can be used to capture

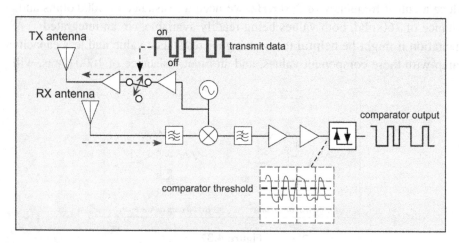

Figure 4.38
Simple conversion between analog and digital domains using a switch and comparator.

the zero crossings from the tag signal, which determine the frequency. Both of these approaches amount to conversion between digital and analog signals with 1-bit resolution.

This one-bit approach to digital radio is adequate for RFID readers, but the use of more capable conversion processes with higher resolution enables the reader to add capabilities through changes in software rather than hardware. On the transmit side, symbols can be smoothed digitally, allowing optimal adaptation to differing data rates and protocols. On the receive side, use of digital filtering allows the bandwidth and center frequency of the received signals to be adjusted to account for different tag data rates and interference conditions. The tradeoff between analog and digital signal processing is complex and subtle, but as computing power increases and software algorithms improve, the balance tilts toward more ***digital signal processing*** (DSP) power and simpler analog circuitry.

In order to exploit the power of DSP, more capable ***analog-to-digital converters*** (ADC) and ***digital-to-analog converters*** (DAC) are needed. There are numerous approaches to performing conversion. We will examine two common approaches.

For converting a digital signal to an analog voltage, a common approach is to use a ***current-steering*** DAC (Figure 4.39). The total current entering the inverting terminal of the op amp is the sum of the current from each of the legs. Since this current must also flow through the feedback resistor R_f, applying Ohm's law we find the output voltage:

$$v_o = R_f \left(\frac{b_1 v_{ref}}{R} + \frac{b_2 v_{ref}}{2R} + \frac{b_3 v_{ref}}{4R} + \frac{b_4 v_{ref}}{8R} \right)$$
$$= \left(\frac{v_{ref} R_f}{R} \right) \left(b_1 + \frac{b_2}{2} + \frac{b_3}{4} + \frac{b_4}{8} \right)$$

(4.28)

where $b_i = 1$ or 0 depending on whether the relevant switch is open or closed.

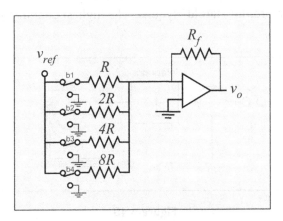

Figure 4.39
Simple 4-bit current-steering digital-to-analog converter.

The switch positions b_i can thus be regarded as the bits of a binary number $(b_1 \ b_2 \ b_3 \ b_4)$, determining the output voltage. The resolution of a current-steering DAC is determined by the number of stages, which is in turn limited by the accuracy with which the resistor values of the ladder can be constructed, and the extent to which the reference voltage can be held constant as the switch configuration varies. If a DAC is used to control the output power of a transmitter, higher resolution will permit more precise smoothing of the output signal and contribute to a narrower output spectrum and thus less interference with readers on neighboring channels.

The corresponding analog-to-digital converter, a *flash* ADC, uses a ladder of 2^n resistors to divide the reference voltage into 2^n slices. Each resistor has a comparator one of whose terminals is connected to the input voltage and the other to the slice of the reference; by recording the location where the comparators switch states, the input voltage is digitized. Flash ADC's are very fast but high resolution requires a large number of accurately-matched resistors and comparators.

A more common analog-to-digital conversion architecture is the *sigma-delta* converter (Figure 4.40). The sigma-delta ADC performs successive approximations to the input voltage, using a DAC; the difference of the input voltage and the DAC estimate is integrated and fed to an ADC. The result is used to improve the DAC estimate. The resolution of the ADC doesn't need to be very high, since its job is only to see if the DAC estimate is too high or too low: a single bit is sufficient. The digital output is the setting of the DAC at the end of some number m of approximation cycles, where m is roughly the resolution of the DAC.

Commercially-available ADC's provide resolutions of 15 bits and sample rate of 100 million samples per second (Msps), greatly in excess of what is needed to digitize the baseband signal for the relatively low data rates used in RFID. To capture a signal of bandwidth BW, one must sample the signal at a frequency of at least $2\ BW$ − one of the many results due to the prolific Harold Nyquist. The highest commonly-encountered

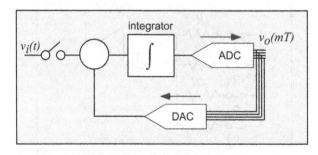

Figure 4.40
Sigma-delta ADC.

frequency, the 3.3 MHz tone used in EPCglobal class 0 tag data, would require a sampling frequency of 6.6 MHz. A resolution of 15 bits means that the smallest resolvable signal is about 90 dB below the largest signal, assuming that noise in the conversion is negligible. One can therefore choose to employ inexpensive, low-power digitization chains with just enough sampling speed to capture the signal, or spend more power and money to sample a broad band with high resolution, enabling the use of more sophisticated and flexible digital processing to provide much of the channel selection filtering and interference rejection.

4.3.6 Circulators and Directional Couplers

In a monostatic configuration (Figure 4.1), a single antenna is employed to simultaneously transmit a CW signal to power the tag, and receive the backscattered signal. If the receiver were simply connected directly to the transmitter, and both antenna and receiver were well-matched, half of the transmitted power would be directed into the receiver. In such a case, not only is half of the transmit power being wasted, expensive in terms of component size and cost and power consumption, but the receiver is subjected to a huge blocking signal (on the order of 1/2 watt) while trying to capture the tag signal (on the order of a nanowatt). It would be helpful if the signal leaving the transmitter were directed only to the antenna, and the returned signal from the antenna were directed only into the receiver, as indicated schematically in Figure 4.1.

There are two basic ways to accomplish this end. The first is to employ a special microwave component known as a *circulator* (Figure 4.41). A circulator is a 3- or 4-port device in which signals introduced into any port come out only at the next port. For a 3-port coupler, an input signal at port 1 appears at port 2, an input signal at port 3, and a signal at port 3 comes out at port 1.

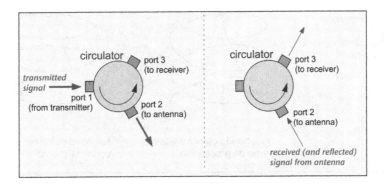

Figure 4.41
3-port circulator operation.

Not only is this a neat trick, it is also a violation of the principle of reciprocity that we used in chapter 3 section 6: if we ignore port 3, putting a signal into port 1 produces a signal at port 2, but a signal at port 2 produces no signal at port 1. Such violations are possible because the circulator contains a bit of *ferrite* — an iron-oxide-based mixture with very high magnetic permeability — placed within a constant magnetic field created by permanent magnets. The electrons in the ferrite act like tiny magnets, and tend to orient themselves along the external magnetic field. If they are prevented from doing so by minor obstacles like conservation of momentum, they will rotate in a plane perpendicular to the field: that is, they *precess*, much like a gyroscope held at an angle to the vertical. Magnetic fields are oriented: a current circulating in a counterclockwise direction in a horizontal plane produces an upwards magnetic field, whereas a clockwise current produces a downward-directed field. So the precession of the electron gyroscopes is preferentially in one direction (counterclockwise if we watch the tip of the magnetic dipole). A right-hand-circularly polarized wave (see chapter 3 section 6) propagating along the field is rotating in the same direction as the electrons are precessing, and will interact differently from a left-hand-circularly polarized wave (Figure 4.42), though the details are rather complex and dependent on the frequency and external field. Thus right-hand and left-hand waves will travel at differing speeds in the ferrite.

Armed with such a non-reciprocal material, we can construct a circulator by placing a pair of ferrite disks on either side of a metal conductor with three conductor lines in a Y-configuration forming the three ports (Figure 4.43).

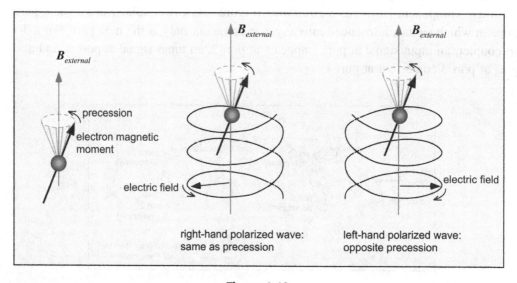

Figure 4.42

Precessing magnetic moment and circularly-polarized waves propagating along the external magnetic field.

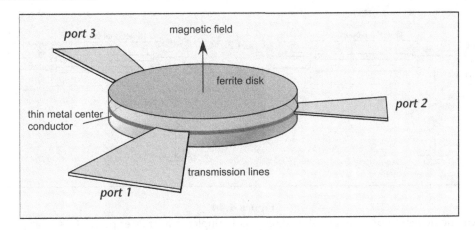

Figure 4.43
Simple 3-port circulator; note that ground planes above and below the disk are suppressed for clarity.

A wave entering one port can be roughly regarded as splitting into right-hand-circular and left-hand circular waves within the disk. Because the disk is a ferrite, and a magnetic field is applied, the propagation velocities need not be equal for the left- and right-going waves. If we can arrange the geometry and magnetic field so that the phase shift of (say) the right-hand circular wave is $2\pi/3$ (60°) from port 1 to port 2, and the left-hand circular wave suffers a phase-shift of $\pi/3$ from port 1 to port 3 (and thus $2\pi/3$ from port 1 to port 2), at port 2 the waves will combine in phase:

$$\phi_{rt}[\text{port } 2] = \frac{2\pi}{3} = \phi_{left}[\text{port } 2] \tag{4.29}$$

At port 3, the phase shift of the right-going wave will be doubled to $4\pi/3$, so we find:

$$\phi_{rt}[\text{port } 3] = \frac{4\pi}{3} = \phi_{left}[\text{port } 2] + \pi \tag{4.30}$$

That is, at port 2 the two waves are π radians (180°) out of phase and cancel. Obviously by symmetry the same applies to the other two ports, so a wave entering at any port comes out at the next clockwise port, as desired.

Circulators can also be constructed using various combinations of microwave coupling devices and ferrite phase shifters (known as **gyrators**). Ferrites can also be used to create **isolators** that pass signals in only one direction.

Practical commercial circulators achieve insertion losses of around 0.5 dB (that is, essentially all the power in comes out at the desired port) and port-to-port isolation of about 20 dB. This level of isolation is not very good — remember that we're trying to see a

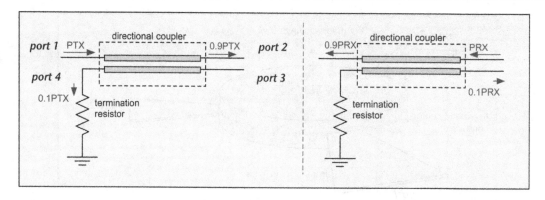

Figure 4.44

Schematic depiction of a coupled-line directional coupler; transmitter power PTX in at port 1, received power PRX from antenna at port 2.

−60 dBm signal in the presence of a +30 dBm CW transmission − but it doesn't really hurt much, as the reflection from the antenna, which cannot be removed by the circulator, is typically only around 15 dB less than the transmitted power. However, circulators are relatively large (typically 2−3 cm in diameter and 1 cm thick), and commercial units are expensive, costing as much as US$200 for a 900-MHz unit.

An alternative approach to separating the transmitted from the reflected signal that is more compact and less expensive is to employ a *directional coupler* (Figure 4.44). A directional coupler extracts a small portion of waves traveling in one direction on a transmission line (e.g. to the right) while being essentially impervious to waves traveling in the other direction (e.g. to the left). Directional couplers can be constructed using two weakly-coupled transmission lines, where coupling is obtained simply by designing the lines to travel near each other on a printed-circuit board. The maximum coupled signal is obtained for a line length of a quarter of a wavelength; shorter couplers can be used to remove only a small portion of the signal.

Commercial couplers of dimensions less than 2 cm on a side and only a few mm thick are available, and can be configured to mount directly to a printed circuit board. Couplers are also relatively inexpensive, on the order of US$10. However, because a coupler is a reciprocal device, if the coupler extracts a large amount of power from the reflected signal, it also extracts a large amount of power from the transmitted signal (though this appears at port 4 and is normally dissipated in the terminating resistor). In order to save transmitted power, it is typical to employ a 10-dB coupler: the coupled signal is 10% of the incident power, so the transmitted signal only encounters about 1 dB of loss, but the reflected signal from the antenna is attenuated by 10 dB in passing through the coupler. This loss in signal power affects sensitivity; we will examine the consequences of this loss in more detail when we consider radio chain analysis in section 5 of this chapter.

4.4 RFID Transmitters

An RFID transmitter has two basic tasks. In the downlink phase, it must provide power to start up passive tags, and modulate its signal so as to send the tags commands and data. In the uplink phase, the transmitter must provide an unmodulated signal that the tag can backscatter in order to return data to the receiver. The transmitter must be able to operate on any of a number of radio channels with accuracy of a few parts per million in frequency, and switch from one channel to another rapidly, in order to meet the requirements of unlicensed use. In order to obtain good read range, the transmitter should provide the maximum output power allowed by the relevant regulatory bodies. In handheld or portable applications, it should do so while consuming as little DC power as possible.

The transmitter's task is rendered significantly more complex by the presence of other radio devices, and in particular other RFID readers. In normal far-field-coupled operation, the transmitter radiates and can interfere with distant radio receivers. Recall from chapter 3 that the radiated spectrum of a modulated signal is much wider than that of an unmodulated signal, and that typical RFID modulations, which are optimized for powering the tag, are very inefficient users of spectrum. Regulatory requirements for radiation out of the allowed bands are often very stringent. It is ordinary practice to provide *guard bands* by refraining from use of channels immediately adjacent to the edge of the allowed band. For example, in the US, a reader may place its lowest channel at 902.75 MHz, so that the region of 902–902.5 MHz is unused, but radiation at frequencies below the lower edge of the ISM band (902 MHz) is reduced. In this fashion, interference with other (possibly licensed) spectrum users is reduced, but no particular benefit is obtained for other users of the unlicensed bands.

In order to reduce interference between unlicensed users, and in particular from one reader to another, it is helpful to minimize the width of the radiated spectrum. Protocols may place specific requirements on the amount of power radiated into neighboring channels by imposing a *spectral mask* requirement; we shall examine the EPCGlobal Generation 2 *dense reader mode* mask in chapter 8. Filtering of the transmitted symbols can reduce the radiated power far from the carrier, as we demonstrated in chapter 3. Alternative modulation techniques, still simple enough to be deciphered by passive tags but more parsimonious in their use of bandwidth, may also be employed, as we will discuss below.

4.4.1 Transmitter Architectures

The simplest transmitter consists of a synthesizer to provide a carrier signal, a switch to turn the signal on and off, and an amplifier to provide sufficient output power (Figure 4.45). However, as is shown schematically in the inset, the resulting spectrum is very broad with substantial radiated power in the adjacent channels, where it may interfere with other readers.

A better approach is to employ a variable attenuator so that the signal may be smoothly switched on and off (Figure 4.46). It is readily apparent that the spectrum resulting from transmission of filtered (smoothed) symbols is much narrower than that from switched symbols. An RF attenuator is a bit more complex than suggested in the figure: if a single variable resistor were used, the signal would be reflected rather than merely absorbed when the resistance value was different from the impedance of the surrounding circuitry. A useful attenuator employs three resistive elements, e.g. in a "π" configuration (parallel / series / parallel); as the series resistor increases, the parallel resistors decrease, so that the attenuator always presents a matched (typically 50 Ω) load. The resistive elements can be p-intrinsic-n (PIN) diodes, or transistors. PIN diode attenuators require additional passive

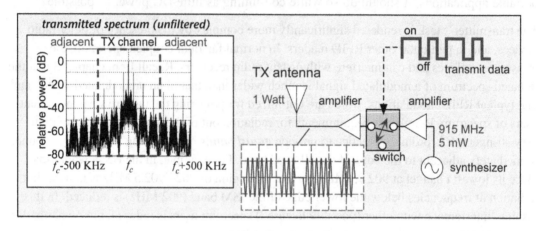

Figure 4.45
Simple switched transmitter architecture; inset shows radiated spectrum (unfiltered PIE modulation).

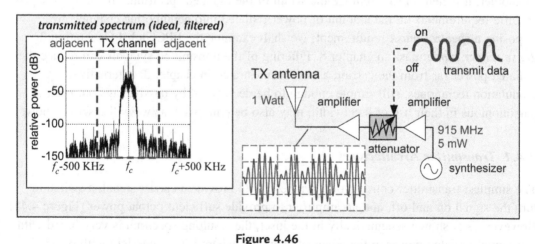

Figure 4.46
Transmitter using variable attenuator and filtered (smoothed) symbols; inset shows radiated spectrum (ideal PIE modulation and filtering).

components for biasing, but transistor-based attenuators may introduce significant additional distortion in the signal, broadening the spectrum and vitiating some of the benefits of filtering. Attenuators are somewhat more expensive than switches, and introduce additional loss, so higher gain amplifiers must be used to compensate. Furthermore, if a linear power amplifier is employed (which is desirable for producing minimal distortion), the amplifier draws power even when the input signal is off.

A more elegant approach to filtering the signal is to vary the DC power supplied to the power amplifier, as depicted in Figure 4.47. By varying the supply power to the output amplifier, the transmitter power can be smoothly changed while minimizing DC power consumption when the pulse amplitude is small. (This is a very old trick, based on the variation of plate voltage of very-high-power tube amplifiers to impose AM on broadcast radio signals.) Reducing the DC voltage may introduce both changes in the delay (which show up as unintended phase or frequency modulation of the output signal) and additional distortion in the signal, so some calibration may be needed to establish a range of voltages in which the amplifier is well-behaved. A similar approach, which may produce less variation in the amplifier characteristics, is to control the bias current in the amplifier. Bipolar transistor amplifiers often use a small reference transistor in a ***current mirror*** configuration to control bias current; it is relatively simple to vary the current in the reference transistor to vary the output power.

The transmitters described above only provide control of the amplitude of the transmitted signal, and don't change the phase of the carrier (except accidentally). The ability to modulate phase as well as amplitude provides a powerful tool, and is widely used in many areas of wireless communications. Phase modulation can improve signal detection in the

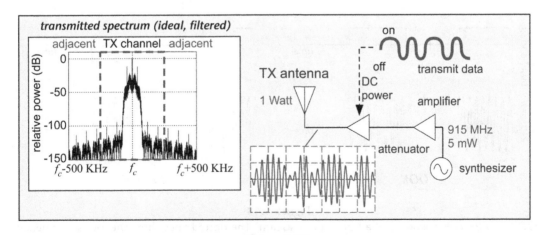

Figure 4.47

Transmitter using variable bias supply to modulate output power; inset shows radiated spectrum (ideal PIE modulation).

presence of noise, and reduce the amount of spectrum required for a given data rate. The use of phase modulation in passive RFID is constrained by the fact that the tags can only extract the amplitude of the reader's signal. Nevertheless, there remain modulation techniques that provide improved spectral efficiency while still allowing simple demodulation. Let's take a look at two of these approaches, and their implications for transmitter architecture.

The first approach is known as ***phase-reversal amplitude shift keying*** (PR-ASK), which can also be regarded as a variant of ***duobinary*** encoding. To implement PR-ASK for a simple binary data stream, we simply invert the phase of each successive binary 1 (Figure 4.48). The highest-frequency OOK data stream is a series of alternating 1-0-1-0...; when such a stream is transmitted using simple on-off keying, even with perfect smoothing, we must have at least a sine wave of frequency equal to half the data rate in order to preserve the information in the signal; this is shown by the dotted line in the figure. When this baseband signal is used to modulate the carrier, the resulting modulated signal spectrum will have energy at least half the data rate above and below the carrier frequency. Using PR-ASK, the envelope of the RF signal is identical, but the essence of the baseband signal can be described using a sine wave of frequency equal to a quarter of the datarate. The modulated signal is half as wide as that for OOK. (Note that real symbols would likely be smoothed (filtered) to minimize excess spectral width due to abrupt transitions between bits, as was discussed in chapter 3, section 3.)

Recall that reader symbols are usually encoded in some fashion that avoids loss of power during long strings of zeroes, such as pulse-interval encoding (chapter 3, section 3). Using PIE, each symbol needs to end with a brief period in which the transmitted power is low or

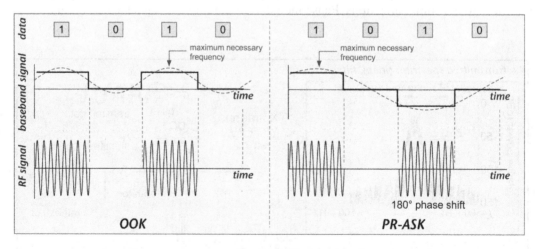

Figure 4.48
OOK and PR-ASK encoding of a binary data stream. The dotted lines show the lowest frequency sine wave that must be included in the spectrum to reproduce the data. Symbols are shown unfiltered for clarity.

zero. We obtain this nice result from PR-ASK modulation if we simply invert the phase of every symbol rather than just every binary 1, since the baseband signal must pass through zero each time it changes sign (Figure 4.49). Note that in the case of PIE, the highest frequency is obtained from a string of binary 0's rather than alternating symbols.

In order to impose this phase inversion on the transmitted signal, we can use a mixer instead of an attenuator or switch to modulate the signal, so that the sign of the baseband voltage is preserved (Figure 4.50). We need to use a balanced mixer like that in Figure 4.21. (We could instead use a pair of switches and a delay element, but this is more complex and because of the delay element is likely to be sensitive to the RF frequency employed.)

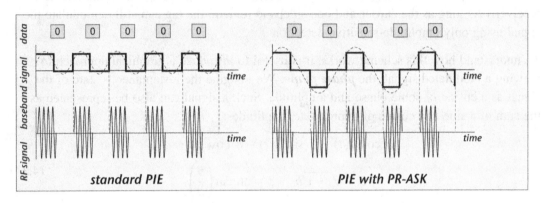

Figure 4.49
String of PIE-binary 0 symbols encoded using amplitude-shift keying (left) and phase-reversal ASK (right); symbols are shown unfiltered for clarity.

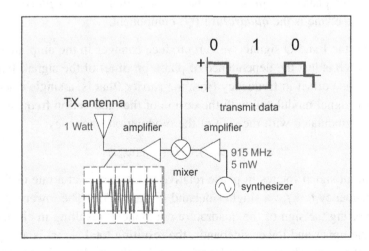

Figure 4.50
Transmitter using (balanced) mixer to allow phase-reversal modulated signal; note the baseband signal is shown unfiltered for clarity.

One other approach to creating a signal that uses minimal spectral width but can still be demodulated by a passive tag is *single sideband* (SSB) or *vestigial sideband* (VSB) modulation. Recall that amplitude modulation of a carrier produces two sidebands, one above and one below the carrier frequency (Figure 3.5 and equation (3.11)). In SSB modulation, only one of these sidebands is preserved, the other being completely removed using techniques to be described below. In VSB, the signal is filtered (typically at a convenient intermediate frequency) to remove most of the upper or lower sideband. (VSB is used in the NTSC television standard, for similar reasons: video images require a lot of bandwidth, but in the 1930's and 1940's, when the standard was being developed, it was simpler to implement filtering on the transmitter, and envelope detection in the television receiver.) As long as the carrier and one sideband remain, the tag can still demodulate the signal using only amplitude-sensitive detection.

To understand how this scheme works, it is useful to introduce a graphical approach to viewing a modulated signal: the *phase plane*. We imagine the instantaneous state of the signal as a cosine of some phase and amplitude. Such a signal can also be represented as the sum of a sine and cosine of appropriate amplitudes:

$$a \cdot \cos(\omega_c t) + b \cdot \sin(\omega_c t) = c \cdot \cos(\omega_c t + \phi)$$

$$c = \sqrt{a^2 + b^2} \quad \phi = \arctan\left(\frac{b}{a}\right) \tag{4.31}$$

If we plot the amplitude of the cosine on the horizontal axis of a graph, and the sine on the vertical axis, the length of the resulting vector is the amplitude of the signal, and the angle from the x-axis is the phase (Figure 4.51). The cosine signal is the *in-phase* or *I* component of the signal, and the sine is the *quadrature* (*Q*) component.

Thus by adjusting the I and Q signals we can produce changes in the amplitude and phase (or frequency, which is just the dependence of phase on time) of the signal. In particular, to produce a single tone offset in frequency from the carrier (that is, a single sideband), we apply an in-phase signal modulated with the cosine of the modulation frequency, and a quadrature signal modulated with the sine of the modulation frequency:

$$a = \cos(\omega_m t); \quad b = \sin(\omega_m t) \tag{4.32}$$

The resulting output signal rotates in phase relative to the carrier at a rate of f_m: that is, it is a cosine with frequency $f_c + f_m$: a single sideband (Figure 4.52). The lower sideband can be produced by inverting the sign of the quadrature component, resulting in clockwise rotation. If we combine the upper and lower sidebands, the resulting total signal is along the real axis: a modulated cosine. The double-sideband signal is the modulated cosine we introduced in chapter 3; the length of the vector to the resulting signal varies with time,

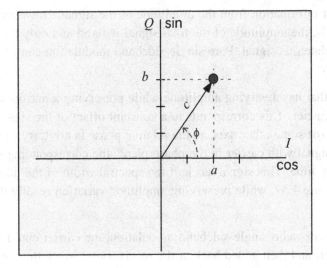

Figure 4.51
Geometric view of addition of *I* and *Q* signals to create a new sinusoidal signal of arbitrary phase.

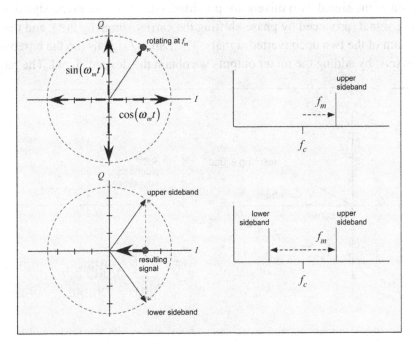

Figure 4.52
Sidebands in the phase plane.

so a tag can extract information from the amplitude of the signal. However, if we remove one of the sidebands, the amplitude of the total signal is fixed and only the phase varies; a tag cannot demodulate this signal. Pure single-sideband modulation can't be used for passive tags.

To create a signal that has a varying amplitude while preserving a narrow spectrum, we need to add some carrier. This corresponds to a constant offset of the single-sideband signal along the real axis (or some other axis, as the relative phase is arbitrary). Figure 4.53 shows a single-sideband signal with carrier in the phase plane, the corresponding spectrum and signal amplitude vs. time. This signal has half the spectral width of the double-sideband signal shown in Figure 4.52, while preserving amplitude variation readily decipherable by a passive tag.

(In traditional amateur-radio single-sideband modulation, the carrier could be suppressed to save transmit power, and then added back at the receiver end, since the human operator could adjust the phase and amplitude of the injected carrier until human-sounding speech was heard.)

The SSB signal has a rather complex relationship between instantaneous amplitude and phase during the RF cycle. The most straightforward and flexible approach to creating such a signal is to use an *I/Q modulation* architecture. This approach corresponds exactly to the phase plane depiction of the signal: two mixers are provided, one for the in-phase signal and one for the quadrature signal (produced by phase-shifting the carrier signal by 90°), and the output signal is the sum of the two upconverted signals. The *I* and *Q* signals are the baseband input to the two mixers; by adding the mixer outputs we obtain the desired signal. The required *I/Q*

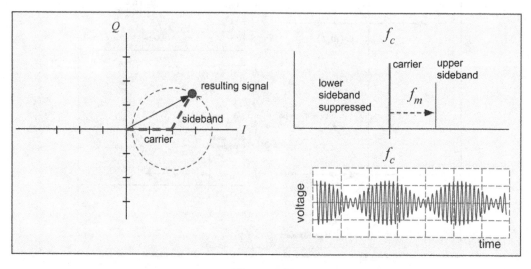

Figure 4.53
A sideband with carrier produces an amplitude-modulated signal with reduced spectral width.

upconverter is shown in Figure 4.54. The two mixers are often combined with the phase shifter into a single component, sometimes known as an *analog quadrature modulator* (AQM), in order to ensure good matching of amplitude and phase between the two branches. To produce a single-sideband output, the input voltage to the *I* mixer contains a DC bias voltage (to produce the carrier contribution) and a cosine at the modulation frequency. The *Q* branch need only receive a sine if the carrier is to be in-phase.

Since any modulated signal can be viewed as a path in the phase plane, the I/Q modulator can be used not just for SSB signals but for more complex signals combining phase and amplitude modulation, not as relevant to passive RFID but widely employed in other areas of wireless communications.

4.4.2 Transmit Power Efficiency

In the United States, an RFID reader can transmit up to 1 watt in the ISM band. (The transmitter may produce a bit more than that to account for losses in the circuitry and cabling.) A substantially larger amount of DC power may be required to produce this single watt of RF output. The output amplifier may be the largest single user of power in the system, and of considerable importance in operating lifetime of portable and handheld devices. It is worth a brief investigation to understand the origin of this inefficiency.

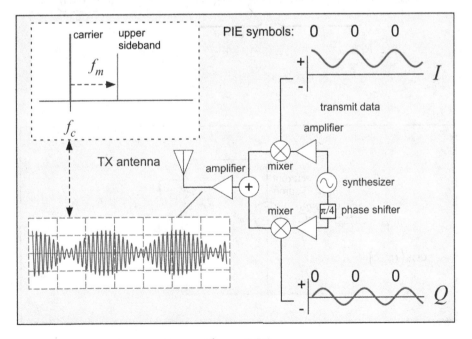

Figure 4.54
I/Q modulator architecture for SSB, with output signal and spectrum.

To convince ourselves that it is possible for an amplifier to efficiently convert DC power into an RF signal, let's examine a greatly simplified ideal current amplifier (Figure 4.55). The amplifier consists of a current source whose output current is linearly proportional to the input voltage:

$$i_{amp} = g_m v_i = g_m(V_b + \cos(\omega_c t)) \tag{4.33}$$

where the constant of proportionality is known as the **transconductance** of the amplifier. The **bias voltage** V_b is imposed to allow the output current to both increase and decrease with the sinusoidal input, since in real amplifiers the current can only flow in one direction. (If there were no bias voltage, the amplifier would act as a rectifier, amplifying only the positive-going part of the input voltage.) Real amplifiers have a maximum current they can provide, I_{max}; in the simplest sort of amplifier (known formally as a **class A** amplifier), the bias voltage is chosen so that the output current when there is no sinusoidal input signal is $I_{max}/2$, and thus the largest possible sinusoidal output signal has amplitude $I_{max}/2$.

In this greatly-simplified circuit all the output current flows through the load resistor R_L. By Ohm's law the voltage across the resistor is the product of the current and resistance. The RF power provided to the load is the product of the variable part of the current and voltage, averaged over a cycle of the signal. For the maximum output signal, we have:

$$
\begin{aligned}
P_{load} &= \left\langle \underbrace{R_L \frac{I_{max}}{2}\cos(\omega t)}_{voltage} \cdot \underbrace{\frac{I_{max}}{2}\cos(\omega t)}_{current} \right\rangle = \frac{R_L I_{max}^{\,2}}{4} \underbrace{\langle \cos^2(\omega t) \rangle}_{\to 1/2} \\
&= \frac{R_L I_{max}^{\,2}}{8}
\end{aligned}
\tag{4.34}
$$

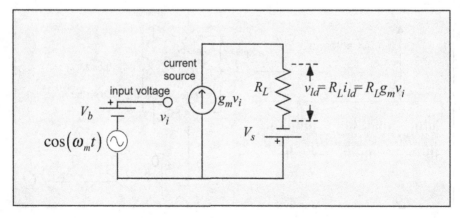

Figure 4.55
Ideal voltage-controlled current amplifier.

The most sensible way to choose the load resistor is to set the value so that when the current is equal to the maximum current, all the supply voltage appears across the load resistance and none across the transistor. Since the DC bias current is chosen to be half the maximum current, in the condition when there is no sinusoidal input (the *quiescent* state), the voltage on the transistor is half the supply voltage, the same as the voltage across the load resistor. The DC power dissipated in the transistor is then:

$$P_{DC} = \frac{R_L I_{max}}{2} \frac{I_{max}}{2} = \frac{R_L I_{max}^2}{4} \tag{4.35}$$

The *efficiency* of the amplifier is:

$$\eta = \frac{P_{RF}}{P_{DC}} = 0.5 \rightarrow 50\% \tag{4.36}$$

That is, it's not unreasonable to hope that half the DC power is converted to RF power in the output amplifier. (In fact, much higher conversion efficiencies can be obtained using more complex amplifier configurations, known as *class B*, *class C*, and on into the alphabet.) It ought to take 2 watts or less of DC power to deliver 1 watt of RF power to the transmitter output.

This optimistic view of the situation falls apart when we recall from section 3.1 above that real amplifiers are not perfectly linear, but distort the input signal. In particular, recall that third-order distortion leads to intermodulation products that are very close to the signal itself, and cannot be filtered out. We wish to use a high data rate in order to count tags rapidly, but when we do so the output spectrum increases in width, even if the symbols are smoothed (e.g. chapter 3, Figure 3.10). The best we can do is to choose a data rate that results in use of the whole channel we are operating in. If we now introduce third-order distortion to the resulting signal, the spectrum grows wider (Figure 4.14), resulting in radiation in the neighboring channel. To keep the amount of power in the adjacent channels – the *adjacent channel power ratio*, ACPR – small, we must back off from the third-order intercept power. Recall that the signal from a tag a few meters from an antenna may be as small as -60 or -70 dBm (Figure 3.30); to keep a nearby reader in an adjacent channel from interfering with the tag, the amount of power it radiates into the adjacent channel should be much lower than that in its own channel. Let's imagine that we wish the ACPR to be (say) 30 dB – that is, the power radiated into the neighbor channel due to distortion should be 1000 times smaller than that in the channel we're operating in. Using the IP3 values (Figure 4.13) as a rough guide, we need to back off from the output intercept by 15 dB from OIP3. If the OIP3 is 10 dB higher than the maximum output power of the amplifier, a fairly typical result, then the distortion-limited output power is 5 dB (3-fold) smaller than the maximum power the amplifier can provide. The calculation is depicted schematically in Figure 4.56. The amplifier efficiency, which we at first hoped would be around 50%, is now $(50/3) = 17\%$. A nominal 1-watt amplifier will provide only

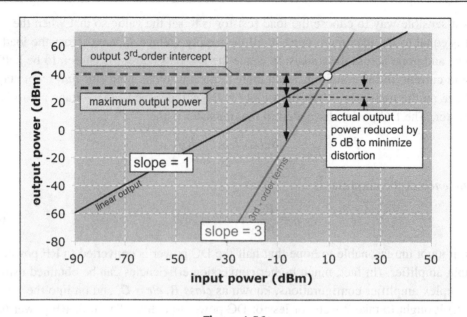

Figure 4.56
Output power is reduced to ensure that 3rd-order distortion is 30 dB smaller than linear power.

330 mW of output power and require 2 watts of DC power; to achieve the desired legal limit of 1 watt, we now need an amplifier with a nominal output power of 3 watts, and DC power consumption of 6 watts.

In practice design requirements vary considerably, and the designer may choose to trade a reduced data rate for reduced power consumption and cost (high power amplifiers at 900 MHz are expensive). At the other extreme, to provide maximum performance and minimize linearity concerns, the designer may choose a power amplifier rated at 10–20 watts maximum output to provide 1–2 watts of actual output power.

4.4.3 Phase and Amplitude Noise

Because an RFID reader for passive tags is transmitting at the same time the receiver listens for the tag response, there is always some leakage from transmitter to receiver. In a monostatic configuration (Figure 4.1), this leakage is dominated by the signal reflected from the antenna back into the receiver, and is typically around 15–20 dB below the transmitted signal. In a bistatic configuration, TX-RX isolation of 30–40 dB is achievable.

There are a couple of basic sources of noise in the transmitted signal. The first is phase noise from the oscillator (section 3.3 above). To get an idea of the magnitude of the phase noise contribution, let's use the example shown in Figure 4.29. The frequency range of interest is that corresponding to the tag signal; for a tag data rate of 100 Kbps we might want to look at frequencies from around 50 KHz to 100 KHz from the carrier, where the phase noise is typically about −95 to −105 dBc/Hz. Very roughly speaking, the bandwidth is around 50 KHz or 45 dB from 1 Hz, so the total noise will be about $(-95 + 45) =$ −50 dBc in this band. Let us imagine that the transmitted signal is 1 watt (30 dBm), of which 15 dBm is reflected from the antenna and sneaks into the receiver. The received phase noise is then approximately $(15 - 50) = -35$ dBm. If on the other hand a bistatic arrangement is used, the injected signal from the transmitter starts at around 0 dBm (or better) and the phase noise is then −50 dBm. (In practice, as we will discuss in the next section, receivers are typically arranged with I and Q channels, and the absolute phase of the leakage signal determines how much of this noise ends up in each channel.)

To convert this phase (or frequency) noise into a voltage in the baseband receiver, we need the output voltage of the mixers to be affected by the frequency of the signal. In general, the mixers are not terribly sensitive to frequency on their own, so if the instantaneous frequency varies for both the local signal and the reflected signal from the antenna, and both are in phase, no variation in the output voltage results. That is, the average value of the product is:

$$\langle V_{mixer} \rangle = \left\langle \underbrace{\cos([\omega + \delta\omega]t)}_{local\ oscillator} \underbrace{\cos([\omega + \delta\omega])}_{antenna\ reflection} \right\rangle = \frac{1}{2} \qquad (4.37)$$

independent of the frequency variation $\delta\omega$. However, in many cases the antenna reflection is not in phase with the local oscillator signal, but is delayed because it needs to travel down some circuit board traces and cables to the antenna and back again (Figure 4.57). This delay leads to an absolute change in the phase of the reflected signal; for some frequency and

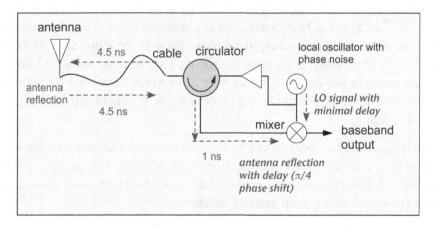

Figure 4.57
Delayed antenna reflection mixes with local oscillator.

delay values, the cosine will be shifted by 90° and become a sine. In this case, the average value of the output voltage of the mixer will be 0:

$$\langle V_{mixer} \rangle = \left\langle \underbrace{\cos(\omega t)}_{local\ oscillator} \underbrace{\cos(\omega[t + \tau])}_{antenna\ reflection} \right\rangle$$

$$= \left\langle \cos(\omega t) \underbrace{\sin(\omega t)}_{when\ \omega\tau = \frac{n\pi}{4},\ n\ odd} \right\rangle = 0 \tag{4.38}$$

When the reflected signal is in quadrature like this, small variations in frequency disturb the perfect null condition and result in a finite average output voltage. The output is no longer independent of frequency:

$$\langle V_{mixer} \rangle = \left\langle \cos([\omega + \delta\omega]t) \underbrace{\sin([\omega + \delta\omega]t + \delta\omega \cdot \tau)}_{\approx \sin([\omega + \delta\omega]t) + \delta\omega \cdot \tau \cos([\omega + \delta\omega]t)} \right\rangle$$

$$\approx \frac{1}{2}\delta\omega \cdot \tau \tag{4.39}$$

(This sort of scheme can be constructed intentionally as a method of demodulating a frequency-modulated signal, where it is known as a *delay line discriminator*.)

A typical cable might be 1 meter long and propagation velocities in coaxial cable are typically around 60% of the velocity of light in vacuum, so the delay (recalling that the signal must travel out and back) is around 10 ns. For the range of frequency offsets from the carrier ($\delta\omega$) we were considering above, 50 to 100 KHz, the sensitivity of the output voltage ($\delta\omega\tau$) is roughly

$$\frac{1}{2}\delta\omega \cdot \tau = \frac{1}{2}(2\pi \cdot 75 \times 10^3) \cdot (10 \times 10^{-9}) \approx 2 \times 10^{-3} \tag{4.40}$$

Since power goes as the square of the voltage, the phase noise power is reduced by a factor of about 5×10^{-6} or 53 dB in being converted to amplitude noise. Thus the equivalent amplitude noise at the receiver resulting from phase noise in the transmitter VCO is around $(-35 - 53) = -88$ dBm. Imagine we need the received signal to be around 13 dB larger than the noise; in this case a signal of -75 dBm would be needed for reliable demodulation, which just happens to be the reader receive threshold shown in figure 3.30 of chapter 3. Go figure.

The phase noise entering the receiver can be reduced by using quieter synthesizers, typically increasing cost and power consumption, or by reducing the transmitter leakage into the receiver. A bistatic reader can expect to achieve 15–25 dB lower phase noise than a monostatic reader, assuming good antenna isolation.

The other source of noise from the transmitter is amplitude variation in the transmitted signal. The desired output of a VCO is a sine wave with no variations in amplitude, so nothing stops us from putting a limiter on the output of the VCO to strip out any amplitude variations (noise) in the oscillator signal. However, the VCO output is typically at around 0 dBm. To get one watt at the transmit antenna, we need 30 dB of gain between the VCO and the output. Recall that an ideal 50-ohm source at room temperature produces -174 dBm/Hz of bandwidth of thermal noise. For the band we're considering (50–100 KHz, or 50 KHz wide), that's -174 dBm $+ 57$ dB $= -117$ dBm. This is the (minimum possible) noise entering the amplifier chain from the matched output of the VCO or limiter. An ideal amplifier would simply amplify this noise by the gain of the amplifier: we'd get about $(-117 + 30) = -87$ dBm of thermal noise at the power amplifier output. Real amplifiers add excess noise, described by the ***noise figure*** NF; if the noise figure were 10 dB, the actual thermal noise is around -77 dBm at the transmitter output. About 3% of this power, roughly -92 dBm, bounces off the antenna and enters the receiver. These estimates are depicted graphically in Figure 4.58.

Just as in the case of phase noise, what happens then depends on the relative phase of this reflected signal and the local oscillator; if the two are in phase (equation (4.37)), the output of the mixer will be linearly dependent on the amplitude of the received signal, so the amplitude noise is converted directly into noise in the baseband of the receiver. (In practice, the receiver has both an I and a Q channel; if by happenstance the I channel is in phase, the Q channel will be in quadrature, so one receiver channel will be dominated by amplitude noise and the other by phase noise.) For our simplified example, we obtain a receiver threshold for reliable demodulation of $(-92$ dBm $+ 13$ dB$) = -79$ dBm: modestly better than the phase-noise-dominated case.

To reduce the amplitude noise from the transmitted signal, we can use lower-noise (more expensive!) amplifiers to reduce the noise figure of the system. If a mixer is used to modulate the output signal instead of a switch, any amplitude noise present in the VCO signal is generally removed since the LO drive level is high enough to cause the mixer to act as nearly an ideal switch. However, in this case any noise on the

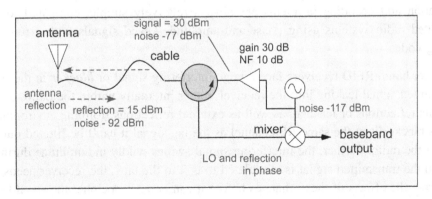

Figure 4.58
Amplitude noise from the VCO is amplified and can leak into the receiver.

low-frequency modulating input (which might arise in a digital-to-analog converter) is amplified and appears at the output. Just as in the case of phase noise, a bistatic configuration will have 15 to 25 dB lower transmit leakage and thus lower amplitude noise in the receiver.

The calculations given above are, of course, very simplified and not precisely representative of any actual reader, but provide rough guidance for the sort of noise performance that can be achieved by a typical homodyne radio system. By reference to section 6 of chapter 3, we can see that for a monostatic system with the noise thresholds we've given above, the read range of a tag is limited to around 15–20 meters by receiver sensitivity. Current passive tags will run out of power to drive the tag IC before they get this far from the receiver, since as we've noted before a passive tag is usually forward-link-limited, but as tag IC's improve receive sensitivity will become more important. For semi-passive tags, the reverse link budget will limit read range, and we can clearly see the benefits of using a bistatic radio in this case; reducing the noise by 20 dB will improve our range by $20/4 = 5$ dB $= 3X$, so that we can expect semi-passive tag reads at around 50 meters for an ideal link, and longer ranges (albeit with fading) in a real environment with reflections.

4.5 RFID Receivers

RFID receivers face a unique set of requirements, quite distinct from those encountered in most other radio systems, particularly when passive tags are used.

In some respects, RFID receivers are easy to build. Because passive tags require so much forward-link power, there is usually not much point in constructing a receiver at the theoretical sensitivity limit, because the tags will have lost power to their IC's by the time they get that far away. The limitations of passive tags also mean that the return link modulation is always some variant of frequency-shift keying, so the demodulation and decoding problems are always relatively simple, compared to more sophisticated radio systems using phase-and-amplitude-keyed signals with error-correcting codes.

On the other hand, RFID receivers face a huge interfering signal or *blocker* in the form of the transmitted signal leaking into the receiver, either internally within the radio, or from the antenna reflections or leakage, as well as external reflections from the environment. Since this blocker is at the same RF channel as the tag signal it can't be filtered out in the RF part of the radio. Further, the interfering signal swings wildly in amplitude during the time when the transmitted signal is modulated to talk to the tags; the receiver needs to recover from the effects of these changes before it can hope to decipher the small tag response. Finally, because (absent Doppler shifts) the received signal and the local oscillator signal are at exactly the same frequency, the absolute phase of the received signal influences the amplitude of the downconverted signal (Figure 4.18), so some sort of phase diversity must be provided to ensure visibility of the tag signal.

4.5.1 Receiver Architectures

The basic receiver architecture for an RFID receiver is that of a direct-conversion I/Q demodulator (Figure 4.59). The received signal is split and directed to two mixers, one excited with the local oscillator signal and the other with the LO signal shifted by 90 degrees (that is, with $\cos(\omega_c t)$ and $\sin(\omega_c t)$ respectively). The received signal is mixed in each branch with the LO signal, and the resulting output is filtered to remove the carrier and harmonics, leaving behind a low-frequency signal containing the tag response.

Recall that the phase of the tag reflection is not predictable or controllable, as it varies by $2\pi = 360°$ each time the distance to the tag changes by $\lambda/2 \approx 16\,\text{cm}$. If the reflected signal from the tag is in quadrature with the LO in the I-branch, it produces zero output voltage from the mixer (Figure 4.18), but in this case the received signal is in phase with the LO signal in the Q branch and produces the maximum possible signal. In the case where the phase of the received signal is in between that of the I and Q LO signals, both branches will generate an intermediate output level.

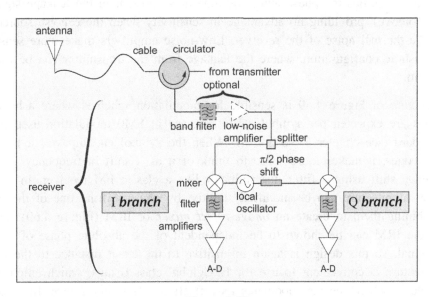

Figure 4.59
Generic I/Q receiver.

Thus, performance can be expected to vary slightly with range to the tag, but in all cases a signal ought to be received.

In conventional radio design, the received signal is almost always directed first to a band filter, to remove interfering signals from outside the band of interest (in this case 902−928 MHz in US operation). In an RFID radio, the band filter may not be very helpful. Because the sensitivity requirements for a radio for passive tags are not very demanding, rejecting low-amplitude out-of-band interferers is less important than in a conventional radio. Recall that the main mechanism by which interferers cause problems is through third-order distortion (Figure 4.15); if mixers with high third-order intercept (section 3.2) are used, the resulting interfering signal amplitude is likely to be too small to concern us in a passive tag receiver. In addition, as we noted previously (Figure 4.35), commercially-available filters for the US ISM band (902−928 MHz) cannot reject the top part of the nearby cellular downlink band (869−894 MHz), so some likely interferers will slip through the filter in any case. A band filter may be omitted in a low-cost compact reader design intended only for monostatic operation, but included in a high-cost portal reader designed for bistatic operation.

A conventional radio also employs a low-noise amplifier (LNA) to increase the signal strength prior to the lossy splitter and mixer. In a monostatic RFID receiver, this LNA may cause more harm than good. The large antenna reflection entering the receiver may exceed the signal level over which the LNA is able to amplify with good linearity, causing reduced sensitivity for the small tag signal, while the amplification provided by the LNA also amplifies the phase and amplitude noise present in the leakage signal (section 4.3 above), providing no advantage in sensitivity when those noise sources dominate the thermal noise of the receiver. Low-noise amplifiers make more sense when used in a bistatic configuration, where the leakage from the transmitter can be kept below 0 dBm.

The architecture of Figure 4.59 is sensible for modulation schemes where a handful of signal edges are expected per symbol, as is the case in FM0 modulation used in the Gen 2 standard (see chapter 3, section 4). When the symbol encompasses a large number of edges, it makes more sense to think of it as a shift in frequency, and detect this frequency shift using a filter discriminator like a classic FM receiver. In this case, the receiver architecture can be modified slightly, by phase-shifting one of the branches and recombining them to create an *image-reject mixer* or IRM (Figure 4.60). The output of the IRM can be shown to be independent of the absolute phase of the received signal, so this design is again insensitive to the exact distance to the tag. The IRM architecture is convenient to use for EPCglobal class 0 tags, which employ an FSK scheme in which the tag modulates its reflection at roughly 2 MHz for a binary 0 and 3 MHz for a binary 1.

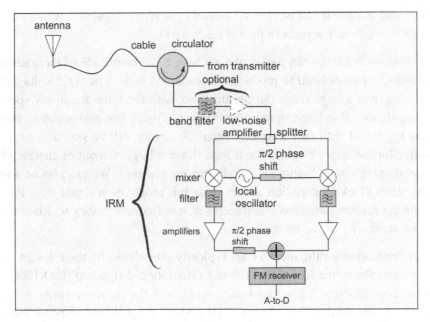

Figure 4.60
Image-reject mixer configuration for frequency-shift-keyed tag signal.

As is apparent from the discussion above, the choice of monostatic or bistatic operation affects radio design and performance. Bistatic radios will generally have superior senstivity to monostatic radios, but when the forward link budget is the main limitation performance may not be strongly affected by receive sensitivity. A bistatic radio is more expensive because of the doubling of antennas, antenna cables, and antenna connectors, though the cost of a circulator within the radio may be avoided in this case. Bistatic configurations are impractical for handheld or portable readers, and less convenient than monostatic radios for many fixed applications. Thus the receiver design choices are affected by the overall system architecture, which in turn is determined by the envisioned usage models for the reader.

4.5.2 DC Offsets and Recovery

An RFID receiver for passive tags is generally a homodyne radio: the received signal is at the same carrier frequency as the local oscillator. Any reflected signal, such as those from the antenna, stationary objects in the antenna field, and internal reflections in the radio, is mixed with the LO and contributes a signal at zero frequency – that is, at DC. The DC voltage, or *DC offset*, generated depends on both the phase and amplitude of the reflected signal (see Figure 4.18). When the transmitter is operating in CW mode, this offset is truly a DC voltage, and as such it can be easily blocked by a series capacitor, which doesn't allow DC or low-frequency signals to pass (Figure 4.61).

However, during the reader-to-tag part of the exchange, the amplitude of the transmitted signal is intentionally modulated to produce the reader symbols. The 'DC' voltage will potentially swing over a large range during this time, with the same frequency spectrum of the reader modulation. This large signal may drive amplifiers into compression, preventing receipt of the tag signal until they have recovered. Recovery will be slow if we use a large value capacitor for the series block, since it then stores a large amount of charge which must leak out into the circuit before DC conditions are restored. We can choose a small value for the series blocking capacitor and exclude this modulation signal from the receiver, but then if the tag response also has components at low frequency, they will be excluded as well, reducing sensitivity of the receiver.

The timing requirements for the receiver are typically established by the relevant tag protocol. For example, in the EPCglobal Class 1 Generation 2 standard (ISO 18000-6C), the delay between completion of a command and initiation of the tag reply varies with the data rate. For a reader average data rate of about 100 Kbps, the tag should respond about 18 microseconds after the reader is finished transmitting (the requirements being somewhat dependent on the tag data rate as well). When the lowest allowed reader data rate is employed, this time is extended to about 75 microseconds. Receiver recovery problems may

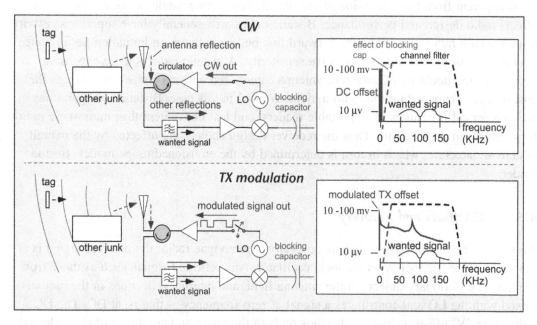

Figure 4.61
Offset problems for direct-conversion RFID receiver.

be invisible at low data rates, but show up as degraded sensitivity when high reader data rates are attempted.

One solution to the receiver recovery problem is to use some sort of adaptive receiver, that changes its configuration depending on whether transmit modulation is active. The simplest approach is to place a series switch in front of the amplifiers, and turn the signal off when transmit modulation is on. Alternatively, one can use adaptive filtering that cuts off the low-frequency modulation signal when the transmitter is modulated, and restores it when the modulation is done and the tag signal is to be received.

4.5.3 Phase and Amplitude Noise and Sensitivity

The sensitivity of the receiver is limited by the noise that enters it. As we discussed in some detail in section 4.2 above, the largest source of noise for an RFID receiver is usually the leakage from the transmitted signal, particularly when a monostatic configuration is employed.

As we saw in our previous discussion, the amount of amplitude noise entering the receiver is maximized when the transmitted leakage from the receiver is in phase with the local oscillator signal, and the amount of phase noise converted to amplitude noise in the receiver is maximized when the transmitted leakage is in quadrature with the local oscillator. Since the receiver normally has two branches, I and Q, in quadrature, the dominant noise source in each branch will depend on the relative phase of the local oscillator and the transmitter leakage, which depends on things like the exact length of the antenna cable and is not usually well-controlled. The conversion of phase noise to amplitude noise is also dependent on the total phase (or delay) associated with the leakage path, since this total phase controls the effect of a change in frequency on the phase relationship between the leakage and the LO. Thus the importance of phase noise depends on cable lengths and other delays.

Let us imagine, for example, that a long antenna cable is in use, and phase noise from the transmitter is the main noise source entering the receiver. Let us further imagine that the leakage happens to be in phase with the LO at the receiver, so that the I-channel is relatively quiet and the Q-channel relatively noisy (Figure 4.62). When the tag is in phase with the LO, the signal on the I-channel is strong and encounters little noise. When the tag is moved an eighth-wavelength farther away, the tag signal will be 90° out of phase with the LO, and the best signal will be found in the Q-channel, where it must compete with converted phase noise. In this case, we can expect that the receiver sensitivity will vary with tag distance, with a period of 1/4 of a wavelength (about 8 cm for ISM-band operation).

Naturally, there is no particular reason why one channel would be preferred: the phase noise could just as easily be mainly in the I channel. Depending on the exact

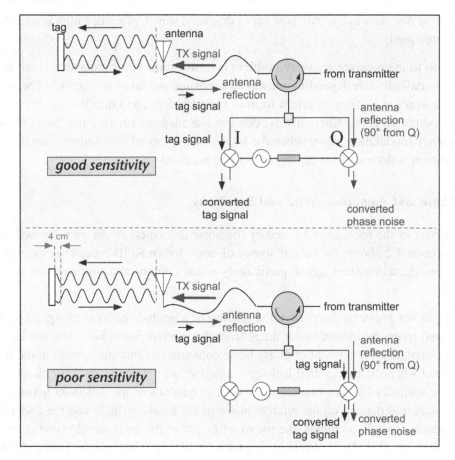

Figure 4.62
Receive sensitivity variation with tag signal phase due to converted phase noise.

characteristics of the radio and cabling, the amplitude or phase noise may dominate, or even be of comparable magnitude, so that there would be little distinction between the I and Q branches and the receiver might be insensitive to absolute phase of the tag signal.

4.5.4 Example Design Calculations

To get a feel for how a design is constructed, let's look at a few example design calculations. We'll start with the question of interference rejection: can we tolerate nearby readers operating when we're trying to receive?

Let's imagine there are two other readers, say 5 and 10 meters distant from ours, all using 6 dBi antennas and transmitting at 1 watt. How much power reaches the receiver from these neighboring transmitters? To arrive at an estimate we employ the Friis equation (equation (3.20)). We find for the 5-meter distance:

$$P_{RX} = P_{TX} G_{TX} G_{RX} \left(\frac{\lambda}{4\pi r} \right)^2 = (1)(4)(4) \left(\frac{0.33}{4\pi r} \right)^2 \tag{4.41}$$
$$\approx 0.45 \text{ mW } (-3.5 \text{ dBm})$$

The 10-meter-distant transmitter will contribute 6 dB less power, or about -9.5 dBm. In both cases the received power is greatly in excess of the tag signal for tags more than a few cm distant, so if either of these two readers happen to be on our channel we will not be able to read tags (unless they are using EPCglobal Generation 2 Dense Interrogator mode; see chapter 8). In US operation, the chance that one of the two readers will hop onto the same channel as our wanted reader is about 1/25. If more potentially interfering readers are present, the chance of co-channel interference increases linearly, and the interfering signal doesn't fall off very rapidly: a reader 50 meters distant will still contribute a received signal of about $(-3.5 - 20) = -23.5$ dBm. When a large number of readers are present in a confined space it is necessary that the duty cycle be low, or it will become very difficult for any of them to work.

The situation is worse if the readers also radiate some power out of their own channel, as is apparent from e.g. Figure 3.12 of chapter 3. Let us imagine the 10-meter-distant reader is not on the same channel as ours but instead on an adjacent channel, but that this adjacent channel contains (say) 1% of the signal energy in the main channel − that is, the power in the adjacent channel is down by 20 dB. The received power is then $(-9.5 - 20) = -29.5$ dBm: still enough to block tags more than 1 or 2 meters away (see equation (3.21)). If the interfering reader is on the next-adjacent channel, the power in the operating channel might be down by as much as 60 dB (this level being compliant with Dense Interrogator operation in the EPCglobal Generation 2 standard, as we shall see in chapter 8). In this case the interfering signal is $(-9.5 - 60) \approx -70$ dBm, and will cause a problem only for tags fairly far from our reader. So while we can have some hope of dealing with interferers two channels away, readers on the adjacent channel to ours will also block our ability to read any but nearby tags. If we consider only next-adjacent channels as non-interfering, we have only 25 non-interfering 'channels' to hop to in typical US operation using 500 KHz channels. With even 3 collocated readers, interference can be expected roughly 8% of the time, and with 25 readers in one facility, reads will be very difficult if all are simultaneously active.

So far we have only considered power radiated into the intended channel due to modulation of the interfering signal. Even if a band filter is present it cannot reject the other reader signals, which are in band. Recall that non-linearity in our receiver can cause intermodulation products to be generated from signals on other channels (Figure 4.15).

These intermodulation products will fall on our wanted channel whenever two interfering readers are equally-spaced; i.e. if our receiver is tuned to channel 1, and the interferers are on channels 3 and 5, or 8 and 15, etc. The intermodulation signal can be found from the magnitude of the interfering signals, and the third-order intercept of the input low-noise amplifier (if present) or mixer (Figure 4.13). Guided by the results above, let us estimate that the interfering signals might be around −10 dBm. In order for the third-order distortion products to be less than −70 dBm, we must be backed off by (60/2) = 30 dB from the input third-order intercept. Thus we require an input device with IIP3 = −10 + 30 = 20 dBm or more. Such performance is achievable from commercially-available components, but at relatively high cost.

How much gain do we need after downconversion (that is, after the mixer)? A tag signal entering a monostatic receiver at −65 dBm might encounter (for example) 2 dB losses in the circulator and switches, 7 dB conversion loss in the mixer, and 1 dB insertion loss in the carrier-reject filter. The signal entering the baseband chain would be about −75 dBm. In a printed-circuit radio, this signal may be in a 50-ohm environment, in which case the associated voltage is around 50 μV. An integrated reader (on an IC) is likely to use higher input impedances in the baseband, resulting in a peak voltage of around 200−400 μV. The amount of gain required to obtain a readable signal depends somewhat on the approach to digital-to-analog conversion, discussed in more detail in the next section. Let us for simplicity assume the final stage is a comparator: that is, a device whose output is either a digital high or digital low level, of roughly 1 V. The required voltage gain is then on the order of 3000−20000, or around 75−85 dB. It is apparent that a few tents of millivolts of transmitter modulation that leaks through to the receiver will readily saturate the receiver, leaving a small tag signal invisible.

4.6 Digital-Analog Conversion and Signal Processing

The general topic of digital signal processing has filled a number of books, and to cover any substantial portion is far beyond the scope of our discussion here. We shall limit ourselves to a few remarks specific to RFID.

Because of the modest computational capabilities of passive tags, only relatively simple modulation schemes are employed for passive RFID, and data rates are low relative to most other communications technologies. When simple ASK symbols are used, the transmitter can be based on a simple switch (equivalent to a 1-bit digital-to-analog converter), with optional analog filtering to smooth transitions and reduce spectral width. Filtering can also be performed in the digital domain using standard techniques such as *finite-impulse-response filters* (FIR) − in this case requiring a more sophisticated DAC with multiple bit resolution. It is important to note that the DAC can then be a source of amplitude noise, and it may be useful to bypass or fix the DAC output during the CW portion of operation, when the receiver is attempting to decipher a possible tag reply.

On the receive side, the optimal approach for signal processing depends on the protocol in use. As we shall see in chapter 8, the EPCglobal class 0 protocol uses subcarrier-modulated frequency-shift keying, with frequencies of 2.2 or 3.3 MHz, for the tag reply to the reader. The relatively high frequency means that one can use a conventional analog FM discriminator, which produces a simple binary output (either 1 or 0) depending on the tone frequency. (Both outputs may occur if a collision is present.) Recovery of the tag clock is not needed since the locations of individual edges in the tag signal don't matter.

Other protocols use approaches in which only one or a few edges (abrupt transitions in the amplitude of the backscattered signal) determine whether a symbol is a binary 0 or binary 1. Since tags normally have only two states, in principle it is sufficient to sample the data once for each change in the tag state, and it is only necessary to provide one bit of information: whether the signal is high or low.

For example, EPCglobal Class 1 Generation 2 uses FM0 signaling, in which a symbol with no transition in the middle is a binary 1, and a symbol with a transition in the middle is a binary 0. In this case, it is vital to be able to determine where the boundaries of each symbol are, which is equivalent to synchronizing the reader to the tag's clock, a task known as *clock recovery*. Clock recovery can be challenging in passive RFID because the tag clocks are not required to be particularly accurate: symbol rates can vary by 10% or more. Consider a tag reply containing a 96-bit electronic product code and 16-bit error check (112 bits): if we obtain synchronization on the first few bits but get the clock timing wrong by 10%, by the end of the message our guess at where the symbol edges lie will be 10 symbols in error! Clearly this is not acceptable for accurate decoding.

One approach to solving this problem employs a block-by-block comparison of the received data with the expected symbols. This approach depends on *oversampling*: the signal is sampled several times per symbol rather than just twice. Fixed blocks of samples are considered, with transition rules that depend on what happened in the previous block. Drift in the clock is implicitly accounted for by these rules, rather than being explicitly applied to vary the sample timing.

For example, let us imagine we extract eight samples of the returned signal amplitude in each nominal symbol time. If the symbol is a binary 1, we expect that all the samples will have the same value (either high or low), whereas if the symbol is a binary 0 the first four samples will have one value, and the last four samples the other. In addition, there are transitions at each symbol edge, so we expect a sign change between adjacent blocks of 8 samples. These expectations are illustrated in sample set (a) of Figure 4.63. In this ideal

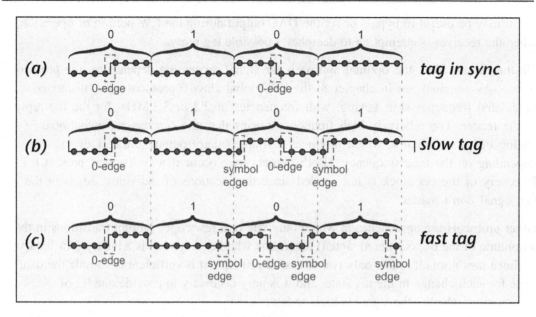

Figure 4.63
Sampling FM0 with tag clock drift.

case it is only necessary to examine whether the first four samples in each block of 8 are the same as the last four samples. Symbol edge transitions always occur between the sample blocks and can be ignored.

Life is more complex when the clock drifts. Sample set (b) illustrates a case where the tag clock is slower than the reader clock. Initially the symbol edges are well-aligned with the 8-sample block and the binary-0 transition is centered. However, as time passes, the symbol edge drifts to the right past a sample time and appears within a block, and the binary-0 transition (when present) becomes off center. Multiple signal transitions are present within a single block of 8 samples. In order to distinguish between symbol edges and meaningful code transitions, we now need to remember what happened in the previous block. By imposing the rule that edges can move at most one sample point to the left or right between successive blocks, we can distinguish between symbol edge transitions and coding transitions (assuming that the distinction was properly made to start with!). We must also account for the case where, for example, the coding transition drifts entirely out of the right side of one block and appears at the left side of the next block (*aliasing*). When such an 8-sample shift is seen, the block in which it occurs will not contain valid data and is ignored: the putative coding edge has moved to the subsequent block (if present).

When the tag clock is faster than the reader clock (sample set (c)), the edges move slowly to the left. Here it is possible to have two coding transitions occur in one block of 8 samples, thus producing two bits from a single block. Proceeding in this fashion, only fixed blocks of samples need be examined, and clock drift is corrected by allowing blocks with no valid resulting bits, and blocks with two resulting bits. Approaches of this type are simple to implement with modest computational requirements, but are somewhat sensitive to noise and offsets: we are relying on the analog circuitry to properly establish the average voltage so that high and low samples accurately reflect the tag's backscattered signal.

An alternative approach is to employ a sliding correlator (also known as a matched filter) for each possible symbol of interest, and assign values to successive symbols based on the resulting correlation scores. Correlation approaches are computationally more demanding (though this approach lends itself to implementation in an ***application-specific integrated circuit*** (ASIC)). A correlator can make efficient use of oversampled data to distinguish between valid symbols and noise or offsets.

It is also possible to transform a set of data into the frequency domain using a ***Fast Fourier Transform*** (FFT). Once in the frequency domain, we can readily filter out high-frequency noise and low-frequency offsets and drifts, and then return to the time domain to extract data bits. This approach is also relatively demanding of computational resources and could benefit from dedicated hardware for FFT's and other operations.

4.7 Packaging and Power

Modern reader radios are generally fabricated by placing discrete components and integrated circuits onto printed circuit boards, the latter being multilayer composites of copper and polymer that provide wiring for interconnections and ground, and to a lesser extent cooling. The boards are normally packaged within metal housings, which provide the dual benefits of isolating the radio's components from interferers in the outside world, and isolating the outside world from the radio. Ideally, the only high-frequency emissions from the reader are at the antenna connector ports.

It is often useful and sometimes indispensable to provide additional isolation between regions of each circuit board, or between neighboring circuit boards. In high-performance circuitry mostly used in military applications, each circuit board resides in its own space milled out of a thick aluminum plate, with shielded interconnections between segments. This sort of approach is prohibitively expensive for most commercial equipment; instead, it is more common to apply covers of formed sheet-metal over sensitive regions of a radio's receiver or synthesizer. Local isolation is also useful to help confine harmonics of the output frequency that may be generated by the power amplifier (see section 3.1 above): a 900-MHz signal, with a wavelength of 33 cm, will not escape very readily from a small hole in the reader housing that is made to allow for a control button or Ethernet connection, but the fourth harmonic at a wavelength of about 8 cm, is much more slippery.

4.8 Capsule Summary

Radio transmitters must be accurate, efficient, and transmit within their allowed frequency band. Receivers must be sensitive (but not to criticism), selective, and detect a huge range of signal strength. Both must be flexible. RFID reader radios usually operate in unlicensed bands and thus must support frequency hopping or other interference-mitigation provisions. RFID readers also have the peculiar problem of being both full- and half-duplex; the use of a bistatic antenna configuration may be beneficial. Because they receive a backscattered signal, RFID receivers are generally configured as homodyne rather than heterodyne radios. The leakage from the transmitter creates offsets which must be filtered or blocked.

Radios are constructed of a number of key components. Amplifiers are characterized by gain, power, bandwidth, noise, and distortion properties, which are often reported in terms of second- and third-order intercepts. Mixers are more complex, and in addition to conversion loss, bandwidth, noise, and distortion, one must consider isolation and a large number of possible spurious output frequencies.

Oscillators are generally constructed using positive feedback through a resonant circuit. Oscillator amplitude noise can readily be removed by limiting the output; phase noise is not so easily dealt with. The resonator quality factor plays a critical role in determining the phase noise of the oscillator. Oscillators use a variable component such as a varactor to adjust the frequency of oscillation.

An oscillator is generally embedded in a phase-locked loop to form a synthesizer, which produces an output signal bearing a controlled relationship to a very stable reference frequency. Integer-N and fractional-N synthesizers can both be used in RFID applications; fractional-N synthesizers are more versatile but more complex to design and can suffer from additional noise.

Filters remove unwanted frequencies from a signal. Filters built of discrete components are usually limited to quality factors of around 10−20 and thus cannot filter very narrow bands; other technologies, such as surface-acoustic-wave devices or dielectric resonators, are used to achieve narrowband high-frequency filtering. Once the signal is converted to baseband, filtering can use discrete components or active filters created by combining an operational amplifier with a frequency-dependent feedback network. The transmitter must be modulated; this may be done with a simple switch, or use a digital-to-analog converter, such as a current-steering circuit. The received signals must at some point be converted from analog voltages to digital data. There are several architectures for performing this operation, including flash and delta-sigma converters.

Monostatic RFID readers also employ specialized microwave components − circulators and directional couplers − that are capable of selecting signals based on their direction of travel.

Transmitter architectures trade efficiency, cost, and transmit bandwidth. Filtering of the transmitted symbols greatly improves the spectral width of the output, but even more improvement can be achieved using phase-reversal amplitude-shift keying, which requires a single mixer for implementation, or single-sideband modulation, which requires a quadrature modulator. Power efficiency is usually dominated by the output power amplifier design; designers must trade off linearity (and thus transmit bandwidth) against efficiency. Phase noise (from the oscillator) and amplitude noise (from the amplifier chain and the DAC) are important because some of the transmitted signal leaks into the receiver and can limit sensitivity. Transmit phase noise is converted to amplitude noise due to the delay-line-discriminator action of the transmitter output, cable, and antenna. The influence of these noise sources is dependent on absolute phase and likely to be different on the in-phase and quadrature channels of the receiver.

RFID receivers are generally homodyne architectures with an I-Q direct downconverter driven by the same oscillator that provides the transmitted signal. For protocols that use pure frequency-shift-keyed signals, an image-reject mixer configuration may be used instead of an I-Q converter. Transmit leakage and other reflected signals will mix to DC, creating offsets that are easily filtered when unchanging, but problematic when they are modulated at rates similar to those employed by the tags. A switch may be used to block the receiver input when the transmitter is modulating. Transmit amplitude and phase leakage will limit sensitivity for monostatic configurations and may still be a problem for bistatic readers.

Once a received signal is digitized, the whole array of digital signal processing techniques becomes available to decipher it. The receiver must somehow sample the signal properly despite inaccurate and inconstant tag clocks. Very simple schemes employ block comparison on a fixed number of oversampled points; more sophisticated approaches use sliding correlators or Fast Fourier Transforms of the data.

Reader radios are constructed of IC's and discrete components assembled on composite circuit boards. The package in which the reader is placed protects the components from the outside world, protects the user from the component voltages, and may play an important role in keeping spurious signals confined to achieve regulatory compliance.

Further Reading

RFIC Design:

The Design of CMOS Radio-Frequency Integrated Circuits, Thomas Lee, Cambridge, 1998: *An encyclopedic introduction to the design of radio components, though the emphasis is much broader than purely CMOS implementation (which was probably added to the title to increase sales). Includes treatments of synthesizer operation, oscillator phase noise, and feedback design.*

Analog-digital conversion:

"Delta-Sigma Data Conversion in Wireless Transceivers", Ian Galton, IEEE Transactions on Microwave Theory and Techniques, vol. 50, #1, p. 302 (2002)

"Analog-to-digital converter survey and analysis", R. Walden, IEEE Journal on Selected Areas in Communications, vol. 17 #4, p. 539 (1999)

Amplifiers:

RF Power Amplifiers for Wireless Communications, Steve C. Cripps, Artech House, 1999: Cripps is bright, opinionated, and brings extensive practical experience to bear on abstruse topics in amplifier design.

Design of Amplifiers and Oscillators by the S-Parameter Method, George Vendelin, Wiley Interscience, 1982: purely microwave-oriented, antedating modern CMOS and SiGe devices, but a useful reference and introduction to matching techniques, low-noise and broadband design.

"A Fully Integrated Integrated 1.9-GHz CMOS Low-Noise Amplifier", C. Kim et. al., IEEE Mcrowave and Guided Wave Letters, Vol. 8, #8, p. 293 (1998)

"On the Use of Multitone Techniques for Assessing RF Component's Intermodulation Distortion", J. Pedro and N. de Carvalho, IEEE Transactions on Microwave Theory and Techniques, vol . 47, p. 2393 (1999)

"Weigh Amplifier Dynamic-Range Requirements", D. Dobkin (that's me!), Walter Strifler and Gleb Klimovitch, Microwaves and RF, December 2001, p. 59

Mixers:

A great deal of useful introductory material on mixers was published over the course of about 15 years by Watkins-Johnson Company as TechNotes. These have been rescued from oblivion (in part by the current author) and are available on the web site of WJ Communications' acquirer, TriQuint Semiconductor, at http://www.triquint.com/products/tech-library/wj-tech-publications.cfm

The material is focused on diode mixers but many issues are generic to all mixer designs. Of particular interest are:

"Mixers, Part 1: Characteristics and Performance," Bert Henderson, volume 8

"Mixers, Part 2: Theory and Technology," Bert Henderson, volume 8

"Predicting Intermodulation Suppression in Double-Balanced Mixers," Bert Henderson, Volume 10

"Image-Reject and Single-Sideband Mixers," Bert Henderson and James Cook, volume 12

"Mixers in Microwave Systems, Part 1," Bert Henderson, volume 17

Reader Architecture and Signal Processing:

"Short-range Radio-telemetry for Electronic Identification using Modulated RF Backscatter", A. Koelle, S. Depp and R. Freyman, Proc IEEE August 1975 p. 1260: *nearly all the elements of a modern UHF RFID system three decades in advance. Plus ça change, plus c'est la meme chose.*

"Design considerations for embedded software-defined RFID readers", Matthew Reynolds and Christopher Weigand, RF Design August 2005 p. 14

"Data Recovery", Nick Sawyer, Xilinx Application Note XAPP224 (v2.5), July 11, 2005

Exercises

RFID configurations:

1. How many antennas are used simultaneously for transmit and receive functions in a monostatic RFID reader?

   _____

2. A bistatic antenna provides 45 dB of isolation from transmit to receive, and is used with a 1/2-watt transmitter. How much transmit power leaks into the receiver?

   _____ dBm

Radio architectures:

3. A superheterodyne receiver (commonly used for conventional radio receivers, though not typical for RFID eqiupment) uses a local oscillator frequency of 800 MHz to receive a signal at 867 MHz. What is the intermediate frequency (IF)? What is the image frequency?

   IF: _____ MHz   image: _____ MHz

4. A direct conversion RFID receiver is intended to successfully receive a tag signal as small as -65 dBm, with a resulting output voltage swing of 0.5 V to drive a comparator with a 300-ohm input resistance. What is the overall gain of the receiver? What voltage would be produced by a -10 dBm input signal?

   Gain: _____ dB   interferer output: _____ volts

Radio components:

5. An amplifier has a gain of 14 dB. An input signal of -10 dBm causes the amplifier to suffer 1 dB of gain compression. What is P1dB?

   P1dB: _____ dBm

6. An RFID receiver operating at 906.5 MHz is also illuminated by two other nearby readers operating at 909 and 912.5 MHz. What spurious frequencies might be produced in the receiver by third-order distortion? Are they likely to interfere with the tag signal at 906.5 MHz?

   low spur: _____ MHz   high spur: _____ MHz

   interference concern?: ___ yes ___ no

7. The reader signals in problem (6) above are both at -15 dBm, at the input into a mixer with an input third-order intercept of 5 dBm. (Here we are defining the intercept as the power at which a spurious output tone is equal to the power of one of the two input tones.) The wanted signal (the one you're trying to hear) is at -50 dBm at the

mixer. What is the level of the output spurious tones from the mixer relative to the wanted signal?

spurious power: _____ dBc

8. A special low-rate, long-range RFID tag can be received by a receiver using only 10 KHz of bandwidth. The room-temperature receiver uses an amplifier with a noise figure of 4 dB, and enough gain so that other sources of noise in the receiver are negligible. What is the smallest signal that can be detected, if a signal:noise ratio of 12 dB is needed?

minimum detectable signal: _____ dBm

9. A mixer is used to detect the tag signal in problem (8) above. Both the tag signal and the local oscillator are at the same frequency, 905 MHz. What are the frequencies of the two output signals from the mixer? Does either signal depend on the absolute phase difference between the tag signal and the LO?

low output: _____ MHz   high output: _____ MHz

10. A local oscillator employs an inductor of 15 nanoHenries (nH) in the resonant feedback circuit. What value of tuning capacitance (picoFarads, pF) must be used to make the oscillator operate at 915 MHz?

C_{res}: _____ pF

11. The local oscillator above is connected to a phase-locked loop, with a compare frequency of 500 KHz. What value of the divisor, N, is needed to produce the requisite 915 MHz output?

N: _____

12. What filter bandwidth is needed to accept the European (ETSI) RFID band at 865–869 MHz? Is such a filter narrower or broader than a filter designed to accept the US ISM band (902–928 MHz)?

bandwidth: _____ %   vs. US ISM: _____

13. An analog-to-digital converter with a maximum input of 1 volt and a resolution of 8 bits is used to digitize the output of an RFID receiver. What is the smallest change in input voltage that can be resolved, neglecting any analog noise in the system?

minimum resolvable signal: _____ V

RFID Transmitters:

14. A transmitter uses a simple switch to turn the signal on or off. Is the output spectrum wider than that of a more elaborate transmitter that uses a filter to smooth the output signal?

____ yes ____ no

15. A receiver is used to capture the output of an RFID transmitter. Upon careful examination, we find that the positive peaks of the RF signal from the (RF-on) part of a symbol occur at the times we would have expected a negative peak based on the previous symbol. What modulation scheme is probably being used?

    amplitude-shift keying (ASK) ___ phase-reversal ASK ___

    single-sideband (SSB) ___

16. A monostatic RFID reader is connected by a 1-meter cable to the antenna. By measurement it is found that the phase noise-amplitude noise conversion efficiency is −40 dB. The antenna is then moved farther away from the reader, so that a 10-meter (low loss!) cable replaces the 1-meter cable. What is the worst-case effect on the phase noise conversion?

    phase-amplitude noise increased by: _____ dB

RFID Receivers:

17. To save money and time, Bob the lazy RF designer[2] constructs an RFID receiver with only a single branch (which we will call the in-phase or I branch). He calls his managers into the lab and demonstrates the ability to read a tag reliably when the tag is 240 cm from the reader antenna, using a fixed frequency of 915 MHz for the demo. How can dedicated designer Amy, witnessing the demo, embarrass her rival and get him fired?

    ___ rotate the tag around its axis

    ___ move the tag 16 cm farther from the antenna

    ___ move the tag 8 cm closer to the antenna

    ___ remove the tag completely and show that it is still being read, since she saw Bob setting up software that fakes reads the previous night

18. Bob also neglected to provide a blocking capacitor between the mixer output and the rest of his receiver, but got lucky in the first demo as the phase of the reflected signal from the antenna happened to be in quadrature with the local oscillator signal and generate no offset voltage. Amy saw the schematic lying on Bob's desk and noted the omission. While the managers are distracted by Blackberry messages regarding SEC investigations of their stock options, she changes out the antenna cable for one slightly longer. How much length does she need to add to put the reflected signal in phase with the LO and swamp the receiver with a huge DC offset, so that no tags can be

---

[2] it will be understood that this is a purely theoretical construct; all real RF designers work so hard that the author is sweating just thinking about it.

detected even with valid software? (Assume the propagation velocity of signals on the cable is 60% of that of light in vacuum.)

___ cm

19. Will Bob realize the error of his ways after losing his job, and actually read this book (which has been sitting on his shelf unopened for several months), or will he become a physical therapist at a Sudoku-addiction treatment clinic? Which path is more likely to provide a lucrative future career?

CHAPTER 5

UHF RFID Tags

5.1 Power and Powerlessness

Tags identify objects. When the objects are very expensive, the cost of the tags is of little consequence, but their endurance is of great import, since expensive objects, and our interest in them, are usually also long-lived. When the objects are cheap, the tags must be cheaper. These are the fundamental dynamics of tag design.

As a consequence, tags that are intended to label long-lived, expensive objects (typically viewed as assets on someone's books) are usually active tags, with their own radio transmitter and receiver, powered by a local battery. Since battery technology has progressed very slowly relative to semiconductor technology, the key issues in designing a passive tag are to minimize the *duty cycle* – the proportion of time during which the tag is doing something other than sleeping – and to minimize the power required both to support the active state and the sleep or idle state. While these are not trivial design challenges, the technology used to fabricate an active tag is substantially similar to that employed in other radios including the reader radio: discrete components and integrated circuits are soldered to a printed circuit board, with the whole attached to a compact antenna and placed within a plastic housing.

Passive tags are mostly meant to identify inexpensive objects, and must thus submit to an economic asceticism that eschews such luxuries. Conventional batteries are far too bulky and expensive to be considered. A conventional radio transmitter or receiver, with the complex and expensive oscillators, mixers and synthesizers over which we labored so diligently in chapter 4, is out of the question. Only inexpensive, low-speed circuitry and simple logic are permitted to us if the tag is to be powered by the pittance of microwatts available at several meters distance from a reader (chapter 3). Instead of a proper transmitter, a switch to change the impedance presented to an antenna must suffice. A single integrated circuit is usually the only electrical component to be placed on the tag, and thus this circuit must be a custom design solely for its specialized application. The expense of creating such an *application-specific integrated circuit* (ASIC) implies that only large volume usage can provide an economic return to the company responsible for it.

Between these extremes lie semi-passive tags, possessed of a battery but bereft of a radio. To date such tags have typically been constructed for specialized applications with moderate volumes, such as auto tolling, and employ fairly conventional fabrication and design approaches, with special attention to duty cycle just as for active tags.

A greatly simplified diagram of the electrical constituents of a passive tag is depicted in Figure 5.1. The radio signal at around 900 MHz from the reader is converted by the antenna in an alternating current, from which the tag must extract both power and information. The tag must then interpret the resulting data, possibly requiring writing data to non-volatile memory, and modulate the load presented to the antenna in order to change the backscattered signal returning to the reader.

In what follows we shall examine a few of the special challenges of designing and manufacturing a passive UHF tag:

- how is power to be extracted from the high-frequency radio signal?
- how can we simultaneously acquire whatever data the reader has sent?
- how do we send back information to the reader?
- how is the resulting chip designed and fabricated?
- how is a completed tag assembled from the chip and other parts?

We shall also briefly examine the corresponding operations for semi-passive (battery-assisted) and active tags, and how to juggle the contrasting requirements of maintaining a communications link with the reader while minimizing the drain on the battery.

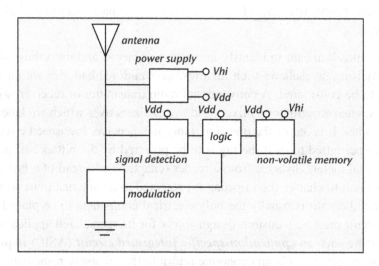

Figure 5.1
Elements of a passive UHF tag.

5.2 RF to DC

To operate, a tag IC needs not just power, but ***direct-current*** (DC) power: a source of voltage that is roughly constant in time, of magnitude from 1 to 3 volts depending on the type of transistors employed in the circuitry, and capable of supplying a few tens of microamps of current. The tag needs to get this DC power from an incoming RF signal whose polarity changes about 900 million times per second, and with the proviso that at a few meters from the reader, a small tag antenna provides an open-circuit voltage of only about 0.1–0.3 V.

To change alternating (AC) voltages to DC, we need an electrical component that treats positive and negative polarities differently: a diode. The left side of Figure 5.2 shows the idealized version of a diode: a component that allows electrical current to flow only in one direction. The right side of the figure shows a more realistic view of a diode's characteristics: in the allowed (forward) direction current turns on slowly until some turn-on voltage is reached, thereafter increasing more rapidly. In the blocked (reverse) direction a small leakage current flows, increasing as reverse voltage is increased.

The actual current flow through a diode is exponential in the voltage; a reasonable approximation is:

$$I = I_0\left(e^{\frac{qV}{kT}} - 1\right) \tag{5.1}$$

where I is current, V is voltage, q the charge on an electron, k is Boltzmann's constant and T the absolute temperature. I_0 is a constant characteristic of the particular type of diode in question. The quantity kT/q is about 0.026 V at room temperature, so if we increase the applied voltage by 1 volt, we increase the current through the diode by:

$$e^{1/0.026} \approx e^{40} \approx 2.3 \cdot 10^{17} \tag{5.2}$$

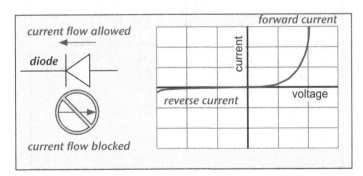

Figure 5.2
Diode schematic symbol (left) and current-voltage characteristic (right).

That is, the forward current increases very rapidly indeed with voltage. For typical values of current flow, we can treat the current as turning on abruptly at some turn-on voltage V_{on}. Two types of diodes are commonly available in standard IC processing: junction diodes, which have saturation currents I_0 around 10^{-10} to $10^{-11}/cm^2$ at room temperature, and Schottky diodes with saturation currents 3−7 orders of magnitude larger, corresponding to a reduction in voltage of about 0.2 to 0.3 V for the same current density. (Schottky diodes are more difficult to fabricate and are not always available, or may increase the processing cost when used.) Rectification can also be accomplished using **_diode-connected transistors_**, in which the drain is shorted to the gate; in this case the turn-on voltage is roughly equal to the transistor threshold voltage.

This idealized version of the current-voltage characteristic is shown in Figure 5.3. The current is zero for all voltages less than the turn-on voltage V_{on}, and can become arbitrarily large when the applied voltage exceeds the turn-on voltage. In this view the diode acts as an ideal switch with an offset voltage. The offset voltage can be estimated from equation (5.1), and varies logarithmically (thus rather slowly) with the DC current required.

Armed with this simplified component model, let us examine the problem of extracting a DC voltage from the RF voltage provided by the antenna to the tag IC. The simplest approach is to place our idealized diode in series with the voltage from the antenna. The result ought to be current flow only in one direction through the diode. We'll use a capacitor to store the current between RF cycles. (Recall from chapter 4 that a capacitor is the analog of a spring. The voltage across the capacitor is proportional to the total amount

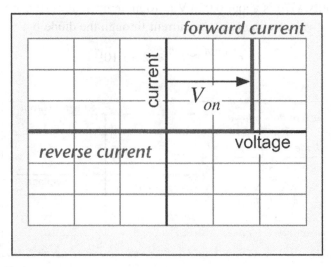

Figure 5.3
Idealized diode current-voltage characteristic.

of current that has flowed into it — the stored charge — analogous to the total extension or compression of a physical spring.) We will represent the remainder of the IC by a load resistor, through which current flows from the capacitor and diode. The whole scheme is shown in Figure 5.4.

Operation of this circuit, treating the diode as an idealized rectifier, is shown in Figure 5.5. When the voltage across the diode is larger than V_{on}, the diode acts like a closed switch, with a voltage offset. A net voltage of $(V_{pk} - V_{on})$ appears across the capacitor and resistor, where V_{pk} is the peak voltage of the signal. During this time a pulse of current flows into the capacitor to charge it up.

When the voltage on the diode falls below V_{pm} the diode turns off. Current now flows out of the capacitor through the resistor, and the voltage across the resistor decreases. The time required for the capacitor to discharge is equal to the product of the capacitance and resistance, *RC*. If this time is long compared to the RF cycle time, the supply voltage will be roughly constant during the RF cycle.

Let's make a simple estimate of the component values required. The RF cycle time is about (1/900 MHz) = 1.1 nanoseconds. Let us assume that the IC consumes about 30 microwatts from a power supply of 1 volt. Since the power dissipated in a resistor is proportional to the square of the voltage and inversely proportional to the resistance we find:

$$P = \frac{V^2}{R} \rightarrow 30 \cdot 10^{-6} = \frac{1}{R} \rightarrow R = 33.3 \ k\Omega \qquad (5.3)$$

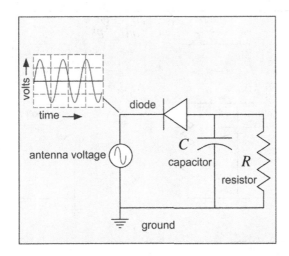

Figure 5.4
Simple rectifier circuit.

To achieve an RC-time constant ten times longer than the RF cycle, we need:

$$RC = 33,300 \cdot C = 11 \cdot 10^{-9}$$
$$\rightarrow C = 3.3 \cdot 10^{-13}\, Fd = 0.33\, pF$$
(5.4)

This is a very modest amount of capacitance, even for an integrated circuit implementation. So far it appears easy to convert incoming RF voltages to DC to power the circuit. However, it isn't sufficient for the DC power to be constant over a single RF cycle: it is necessary that the tag still be powered even when the RF power is briefly switched off to send data to the tag (see chapter 3). For plausible data rates the power could be off for around 10 microseconds. To achieve this RC time constant we need a capacitance of:

$$RC = 33,300 \cdot C = 10 \cdot 10^{-6}$$
$$\rightarrow C = 3 \cdot 10^{-10}\, Fd = 30\, pF$$
(5.5)

To achieve a reasonably constant supply voltage over the course of an RF cycle we naturally need much more storage capacitance: on the order of 300 pF to keep the variation in supply voltage small. This is a substantial amount of capacitance and will require about 40,000 to 60,000 square microns of the IC (whose total area is typically 500,000 to 1,000,000 square microns). In addition to just storing enough charge, we also need to distribute the stored charge over the circuit so that those locations that need power at any given moment have it available; otherwise one transistor switching on will tend to reduce the power supplied to its neighboring transistors, leading to crosstalk and logic errors.

In addition to the problem of providing enough capacitance, we need to provide the full operating voltage of the integrated circuit from the available RF voltage from the antenna.

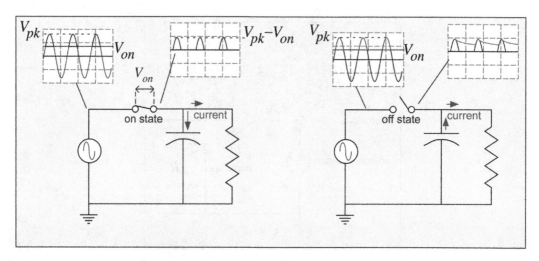

Figure 5.5
Rectifier circuit in on state (left) and off state (right).

This is also challenging, because by reference to Figure 5.5 we can see that the output voltage of the simple rectifier is not the peak voltage of the input, but the difference between the peak voltage and the turn-on voltage of the diode. If the incoming peak voltage is less than the diode turn-on voltage, the diode will never turn on and no power will be delivered to the circuit. Even when the incoming voltage is large enough to get through the diodes, the sacrifice of V_{on} is painful: an IC needs 1 or 2 volts to run, and the turn-on voltage of a typical diode at the relevant current densities might be around 0.5 V (rather dependent on how much diode area we are willing to devote). All this has to be squeezed out of an antenna that is itself providing only about 0.2 volts at a distance of a few meters from the reader antenna. How are we to get enough voltage to run the chip?

The first tool we can make use of is reactive matching. The voltage provided by the antenna is associated with a specific source resistance and reactance; typically, the source resistance varies from a few tens to a few hundred ohms depending on the antenna configuration. The IC draws a few microamps at a volt or two, so the dissipative part of the IC appears as a rather larger resistance (typically 1 to 10 kohms). By using inductors or capacitors to match the source and load, we can theoretically gain an increased voltage proportional to the square root of the ratio of these impedances. However, it is not practical to extent this approach indefinitely. Real matching elements have finite loss, and a very high Q also results in narrow bandwidth (see section 3.4 of chapter 4). For example, let us assume that wish a tag to operate over the whole region of frequencies in use worldwide, that is from 860 to 960 MHz. The relative bandwidth ought to be around $(100/900) = 11\%$, so the matching network should have a quality factor around $1/0.11 = 9$ or 10. Therefore we can achieve an increase of $5-10$ fold in the antenna voltage using reactive matching.

A very common approach to obtaining higher voltages from a rectifier is the use of a *charge pump*: a number of diodes connected in series so that the output voltage of the array is increased. The simplest sort of charge pump, a *voltage doubler*, is shown in Figure 5.6. Two diodes are connected in series, oriented so that forward current must flow from the

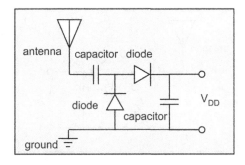

Figure 5.6
Voltage doubler schematic.

ground potential to the positive terminal of the output voltage V_{DD}. A capacitor prevents DC current from flowing between the antenna and the diodes, but stores charge and thus permits high-frequency currents to flow. A second capacitor stores the resulting charge to smooth the output voltage.

When the RF voltage is negative and larger than the turn-on voltage, the first diode is on (Figure 5.7). Current flows from the ground node through the diode, causing charge to accumulate on the input capacitor. At the negative peak, the voltage across the capacitor is the difference between the negative peak voltage and the voltage on the top of the diode. The output (right) plate of the capacitor is $(V_{pk} - V_{on})$ more positive than the RF input.

When the RF input becomes positive, the first diode turns off and the second (output) diode turns on (Figure 5.8). The charge that was collected on the input capacitor travels through the output diode to the output capacitor. The peak voltage that can be achieved is found by adding the voltage across the input capacitor, which we found above, to the peak positive RF voltage, and subtracting the turn-on voltage of the output diode:

$$V_{DD} = V_{pk} + (V_{pk} - V_{on}) - V_{on} = 2(V_{pk} - V_{on}) \tag{5.6}$$

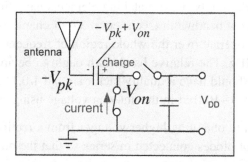

Figure 5.7
Charge pump during negative part of RF cycle.

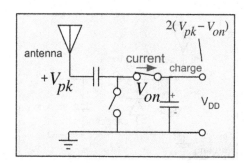

Figure 5.8
Charge pump during positive part of RF cycle.

In the limit where the turn-on voltage can be ignored (e.g. when the input voltage is very large), the output DC voltage is double the peak voltage of the RF signal, from which fact the circuit derives its name. The actual output voltage depends on the amount of current drawn out of the storage capacitor during each cycle, that is on the value of the load resistance (not shown here).

To produce higher output voltages we can provide additional stages of multiplication, to produce a *Dickson charge pump*. A two-stage configuration is shown in Figure 5.9; in the case of ideal diodes with negligible turn-on voltage, the output would be four times larger than the peak RF input voltage. In general for N stages we find:

$$V_{DD} = 2N(V_{pk} - V_{on}) \tag{5.7}$$

It is tempting to imagine that one could continue to add as many stages as required to convert even the most modest input voltage into something adequate to power the IC, but as we add stages we encounter diminishing returns. A very simple analysis of the problem is shown in Figure 5.10. All the DC current must flow through all the diodes in series, so as we add more stages, we waste more and more power in the turn-on voltage of the diodes:

$$P_{load} = V_{DD}I_{load}; \quad P_{diodes} = 2NV_{on}I_{load} \tag{5.8}$$

The power efficiency of the charge pump thus decreases as the number of stages increases for a given turnon voltage and output voltage:

$$\eta_{cp} = \frac{P_{load}}{P_{total}} = \frac{V_{DD}I_{load}}{V_{DD}I_{load} + 2NV_{on}I_{load}} = \frac{V_{DD}}{V_{DD} + 2NV_{on}} \tag{5.9}$$

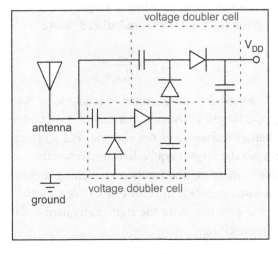

Figure 5.9
Two-stage Dickson charge pump.

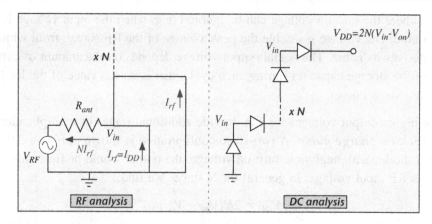

Figure 5.10
Charge pump output analysis. The RF analysis is shown at the time when vertical diodes are on; the horizontal diodes produce a different topology but the same result.

(This analysis turns out to be a bit optimistic for the single- and two-stage cases if substrate loss — current flowing into the bulk of the silicon wafer due to the capacitance of the diode to the substrate — is significant.)

The resulting behavior is shown in Figure 5.11. We can see that the more stages we add (to enable the IC to run with a smaller RF power and thus extend the nominal range) the less efficient the charge pump becomes.

A reasonable approach to estimating the number of stages is to extract an equivalent resistance from the load, given the total power calculated above:

$$R_{eq} = \frac{V_{in}^2}{2P_{total}} \tag{5.10}$$

where the input voltage is adjusted to produce the requisite load voltage from equation (5.7). Roughly speaking, the largest resistance that can be matched to the antenna is Q^2 times larger than the radiation resistance of the antenna. For a typical dipole-type antenna this value is 10–50 ohms, so the largest equivalent resistance that can be optimally matched is around 5 kohms, assuming the limits on matching mentioned above. (Higher values can be used but at some sacrifice in bandwidth.) We can thus adjust the number of stages in the charge pump to provide about the right equivalent resistance for the value of Q we expect to achieve in matching.

Several weaker but non-negligible effects are important in arriving at a final design. The area of the diodes has a weak (logarithmic) effect on the turn-on voltage and thus on the

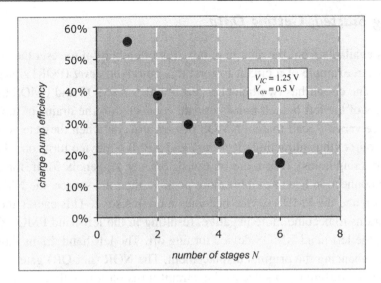

Figure 5.11
Power efficiency vs. number of charge pump stages from equation (5.9).

efficiency, so one is tempted to make the diodes large. However, the diode capacitance grows linearly with the diode area, and since the equivalent resistance of the charge pump is fixed (by the power and voltage targets, as described above), as the diodes are made larger the capacitor starts to draw a substantial reactive current. The capacitance is also voltage-dependent (increasing noticeably as the diodes are turned on), and the variation in capacitance degrades the performance of the matching network, particularly for narrowband, high-Q networks. A charge pump with more stages has smaller capacitance variations because the peak voltage across each diode is closer to the turnon voltage; a typical change for a junction diode is around 25−30% of the zero-bias capacitance. Larger diodes contribute more capacitance but need less peak forward voltage for the same current, so the capacitance variation grows rather more slowly than the capacitance itself. A plausible guideline is that the change in reactance of the equivalent input capacitance be comparable to the equivalent resistance of the load, leading to an input capacitance of around 0.25 to 0.5 pF for typical parameter values. The exact results are of course sensitive to the details of the process technology used.

Even from this rough modeling approach, it is clear that key leverage in operating at higher efficiencies and lower power lies in reducing the turnon voltage of the diodes comprising the charge pump (ideally without excessive increases in diode capacitance), and it can be expected that progress along those lines will continue to improve passive chip performance.

5.3 Getting Started, Getting Data

Once power is available from the charge pump, it is often helpful to reset the state of the logic circuitry. An example of a circuit to send this ***power-on reset*** (POR) signal is depicted in Figure 5.12. The circuit has two branches, each with an NMOS and PMOS transistor in series. The gates of the left branch transistors are connected to the drains of the right branch, and vice versa. Recall that an NMOS device turns on when the gate is positive with respect to the source (thus attracting electrons) and a PMOS device turns on when the gate is negative (attracting holes). (For more on transistors, see Appendix 3). If, for example, the right side gate connection becomes a bit more positive than the source, the NMOS device begins to turn on and the PMOS device becomes more resistive. This causes the voltage on the common drains to become more negative, resulting in the left-hand PMOS device turning on and the left-hand NMOS device turning off. The left-hand drains thus become more positive, enhancing the original displacement. The NOR (not-OR) gate output is high when the inputs are different, so since as this circuit turns on it rapidly assumes opposite states on the two branches, the NOR gate launches a positive-going pulse to reset the remainder of the circuit (POR).

As we noted in chapter 3, the signal from an RFID reader is amplitude-modulated to convey information to the tags. In order to demodulate the signal, we need to create a baseband voltage whose magnitude is proportional to the peak voltage of the reader signal: the ***envelope*** of the RF signal. We have already created a circuit that will do the trick (Figure 5.4), if we choose a storage capacitance value C that provides a storage time long compared to the RF cycle (around 1.1 nanoseconds) but short compared to the length of the

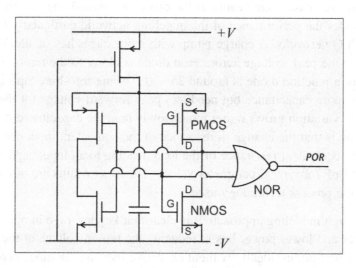

Figure 5.12
Power-on reset circuit; after Curty et al.

data-carrying modulation pulses (one to several microseconds). If necessary (if the antenna is a DC-open or is capacitively coupled) we may provide a resistor to provide a discharge path for the storage capacitance; the values of the capacitor and resistor are chosen to provide a time constant RC of a few nanoseconds. To provide a larger signal voltage for a given RF peak voltage, we can construct a multi-stage charge pump like that of Figure 5.9, though again happily the requirements are much less demanding, since we need deliver only enough voltage to allow the subsequent circuitry to distinguish between the RF power-high and RF power-low conditions of the reader signal. It may also be useful to construct a similar circuit with a longer time constant, to extract a voltage corresponding to the average RF power over many symbols. A simple circuit suitable for these purposes is shown in Figure 5.13.

Once we have obtained a low-frequency voltage proportional to the RF power, we need to extract information from it. If the data is (for example) PIE coded, what we need to figure out is how long the signal remains high between pulses: a long RF-high period represents a binary-1 and a short RF-high period represents a binary-0 (see for example Figure 3.7). One simple approach is shown in Figure 5.14. The output of the envelope detect circuit is directed to a trigger circuit, with an optional current source or other provisions to set the threshold for the trigger. The trigger thus changes state at the beginning of an RF-on pulse, and resets an integrator, which then starts accumulating charge while the voltage is high. The output of the integrator grows linearly with time as long as a signal is applied to its input; on the next rising edge of the RF signal, the integrator output is large if the power was high for a long time, or smaller if the RF-high time was short. That output is applied to a discriminator, which then outputs a 1 or 0 depending on the duration of the RF-high period, just as desired.

Recall from our discussion of chapter 4 that a radio receiver typically includes filtering to select the desired channel, and minimize noise and interference from other transmitters.

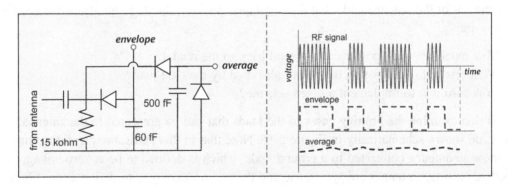

Figure 5.13
Circuit for extracting RF envelope and average power from incoming RF signal; after Curty et al.

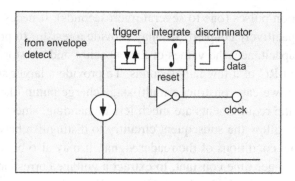

Figure 5.14
Simple demodulation circuit for pulse-interval-encoded data; after Karthaus and Fischer.

A passive tag has no such luxurious amenities. Since there is no local RF oscillator or mixer, there is no conversion operation to allow for channel filtering: all received signals at any RF frequency are converted to baseband by the charge pump. As we will discuss in somewhat more detail in chapter 7, the tag antennas provide some frequency selectivity, but will generally cover at least the 902−928 MHz ISM band for US operation. Thus any transmission in this band, including not only other RFID readers but cordless phones, wireless networks, cell phones and cell basestations, alarm systems, badly grounded spark plug wires, and any other nearby RF radiators all blast right into the IC's receiver.

5.4 Talking Back

Passive tags use modulation of the power scattered by the tag antenna to reply to the reader. In our earlier discussions of backscatter modulation, we imagined a very simple scheme in which the IC simply interrupts current flow through the antenna to modulate the scattered power (see for example Figure 3.15). Let's take a closer look at how one might modulate the behavior of the antenna and what the consequences are. The three questions we need to address are:

- How much scattered power can we send back to the reader?
- What effect do we have on the power absorbed by the (IC) load?
- How hard is it to implement a given scheme?

Let us first examine the limiting cases of the loads that can be presented to the antenna. These are shown schematically in Figure 5.15. Note that in this and subsequent diagrams we show an antenna connected to a ground node, which is defined to be at zero voltage. In practice, most tag antennas are symmetric and there is no easy way to define a true "zero" voltage, but the principles are the same, and it is much easier to discuss the problem in this *single-ended* configuration, rather than the actual *balanced* or *differential* connection.

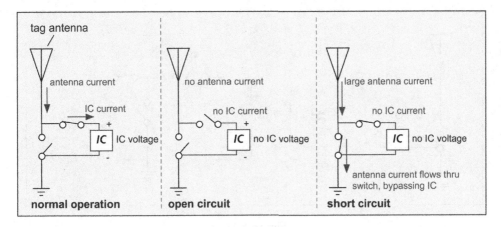

Figure 5.15
Elements of simple backscatter modulation schemes. Single-ended antenna connections are shown for simplicity.

In normal operation we shall assume that the integrated circuit is **matched** to the antenna: that is, that the antenna and IC have been adjusted so that the largest possible power is delivered to the IC. We shall momentarily specify in more detail what this implies about the behavior of the system.

When an open circuit is presented to the antenna, the path to ground is blocked and no current flows in the antenna; since there is no current flowing, no power is radiated (backscattered) in this state. (Real antennas are not quite so simple, and do scatter some power even when they are presented with an open circuit load, and real antenna designs don't quite correspond to the configuration shown here. We shall defer a discussion of these subtleties to chapter 7.)

When a short circuit is presented to the antenna, current flows readily to ground without encountering any resistance or creating any voltage. It may be inferred that in this case a large antenna current flows and substantial scattered power results.

In order to examine the use of these three states for modulation in a quantitative fashion, we need to construct an **equivalent circuit** for an antenna. We'll use the very simplified circuit in Figure 5.16. The antenna is represented by a voltage source V_{ant}, arising from the impinging RF electric field from the reader, and a **radiation resistance** R_{rad}, so named because it arises not from the electrical resistance of the metal of which the antenna is constructed, but from the power lost in the scattered waves that result when current flows in the antenna. The antenna current must flow also through a load consisting of the (constant) load resistance due to the IC's power supply (the charge pumps we discussed in sections 2 and 3 above), with provisions for opening or closing switches to present either an open or a short to the antenna.

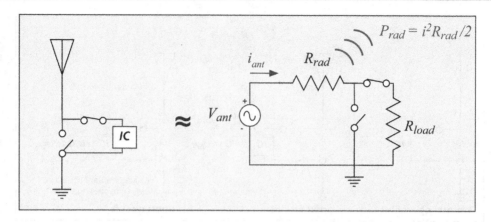

Figure 5.16
Equivalent circuit for an antenna with an IC load, with provisions for switched modulation of the load.

The power delivered to the radiation resistance (and thus scattered back into the world) is proportional to the product of the square of the current and the resistance, like any other resistor (see section 2 of chapter 3). When the switches are in the default configuration shown, the same current must flow through the load resistor. The total current is readily found from Ohm's law (see Appendix 3):

$$i_{ant} = \frac{V_{ant}}{R_{rad} + R_{load}} \qquad (5.11)$$

When the load resistance is very small compared to the radiation resistance, there isn't much voltage across the load and little power is delivered to the load. When the load resistance is much larger than the radiation resistance, all the voltage appears across the load, but little current flows, so again not much power is delivered to the load. It is easy to show that the optimum power transfer — the matched condition — occurs when the values of the source and load resistance are equal. This is the state to which all our previous calculations of the power available from an antenna (section 5 of chapter 3) referred. Notice that as a consequence, in the matched condition, the power dissipated in the load is equal to the power dissipated in the radiation resistance — that is, a matched antenna *scatters as much power as it receives*. Let us denote this power P_{av}, the *available power*. This is the baseline backscattered signal of the unmodulated, matched antenna. We must now examine how this signal changes when we change the load.

Let us first examine the use of an open circuit load for modulation (Figure 5.17). In order to transmit information back to the reader, we can switch the antenna between state 1, in which the antenna sees the matched load of the IC, and state 2, in which an open circuit is

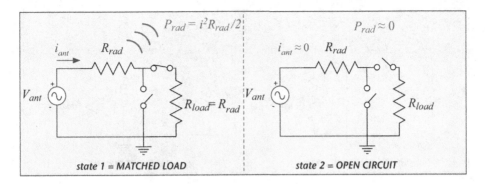

Figure 5.17
Modulating the backscattered signal by switching between a matched load and an open circuit load.

presented to the antenna. As we noted in chapter 3, typical tag modulation approaches, such as FM0, switch symmetrically between the possible tag states, so the tag will spend an equal amount of time in state 1 and state 2.

The signal power is the result of the difference between the current in the radiation resistor corresponding to state 1 (which is equal to the absorbed power in the normal state, P_{av}, as we noted above), and that in state 2, which is equal to 0. Therefore the average backscattered signal power, assuming equal frequency for both states, is:

$$P_{BSC} = P_{av} \tag{5.12}$$

The power delivered to the IC during modulation is the average of that in state 1 (the normal power) and that in state 2 (in which the IC gets no current and so no power):

$$P_{IC} = \frac{P_{av}}{2} \tag{5.13}$$

Another possible approach to modulation is to use a short circuit on the antenna as state 2 (Figure 5.18). In the short circuit condition no power is delivered to the IC, because there is no voltage across it. However, the current in the antenna is doubled relative to the matched case: $i_{ant} = V_{ant}/ R_{rad}$. The signal power here is found from the difference in currents, and is:

$$P_{BSC} = P_{av} \tag{5.14}$$

That is, the modulation is the same as before even though the peak scattered power is larger. The power delivered to the IC is the same as it was in the previous case:

$$P_{IC} = \frac{P_{av}}{2} \tag{5.15}$$

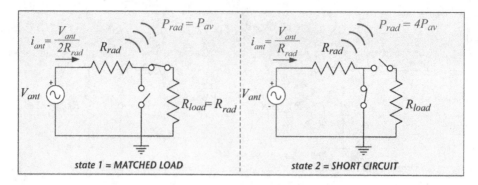

Figure 5.18
Modulating the backscattered signal by switching between a matched load and a short circuit.

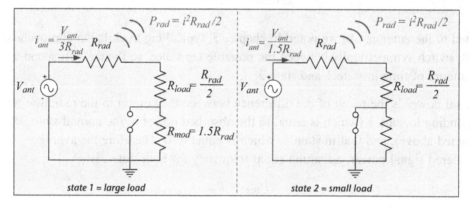

Figure 5.19
Modulation using a resistive circuit in place of a short (resistive amplitude-shift keying).

It is obviously possible to obtain a higher backscattered signal power $4P_{av}$ by switching between an open circuit and a short circuit — but in this case no power at all is delivered to the IC! Such a configuration can be considered for a semi-passive tag.

Note also that in both cases examined above, the IC power is reduced during modulation. This is an important practical problem. In early tag designs, it was not uncommon for a tag to start its reply to a reader and then run out of juice in the middle of transmitting its ID. Since tags may often be found to be forward-link-limited (section 5 of chapter 3), it is helpful to consider trading off some backscattered signal for more tag power (though in the real case the backscattered signal may be lower than the idealized values obtained here; see chapter 7). One approach we might consider is to place a resistance in series with the load; we will choose some specific values so that the total power delivered from the antenna is the same in both modulation states (Figure 5.19).

The total power delivered to the (load + modulation resistor) by the antenna is the product of the current and the total resistance:

$$P_{del,1} = \frac{1}{2}\left(\frac{V_{ant}}{3R_{rad}}\right)^2 2R_{rad} = \frac{1}{9}\frac{V_{ant}^2}{R_{rad}}$$

$$P_{del,2} = \frac{1}{2}\left(\frac{V_{ant}}{(3/2)R_{rad}}\right)^2 \frac{R_{rad}}{2} = \frac{1}{9}\frac{V_{ant}^2}{R_{rad}}$$

(5.16)

However, the power delivered to the load resistor (representing the useful power for the IC) is NOT the same:

$$P_{load,1} = \frac{1}{2}\left(\frac{V_{ant}}{3R_{rad}}\right)^2 \frac{R_{rad}}{2} = \frac{1}{36}\frac{V_{ant}^2}{R_{rad}}$$

$$P_{load,2} = \frac{1}{2}\left(\frac{V_{ant}}{(3/2)R_{rad}}\right)^2 \frac{R_{rad}}{2} = \frac{1}{9}\frac{V_{ant}^2}{R_{rad}}$$

(5.17)

The average power delivered to the IC is thus:

$$P_{IC} = \frac{2.5}{36}\frac{V_{ant}^2}{R_{rad}} \approx 0.55 P_{av}$$

(5.18)

which is about 0.5 dB better than our simplistic modulation schemes above. The backscattered power is determined by the difference in the currents:

$$P_{BSC} = \frac{1}{2}\left(\frac{V_{ant}}{(3/2)R_{rad}} - \frac{V_{ant}}{3R_{rad}}\right)^2 R_{rad}$$

$$= \frac{1}{2}\frac{V_{ant}^2}{R_{rad}}\left(\frac{1}{9}\right) = \frac{4}{9}P_{av} \approx 0.44 P_{av}$$

(5.19)

which is about 3.5 dB worse than in the case of the short or open circuit. So modulating the load resistor doesn't really seem get us much advantage in power to the IC, and gives up some backscattered power.

What about the use of a capacitor instead of a resistor to modulate the load? This change in the reactive properties of the load will result in a change in the phase as well as amplitude of the current flowing through the antenna, and thus is ***phase-shift keying*** of the tag. An example scheme is shown in Figure 5.20. It will be understood that the positive and negative reactances are likely to be implemented by changing the value of a capacitance. (Recall that there is a capacitance in parallel with the resistive load of the IC, which we have presumed to be removed by the matching network.)

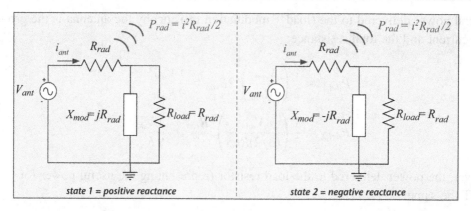

Figure 5.20
Modulation using reactance rather than resistance.

The computation of the currents and powers is rather more involved in this case, but roughly what is going on is that the capacitance changes the phase of the current without changing the amplitude very much. As a consequence, the power delivered to the load resistor is almost the same as in the unmodulated case, but the backscattered power is substantial. The process is depicted graphically in Figure 5.21. The currents are:

$$I_{ant,1} = \frac{V_{ant}}{R_{rad} + \dfrac{jR_{rad}^2}{R_{rad} + jR_{rad}}} = \frac{V_{ant}}{R_{rad}} \frac{1+j}{1+2j}$$

$$I_{ant,1} = \frac{V_{ant}}{R_{rad} + \dfrac{-jR_{rad}^2}{R_{rad} - jR_{rad}}} = \frac{V_{ant}}{R_{rad}} \frac{1-j}{1-2j}$$

(5.20)

The backscattered signal power is due to the difference between these currents:

$$P_{BSC} = \frac{1}{2}|I_{ant,1} - I_{ant,2}|^2 R_{rad} = \frac{1}{2}\frac{V_{ant}^2}{R_{rad}}\left|\frac{1+j}{1+2j} - \frac{1-j}{1-2j}\right|^2$$

$$= \frac{1}{2}\frac{V_{ant}^2}{R_{rad}}\left|\frac{3-j}{5} - \frac{3+j}{5}\right|^2 = \frac{1}{2}\frac{V_{ant}^2}{R_{rad}}\frac{4}{25} = \frac{16}{25}P_{av} \approx 0.6 P_{av}$$

(5.21)

The voltage across the load can be found by subtracting the voltage across the radiation resistance from the antenna voltage; it has the same magnitude for both states, and is:

$$V_{load} = V_{ant}\frac{\pm j}{1 \pm 2j}; \quad |V_{load}|^2 = \frac{V_{ant}^2}{5}$$

(5.22)

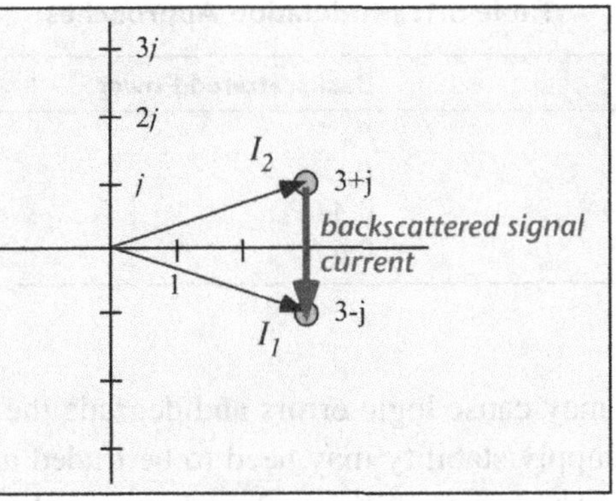

Figure 5.21
PSK current values and difference, depicted as vectors in the complex plane. (Current values scaled by a factor of 5 for clarity.)

and thus the power dissipated in the load resistor (the IC) is the same in both states:

$$P_{IC} = \frac{1}{2}\frac{V_{ant}^2}{5R_{rad}} = \frac{V_{ant}^2}{10R_{rad}} = 0.8P_{av} \qquad (5.23)$$

The performance of these variant modulation schemes is summarized in Table 5.1. Clearly the best compromise between power delivered to the IC and backscattered power is produced by phase-shift keying. The distinctions are on the order of 1−3 dB. For forward-link limited performance, recall that a 2 dB improvement in power to the IC will only add about 1 dB (10%) to the read range. Similarly, a 3 dB improvement in backscattered power will only provide about 0.8 dB (8%) improvement in reverse-link-limited range, since the backscattered power depends on the fourth power of distance. So the distinctions between these approaches are modest in terms of overall tag performance.

We have used very specific examples of ASK and PSK component values for simplicity. In the general case where the component values are allowed to vary from those corresponding to a very small modulated signal to the limit of open and short circuits, it is found that the general conclusions we have arrived at are still applicable: at moderate backscattered power, PSK provides substantially better power delivery to the IC load than ASK, but the backscattered power is always lower than in the extreme case where the load is varied from an open to a short circuit.

It is also worth noting that in the case of the three amplitude-shift-keyed approaches, the power delivered to the load − the rest of the IC − is very dependent on the modulation state, which creates a challenge for the regulation of logic power on the chip, the effect being larger as the modulation efficiency increases. If insufficient regulation and storage are

Table 5.1: Modulation Approaches

Approach	Backscattered Power	IC Power
Match <=> Open	P_{av}	$P_{av}/2$
Match <=> Short	P_{av}	$P_{av}/2$
Resistive ASK	$0.44\, P_{av}$	$0.55\, P_{av}$
Reactive PSK	$0.6\, P_{av}$	$0.8\, P_{av}$

provided, these variations may cause logic errors and degrade the ability of the chip to function properly. Power supply stability may need to be traded against backscatter efficiency and complexity. The use of PSK has the advantage that power delivery to the chip is almost independent of the modulation state, simplifying power management. On the other hand, the effectiveness of all these modulations in producing backscattered power is dependent on the matching approach and antenna parameters. As we will see in chapter 7, adding a matching network doesn't change the qualitative results above, but the matching network has a complex effect on the way the signal is created, so that ASK at the tag can become PSK at the antenna and vice versa.

In order to create any sort of useful modulation, the tag needs some sort of clock to tell it what the difference is between a binary '1' and '0', as well as to synchronize the tag's logic circuitry. The clock is usually implemented as an oscillator circuit, possibly with some provisions for calibrating the clock speed based on the reader preamble. An example low-power oscillator circuit is shown in Figure 5.22. The circuit begins to oscillate when the enable connection is pulled high (to positive voltage). Operation of the oscillator assumes the availability of reference voltages for PMOS and NMOS devices that ensure that a certain fixed reference current per micron of gate width flows through transistors biased with the cited voltages. The reference voltages are created by tapping off the gate voltage of similar NMOS and PMOS transistors adjusted in a current-mirror configuration to keep their output current constant.

The circuit alternates between the two states depicted in Figure 5.22. In state (a) the output is high. The capacitor is positive, thus holding T2 off and T3 on. The output voltage is also fed back to T1, which connects the NMOS reference transistor to the negative supply voltage. This reference current discharges the positive voltage stored on capacitor C. When the capacitor voltages falls sufficiently, the circuit switches to state (b). T2 turns on and T3 turns off; the output voltage is pulled negative, and fed back to turn T1 off and T4 on. The NMOS reference current terminates and the PMOS reference current charges the capacitor. Thus the oscillation frequency is set by the capacitor size, transistor threshold voltages, and the reference current.

Note that operation of this circuit requires that five transistors operate in series from the difference between +V and −V. This circuit was designed for implementation in a

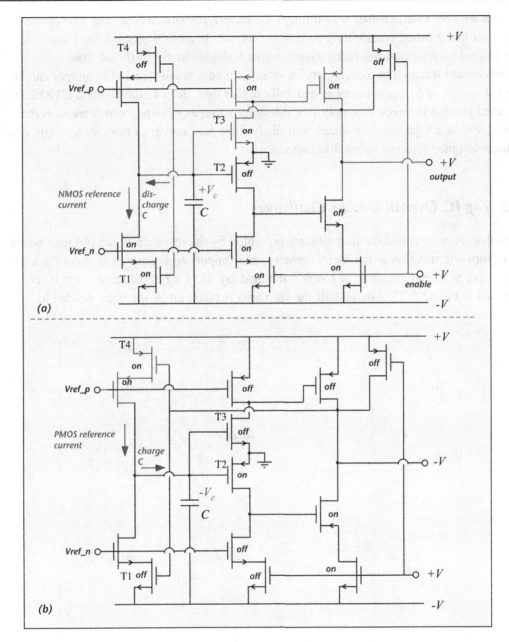

Figure 5.22
Tag oscillator circuit, after Curty et al.

low-threshold-voltage, *silicon-on-insulator* (SOI) process. A conventional field-effect transistor is manufactured by placing a gate electrode in close proximity to a doped region in a bulk silicon wafer, separated by a thin oxide insulator. The transistor is turned off by adjusting the voltage on the gate to repel the channel carriers (electrons or holes), but some carriers are able to make their way through the underlying silicon from one side of the transistor to the other, contributing to leakage current. In an SOI process, the channel is constructed on top of an insulating layer, so that leakage current in the off state (*subthreshold* leakage) is reduced, and devices with very small threshold voltages can be used. The cost of SOI processing is generally higher than that of conventional CMOS. In a standard process the need to supply five threshold voltages in series, which requires the availability of a high supply voltage, will likely limit read range, so this circuit might need to be redesigned for conventional implementation.

5.5 Tag IC Overall Design Challenges

Now that we've touched on the problems presented by the interface to the physical world, let us turn our attention to the logic, memory, and supporting systems that make the tag responsive to its environment. A rough functional layout of a typical passive tag IC is depicted in Figure 5.23. Around half the chip area is taken up by the logic needed to

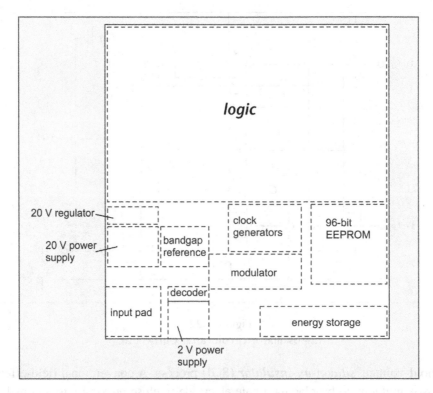

Figure 5.23
Functional layout of a tag IC (after Stewart).

implement the relevant protocol: about 50,000 transistors for an 18000-6C (EPCglobal Class 1 Generation 2) IC.

We've looked at the key RF-related challenges in sections 2−4 above. The remainder of the chip operates at baseband frequencies and is generally similar to conventional mixed-signal design. However, there are some special challenges peculiar to the RFID world.

The first challenge is, of course, that of cost. The cost of a chip is dominated by its size if yield is reasonably good. Modern IC manufacturing facilities use 200-mm- or 300-mm-diameter wafers. A standard 200-mm-diameter silicon wafer offers a useful area of about 30,000 square millimeters. (This is a bit less than the total surface area: the region within about 3 mm of the wafer edge is usually not useful for processing.) If a single IC has a useful area of around 1 mm^2, we can get about 22,000 chips from a wafer assuming that 90% of the chips are good (that is, that the yield is 90%). In small volumes, it costs about $1000 to purchase a processed wafer, so the cost of these ICs would be roughly $0.05 per chip. The use of 300 mm wafers increases the initial cost for masks, but in high volume the ongoing cost is reduced by 30−40% vs. 200 mm wafer. The actual numbers are influenced by such commercial issues as volume pricing − I don't pay $1000/wafer if I buy several hundred wafers! − but it should be apparent that at 1 square millimeter, the chip cost is a substantial fraction of the $0.05 tag cost goal promulgated by such organizations as EPCglobal. It is imperative to keep the chip as small as possible to minimize IC cost.

Traditionally, the size of digital integrated circuits has been strongly influenced by *scaling*: the reduction in the size of transistors due to improvements in lithography and processing, which results in a reduction in the size of the chip for the same number of transistors. Technologies are usually named for the smallest feature size used in the process: for example, a typical high-speed fabrication process might make use of an 0.13 μm line to form the transistor gate, and would generally be referred to as an 0.13 μm process. For many years scaling the size of the transistor down also resulted in reduced power consumption per transistor. However, there are some obstacles to achieving the benefits of scaling in RFID chips over the next several years. First of all, the most advanced process technology is always expensive. It is much cheaper to purchase masks and wafers for 0.18 μm processing than for 0.13 μm processing. Secondly, the benefits of scaling are decreasing as fundamental limits in process technology are reached. For example, silicon dioxide films, critical for forming MOS transistors, have reached thicknesses equivalent to only about 3 molecular layers, and can't be reduced much more. Efforts to replace silicon dioxide in this role have so far been unfruitful. Because of leakage through these thin oxide films, power consumption in very small devices is also not as small as one might have expected from extrapolation from older technologies.

A substantial fraction of an RFID IC consists of analog functional blocks: RF rectifiers for power supplies, capacitors for energy storage, and circuitry for decoding and modulation. The size of analog blocks doesn't necessarily change just because the minimum feature size is reduced. For example, the area required for a storage capacitor is set by the total power consumption of all the chip features, so if power consumption doesn't go down the capacitor must remain the same size. A diode's size is set by the parasitic resistance and capacitance associated with it, which determine the frequency it can operate at and the impact it has on the antenna match. A modulation capacitor's size is set by the characteristics of the antenna and the modulation efficiency we seek to achieve. Protection circuitry requirements are determined by the largest voltage the tag expects to see, not by the smallest voltage its transistors could operate at. While in general the size of the analog blocks can shrink if the power consumption of the logic they support is reduced, analog functions usually don't scale down in size nearly as readily as the corresponding digital circuitry.

So we can't expect scaling alone to magically reduce IC costs. It is also important to exploit every possible measure to reduce size and power consumption of the logic blocks. Automatic routing of wires between transistors is fast and convenient, but clever human designers can squeeze space out of the design by (laborious) manual optimization. Power consumption in the logic circuitry can be reduced by operating the devices near the threshold voltage (the minimum voltage to turn the transistors on), but threshold voltage differs from one chip to another due to variations in manufacturing and due to temperature variations in operation. It is possible to use onboard non-volatile storage to adapt the operation of each chip to its conditions. Such threshold adjustment techniques may also allow the use of MOSFETs instead of junction diodes or expensive Schottky diodes for rectification.

Finally, a tag IC has a number of logic blocks, all running from a very high-impedance antenna (that is, an antenna that has a hard time supplying much current). Each time a logic gate switches its state, a transient current flows from the local power supply connection. If the local supply voltage fluctuates as a consequence − that is, if *decoupling* is insufficient − the change could be interpreted by a nearby gate as a logical input, leading to errors in operation of the circuit. The challenge of properly isolating the individual gates, and the segments of the circuit, is much larger than in conventional circuit design, where power at a reasonably fixed voltage is available from a battery or power supply.

5.6 Packaging: No Small Matter

So far we've focused on the electrical guts of a passive tag, but the physical construction is also of great importance. The integrated circuit must be connected either to an intermediate *strap* or the substrate itself, the latter formed of plastic or paper, at the same time making electrical contact to a separately-created antenna structure. At this stage the tag is often called an *inlay*. The inlay can be used on its own, either as is or coated with adhesive to be attached to an object. Alternatively, the inlay can be laminated between sheets of paper or plastic to form a *smart label* or a *smart card*. These alternative fates are depicted in Figure 5.24

As a consequence, the overall manufacturing process for a typical smart label containing a passive tag looks something like that depicted in Figure 5.25. In the remainder of this section we will briefly examine each of these steps.

More has been written about silicon CMOS fabrication than is prudent to contemplate, and we shall not attempt to recapitulate such an extensive field in this brief aside. We shall content ourselves with re-emphasizing the role of IC size in determining IC cost: the total cost of a completed wafer is roughly independent of what is on it, so the smaller an IC chip is, the more you get per wafer as long as most of them work. High yield of good devices is also very helpful in minimizing test costs. If a large percentage of chips fails, it is necessary to test carefully at the beginning of the process to minimize the labor and materials wasted on bad devices, whereas if only an occasional failure is encountered, testing can be postponed to the end, reducing total cost. Yield is influenced by a plethora of factors, including process design and equipment maintenance in the wafer fabrication facility, and thoughtful design practices.

Historically, a completed integrated circuit connects to the outside world by providing square pads 100 microns or so on a side at the edges of the chip. The chip is placed in a plastic or ceramic package, and electrical connections are made by *wire bonding*: one end of a gold or aluminum wire around 25 microns (0.001") in diameter is bonded to the IC pad

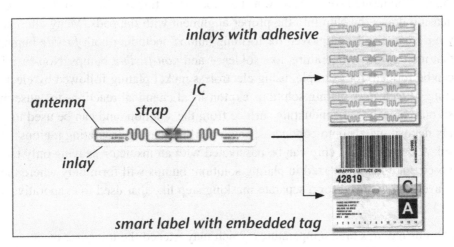

Figure 5.24
Components of a tag and inlay or smart label final forms.

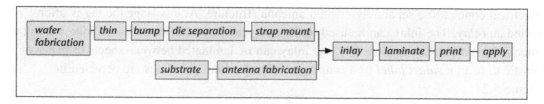

Figure 5.25
Smart-label manufacturing process.

using ultrasonic excitation to scrub away surface layers, and the other end to a similar contact pad on the package. Wire bonding is highly automated, fast, and reliable. However, it is intrinsically a serial process, with each bond made in sequence, and thus relatively expensive when the target is a $0.01 chip.

Fortunately, an alternative means of connecting ICs to the outside world is available. This approach, variously known as *flip-chip* or *chip-scale packaging*, involves the formation of thick metal *bumps* on top of the contact pads on the IC. The chip is inverted, and the bumps are placed onto corresponding conducting regions in a package, strap, or tag antenna.

Much early work in this area was performed by IBM, and made use of evaporated lead and tin to create conventional solder bumps. (Solder, a tin-lead allow, melts at around 220°C, and has historically been widely used for wiring and interconnections, although concerns about the environmental impact of lead have resulted in increasing use of lead-free alloys.) Solder can also be electroplated. The solder-bumped chip is placed over an array of plated or freshly-cleaned copper pads and heated to melt the solder, which then forms a ball (Figure 5.26). The molten solder balls wet the exposed plating or copper surface, and surface tension draws the chip into the proper alignment with the pads. Many alternative metal systems can also be employed for forming bumps, including both *fusible* bumps – bumps that melt at low temperature, like solder – and *non-fusible* bumps. Non-fusible bumps can be formed, for example, using electroless nickel plating followed by electroless gold plating. (*Electroless* plating solutions exploit local chemical reactions to cause metal to deposit onto an exposed conducting surface from the solution, and can be used to selectively deposit metals onto conductive regions while leaving insulating regions untouched. A completed IC chip can be passivated with an insulator, leaving only the bond pads exposed, and then immersed in plating solution; bumps will form only where the bond pads are, avoiding the need for a separate masking step like that used in evaporative solder definition.)

Melting bumps requires high temperatures which may exceed the tolerance of the inexpensive plastics used for passive tag assembly. A popular alternative is to employ non-fusible bumps and conductive pastes, typically fabricated by mixing fine metal

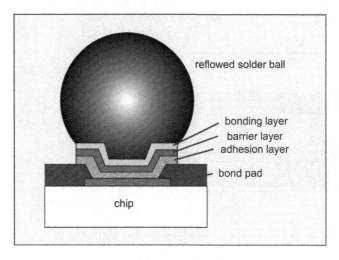

Figure 5.26
Cross-sectional view of solder bump after reflow; ball diameter typically 100−200 microns.

particles, often of highly-conductive silver, with a polymer binder, typically an epoxy resin. It is possible to replace the metal bumps with conductive paste, applied by screen printing, or to use conductive pastes to make contact between bumps and the antenna or strap pads. A perhaps more useful approach is to employ metal bumps in conjunction with ***anisotropic conductive adhesives*** (ACAs, Figure 5.27). An ACA is a conductive polymer with carefully adjusted proportions of metal and binder, so that the material is conductive when compressed but not in its default state. If a layer of ACA is placed between a bumped chip and contact pads and the chip pressed against the adhesive during cure, the compressed regions under the bumps will become conductive and the remainder of the adhesive will be insulating, forming a self-aligned conductive contact to the underlying pads. Such a scheme avoids the need to accurately align the bumps, conductive paste, and pads. ACAs are used in flat-panel display assembly for the same reason. Isotropic conductive adhesives are less expensive and may use lower cure temperatures, but require more accurate assembly, and are more likely to be employed in attaching the strap to the inlay, where alignment tolerances are less crucial.

Because tag ICs must be physically small, separating them becomes an important issue. Large ICs, such as microprocessors or memory chips, are separated from their parent wafer using a saw with a narrow blade to slice the wafer up into rectangular blocks, each containing a chip. ***Kerf loss***, the amount of silicon removed by a conventional wafer saw, is around 100 microns. This is equivalent to expanding the chip on all sides by 1/2 of the saw kerf: a 1×1 mm chip becomes a 1.1×1.1 mm chip, with an area of 1.2 mm^2: a 20% increase in the cost of the chip. Kerf loss can be reduced if the wafer is reduced in thickness, typically by grinding the back side, before the individual dice are separated.

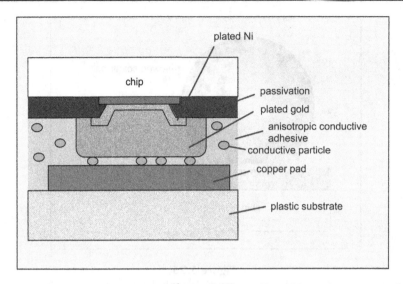

Figure 5.27
Assembly to flexible substrate using plated non-fusible bumps and anisotropic conductive adhesive (not to scale).

Alternative processes for die separation can also be considered. Chips made of gallium arsenide and other compound semiconductors are often separated by *scribing*: a diamond-tipped stylus cuts a linear notch between each row or column of chips, and then the wafer is bent or stretched slightly to cause it to break along the scribed lines. This process works extremely well for small ICs on these materials, because they are more delicate than silicon, and also break readily in certain directions (known as cleavage planes). Scribe and break can also be used for silicon devices, but silicon is much stronger than gallium arsenide or indium phosphide, and thinning the wafers is indispensable to achieve good separation. Scribing also tends to cause mechanical damage and particle problems.

Another alternative is to use a corrosive liquid to etch the material between the ICs away. Etching produces smooth surfaces and no mechanical damage, but requires some sort of *mask* to protect the ICs themselves from the etchant: an additional step with additional labor and materials costs. Etching solutions can be either *isotropic* (etching at an equal rate in all directions) or *anisotropic* (etching most rapidly in certain directions with respect to the underlying silicon crystal lattice). The kerf loss due to an isotropic etch used for die separation is about twice the thickness of the material being etched, since the etchant proceeds sideways at the same rate it proceeds downwards, so again it is useful to thin the wafers before attempting to separate the dice. Anisotropic etchants reveal specific crystal planes, and may allow a more efficient use of the wafer area, but are usually slower than isotropic etches.

Once the individual ICs are separated they may be assembled onto an intermediate strap prior to attachment to the inlay (Figure 5.28). In this fashion the relatively high-precision

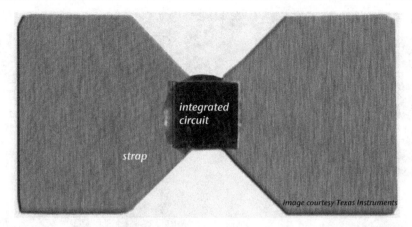

Figure 5.28
Integrated circuit attached to strap in preparation for mounting on antenna.

strap attach can be done in a specialized facility, and the lower-precision strap-to-tag attach, which is unique to each tag design, can be performed separately, using isotropic conductive adhesives and standard assembly techniques.

Alien Technology has pioneered the use of ***fluidic self-assembly*** to place large numbers of chips precisely onto their straps. Because the chips are not handled mechanically, the kerf between chips is limited by the separation process only, and can be as small as 20 microns using anisotropic etching of a thinned wafer. The separated chips are released into a fluid carrier from which they self-locate onto precision-etched openings in an intermediate support structure, a strap web. The configuration of the chips and openings ensures that the chips lie right-side-up. The chip-support structure is then laminated and openings are cut in the laminate to permit contact using a printed conductive material. The resulting conductive straps can couple to a nearby antenna to permit non-contact preliminary testing of the parts before assembly. Other vendors use more conventional assembly techniques employing automated pick-and-place equipment.

Once the chip is placed on a strap, it must be assembled to make an inlay. The inlay is usually constructed from plastic; a common material polyethylene terepthalate, PET. PET is an inexpensive, mechanically robust plastic with good resistance to most common chemicals and low dielectric constant (which we'll see in chapter 7 helps in antenna design). It is widely used in textiles, capacitors, recording tapes, and other applications. Polyimides are also used.

An antenna must be constructed on the inlay. The standard means for producing patterned conductive materials on plastics, taken from the printed circuit technology in wide use in nearly all electronic products, involves electroplating a thin copper layer onto the plastic,

Figure 5.29
Continuous parallel processing is used to minimize inlay manufacturing cost.
Image courtesy Motorola.

and then applying a mask to protect the regions where the metal is to remain and removing the undesired copper with a liquid etchant. This type of subtractive-etching process is mature and robust, and produces films with excellent conductivity. Copper thicknesses of 10–40 microns are readily achieved, corresponding to sheet resistances of less than 1 milliohm/square. (*Sheet resistance* is the resistance of a square piece of a thin material with contacts made to two opposite sides of the square. It is independent of the size and depends only on the material and thickness, and is numerically equal to the *resistivity* of the film divided by its thickness. The resistance of a line may be estimated by multiplying the sheet resistance by the aspect ratio of the line; for example, a line 5 cm long and 1 mm wide has an aspect ratio of 50, and using material with a sheet resistance of 1 mΩ/square would have a DC resistance of 50 mΩ.) Metal may also be patterned by stamping – that is, cutting the requisite pattern out of a foil with a sharp-edged tool – and adhering to the plastic.

A significantly simpler and less expensive approach is to employ a conductive paste to form the antenna. In order to get good performance, high conductivity materials must be used. Modern conductive inks, made using silver particles embedded in a specialized polymer matrix, achieve bulk resistivity of 30–60 $\mu\Omega$-cm, about 10 times higher than solid silver or copper but acceptable for tag applications. Corresponding sheet resistances are around 12–20 mΩ/square. A typical tag antenna segment with an aspect ratio of 20 will thus add a DC resistance of around 0.4 Ω, negligible for most antenna designs. Surface roughness is also important, since at UHF frequencies most of the electrical current flows in a layer a few microns thick near the surface of the film. Recent films have shown improved surface roughness and better RF performance. Challenges for conductive ink assembly include susceptibility to corrosion and oxidation, and problems achieving reliable attachment of the IC.

Whichever approach is used, it is necessary to use high-speed volume manufacturing techniques to minimize cost. The inlays are generally fabricated on long continuous rolls of plastic, using specialized high-speed patterning and assembly equipment (Figure 5.29).

Once the IC or strap is assembled onto the inlay/antenna structure, polymer coatings may be applied to protect the IC and antenna. If the inlay is to be used on its own, it may receive an adhesive coating on the backside and be laminated onto a paper backing. Inlays are also often laminated into a conventional paper or plastic adhesive-backed label; this process is often known as *label conversion*, with a *smart label* being the result. Such labels are usable in specialized printers that both print the human-readable printed labeling, and write and verify the ID of the encapsulated tag. The resulting labels may be automatically applied using a label applicator, or applied by hand.

Inlays incorporated into labels or rolls must be able to tolerate bending, rolling, and compression during the printing and application processes, as well as electrostatic discharge resulting removal of protective layers for adhesive application, and from general handling.

5.7 Other Passive Ways

The current author has never tried skinning a cat, and has no idea if a preferred method exists or the advantages accruing to such relative to alternative approaches, or whether the latter would be more common in other universes with slightly different physical laws. However, there are unquestionably alternatives to fabricating a UHF RFID tag using a silicon integrated circuit at its heart.

The most extensively explored of these approaches is the *surface acoustic wave* (SAW) tag. You may recall that we have already briefly introduced SAW devices in connection with filtering RF signals (see chapter 4, section 3.4). In such a device, an electrical signal (from the antenna) is converted into a sound wave, using a set of periodic metal electrodes known collectively as an *interdigitated transducer* (IDT), constructed on a thin slice of a piezoelectric material such as quartz or lithium niobate. The sound wave propagates much more slowly than the speed of light, so at UHF frequencies wavelengths are a few tens of microns, making it straightforward to construct arrays of electrodes spaced at integer half-wavelengths. Once the wave is launched away from the transducer, it travels away to the far end of the slice, not terribly interesting by itself. However, if we place an additional electrode along the propagation path, the acoustic properties of the near-surface region are slightly modified, and a portion of the wave will be reflected, just as a water wave can be partially reflected by an object floating on the surface of the water (Figure 5.30). When the reflected wave reaches the IDT, it creates a small voltage in the antenna, and thus launches a tiny delayed scattered pulse. By adjusting the pattern of these electrodes, a coded message can be sent, containing (for example) the unique ID of the SAW tag.

In order to detect the delay in the pulse, the reader signal must typically be time-dependent. The reader may transmit a pulse, or a signal whose frequency increases or decreases with time (a *chirped* signal) may be employed. Let's examine the case where a pulse is used. The length of the SAW chip (which, just like a silicon IC, has an important influence on its cost of manufacture) is determined by the time delay it must support: the longer the chip, the more time it takes for the sound wave to get to the end and the more room there is for adding reflectors. The number of reflectors is determined by the requirement that they be separated in time by something comparable to a pulse width: if a pulse is (say) 100 nsec long, successive reflectors ought to be separated by a distance that takes around 100 nsec for the sound wave to cross. The more reflectors we have, the more data we can store on the chip. So short pulses are best.

If we build a chip with a round-trip delay of 1 microsecond, and use 100 nsec pulses, there is room for roughly 10 reflector electrodes (perhaps fewer, since we may need to use one to mark the beginning of the pulse train); if each electrode position encodes 1 bit through an electrode being present or absent there, we have a 10-bit ID, hardly adequate for most applications. To get to a full 96-bit ID with 16-bit error check we need a delay of 11 microseconds. For a typical sound velocity of around 4000 m/s, the round-trip delay of 11 microseconds corresponds to a chip length of 2.2 cm — substantially larger than the Si IC sizes discussed in the previous section. In SAW tags, the size of the ID space is directly traded against the size (and thus cost) of the chip.

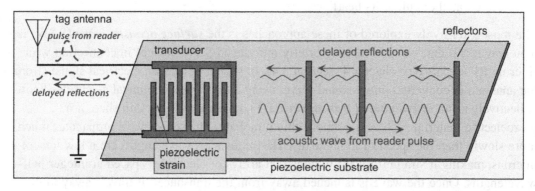

Figure 5.30
Tag employing a surface-acoustic-wave (SAW) active element.

To mitigate the situation, we can use shorter pulses or encode more bits in each pulse. The problem with shorter pulses is that short pulses use bandwidth. Recall that in chapter 3, section 3, we noted that the faster the modulation, the more bandwidth is used. To produce a pulse that lasts 100 nsec, we need around 10 MHz of bandwidth. But the amount of bandwidth is limited by regulation: in the United States, only 28 MHz is available in the 902−928 MHz ISM band, and much less is available in other jurisdictions. To get better performance, it makes more sense to operate at 2.4 GHz, where about 80 MHz is available for unlicensed use in the US and tens of MHz in most other areas of the world. To encode more bits per reflector, we can use pulse-position modulation, in which we try to resolve small displacements in time of the successive reflections. A more powerful but more complex approach is to detect not only the amplitude and timing of a reflected pulse, but also the relative phase of the RF signal. Phase detection is equivalent to detecting the location of the reflector to within a fraction of an RF cycle. At 4000 m/s, one 2.4-GHz RF cycle is equivalent to 1.7 microns, so by using phase detection we acquire a very high-resolution view of the electrode position, at the cost of a very stringent requirement on the position of each electrode on the chip.

The great advantage of a SAW tag is that there is no logic to power on the tag chip, so the read range is limited by the reverse link budget only. The mature techniques of pulsed radar design can be applied to the reader, providing good sensitivity in the presence of high-amplitude pulses. Long reverse-link-limited read ranges can be obtained. SAW tags can be placed close to metal objects or aqueous fluids (conditions where, as we will discuss in more detail in chapter 7, the available electric field from the reader is reduced) and still be read, because there is no requirement for a minimum power at the tag.

On the other hand, SAW tags have no logical capabilities, and can only reply with a stereotyped pulse string. All intelligence must be incorporated into the reader. The lack of responsiveness becomes a challenge when multiple tags are present in the region being examined by the reader. Since the tags can't receive and interpret instructions, all will respond, generally leading to an incomprehensible *collision* between the tag responses. Various approaches to solving the collision problem are available: the reader can take the strongest tag first, decipher the signal (if there aren't too many others), and then subtract that signal from the whole and go after the next strongest. Directional antennas can be used to limit the physical region being interrogated, as we will discuss in more detail in chapter 6; this approach requires use of 2.4 GHz or higher frequencies, as highly directional antennas at 900 MHz are quite large and unlikely to be practical for indoor use. Tags can be designed with built-in fixed delays, so that they respond at different times and don't overlap, but this approach sacrifices code space and thus raises tag cost for the same number of bits of information. Combinations of all these approaches are also possible. It is apparent that when only a few tags are present, collisions will be readily dealt with by a combination of the techniques above with appropriate procedures, but that if tens or hundreds of tags are to be read, an IC-based approach is likely to be superior.

A second alternative to which we shall devote some brief consideration is the use of organic materials to construct an electronic circuit. Semiconductor devices, upon which all silicon integrated circuits are based, exploit the fact that silicon (and certain other materials) can exhibit substantial electrical conductivity when a pure sample is *doped* with an appropriate alloying element. These dopants either contribute an electron to the crystal lattice, or extract an electron from it (leaving behind a positively-charged *hole*); the two kinds of dopants are known as n-type and p-type, respectively. The free electrons and holes that are created by the dopants can move through the silicon, allowing electric currents to flow, and can be reversibly driven away from some regions of the material through the use of electric fields, allowing a voltage to control a current flow and thus enabling the operation of transistors. To make an integrated circuit rather than just a transistor requires the additional capability to fabricate basic electrical components: wires, resistors, capacitors, and (less often) inductors. Since the invention of the transistor in 1947, and more significantly of the planar integrated circuit in the early 1960's, hundreds of billions of dollars have been invested in the technology to design, fabricate, test, and employ silicon integrated circuits. Any alternative technology must surmount the very substantial competitive obstacles presented by the need to functionally replicate this vast infrastructure.

The idea of replacing the silicon integrated circuit (and possibly the antenna as well) in an RFID tag by conductive organic materials is of interest because of the belief that such an approach will eventually enable the use of very high-volume, very low cost processes such as printing and lamination to create RFID tags. Most plastics are insulators, being made of exclusively covalent bonds; but it has long been known that organic materials can provide substantial electrical conductivity when structures encouraging the formation of delocalized electron states are used; graphite (pure carbon organized in a planar structure) is the archetypal example. Graphite is, however, mechanically awkward, and a poor choice for wiring or circuitry. Polymeric materials with unsaturated or delocalized bonding are more plausible candidates for materials with substantial electrical conductivity and the desirable properties of plastics. Organic conductors have been under active development since the 1960's.

Certainly a great deal of progress has been made in organic conductors since the current author laboriously grew miniscule crystals of TTF-TCNQ (the stylish candidate of that ancient day) for his undergraduate thesis. Modern materials, such as vacuum-evaporated pentacene (a block of five hexagonal benzene rings sewn together at the edge, for those who care) display electron mobility − the response of an electron to an imposed electric field − of 1 to 10 cm^2/V sec, much less than achieved in bulk silicon (where mobilities in the hundreds are the norm), but quite comparable to the performance of amorphous silicon materials that have seen wide commercial use in active-matrix flat panel displays. Doping is difficult but possible. Transistors can be fabricated using inkjet printing or spin casting techniques, though vacuum sublimation produces better material properties. Working tags at LF (125 KHz) have been demonstrated, though at the time of this writing it appears that only certain components of a 13.56 MHz tag have been constructed. Rectification (needed to generate power for the IC – see section 2 of this chapter) has been demonstrated at up to a few 10's of MHz but not at UHF frequencies.

There are very substantial obstacles to commercial deployment of these techniques. Operating voltages are typically 10–20 V, much higher than the 1 to 3 V used in silicon ICs. To achieve even such voltages requires that feature sizes around 3 microns be resolved in the structure. Such resolution, roughly equivalent to 8500 dots per inch, is much finer than that normally achieved in low-cost printing processes. The electrical properties of the polymers used today are often unstable in use and on extended exposure to air. The best semiconductor layers are formed by vacuum techniques at a rate of around 0.1 molecular layer per second, not conducive to high throughput or low cost.

Reliable production equipment and processes must be developed for high volume fabrication, a very considerable investment if quantities of hundreds of millions to billions of tags are envisioned.

Even if all these technical hurdles are overcome, challenging economic hurdles remain. At the time of this writing, a 512 MB DRAM chip costs around $6, which is equivalent to around 11 nanodollars (!) per bit or very roughly 0.25 nanodollar per printed 'dot', the actual value varying somewhat depending on the memory structure used. A tabloid printed in moderate volume on 18″ × 12″ (47 × 31 cm) paper at 200 dpi (8 dots/mm) costs around $0.20/copy, resulting in a cost of 3 nanodollars/dot. Larger-volume printing costs are about 10 times lower, or 0.3 nanodollars/dot. That is, the cost of conventional printing with inks whose basic formulation is centuries old is if anything more expensive per feature than the cost of modern silicon IC fabrication. It is very unlikely that the exotic materials employed in making a low-voltage printed organic IC would achieve these cost levels in the near term. Printing an IC only seems cheap if we assume that the IC complexity is comparable to the complexity of printing a bar code. When we include the magnitude of the actual task to be undertaken if organic circuitry is to provide functionality comparable to its silicon counterpart, it is no longer very clear that printing offers a substantial cost advantage.

Some work has also been done in identifying objects using conductive materials or fibers with no circuitry whatsoever, relying on the frequency-dependent behavior of radio reflection from these antenna-like strands to distinguish one object from another. Such an approach promises a very low cost for the 'tag' structures, since they contain no circuitry of any kind. However, a sophisticated reading device employing millimeter-wave frequencies (>10 GHz) is needed, and since the 'tags' have no logic capabilities, use models are constrained. At these high frequencies, diffraction is much reduced compared to the 900 MHz band, and the advantages of non-line-of-sight operation are lessened. Such circuit-less tag technologies may find special niches where their very low ongoing cost makes up for complexity in implementation. Non-RF-based techniques for authentication, using chemical compounds with unique optical or other analytical signatures, are also used and may have advantages over RFID in this type of application, though such approaches hardly seem adaptable to unique item serialization.

5.8 Assault of the Battery

When one's patience with the pittance of power provided by the petulant passive path is penultimately played out, it's finally time to add a battery to the tag (and stop overusing words starting with 'p'). We found in chapter 3 (Figure 3.36) that merely removing the need to power the logic with received RF — that is, replacing a passive tag with a semi-passive or battery-assisted tag — expanded our line-of-sight range to the order of 100 meters, limited more by the ability of the reader to hear the backscattered signal than by the tag's sensitivity. If we include a transmitter at the tag — an active tag — the range becomes limited mainly by real-world obstructions, absorbers, and interferers rather than tag or reader sensitivity (Figure 3.37). The reader is no doubt familiar with the ability of a sophisticated, expensive, battery-powered cellular handset to reach a network from substantial distances with no line-of-sight path in evidence. Batteries turn a dumb hardly visible tag into a powerhouse of communication and information processing.

However, pretty much every battery-powered device faces the same problem: the battery holds a finite amount of energy. The device it powers must either:

- operate for its expected lifetime on a single charge received at manufacture or assembly;
- recharge or replace the battery when it is exhausted; or
- fall back to another power source when the battery is exhausted.

Each of these approaches is an acceptable solution for some applications. An RFID tag might be used to keep track of people — to help find your kids at a theme park, for example. If a visitor receives the tag when entering the park, it is already handled once at each use; a charging operation or battery swap may be added with minimal additional cost. Kids are valuable — or at least, they're certainly expensive — so a few extra pennies will not impact the utility of the tags.

On the other hand, a tag used to mark an asset that becomes inaccessible or nearly so should last at least until the next time the tag can be reached for maintenance. Active tags can be used to track shipping containers, which are typically 12 meters (40 feet) long and 3 meters high. Containers are often piled 3 or 4 high in storage at ports, and on the huge ships that take them across an ocean. If the reader wishes to climb one of these piles to replace a tag battery, the author will be happy to watch and offer encouragement; it isn't impossible but it isn't practical. It might be OK to spend a dollar, but the tag battery needs to last months or years in this application.

Finally, in a classic supply-chain application like tracking a consumer product from manufacturer to retailer, adding a battery may greatly improve performance, but cost constraints are very severe. The battery must last from the time the product is marked at manufacture until it is sold to the consumer, and if it fails, the tag must still be readable by some means.

To make a battery last longer, you can use a big battery, draw very little current from it, or turn your device off when not in use. The cost constraints in most RFID applications preclude use of a big battery to solve endurance problems. Modern integrated circuits consume very little power when inactive, but if endurance is measured in years, very little is still a lot. To get a feel for the magnitude of the problem, let's consider a button battery, like those used in wristwatches and other small portable devices. A typical alkaline battery of this type has a capacity of 150 mA-hours at about 1.2 V (and is already awfully expensive for many passive-tag-like applications). A tag IC that uses 20 microwatts (17 μA at 1.2 V) will last about 1 year under ideal conditions. Real-life service is likely to be shorter, and there is a strong economic incentive to reduce the battery size.

To make a tag that lasts a reasonably long time on a small, cheap battery, we must exploit a key fact about RFID tags: they are hardly ever read. Read frequency varies from one application to another, but in general tags are read at most once every few minutes. Tags in storage in a warehouse may be read only when they are moved or pulled for shipment. Tags on shipping containers may be read a few times a day. All the rest of the time, power drawn from the battery is wasted. Thus, a battery-powered tag is almost always designed to operate in a minimum-power sleep-like mode, and wake up only when it needs to be read.

A simple approach to constructing such a tag is to start with a passive tag and add a battery to power the logic circuitry, while preserving the backscatter modulator and diode detection. The battery is disconnected from the logic circuitry by a switch until the tag detects substantial RF power, indicating that a reader is illuminating it. An example of this sort of architecture is depicted in Figure 5.31. A switch is interposed between the battery and the tag's logic circuitry. The switch can be turned on by a wake-up circuit, which monitors the output of a charge pump. The charge pump rectifies any RF signals from the antenna. When the RF signal rises high enough, a comparator in the wake-up circuit trips to turn the battery switch on. The battery then powers the tag's baseband circuitry, which can decode the reader's commands and respond using a backscatter modulator, just as a passive tag would do. When the switch is off, the battery uses very little power: just a few tens of nanoamps to cover the battery's internal leakage and the leakage of the switch. When the

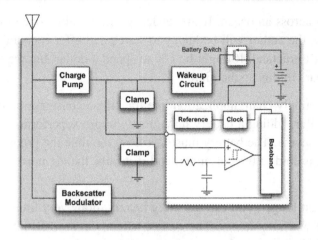

Figure 5.31
Simple battery-assisted semi-passive tag. *After Che et al.*

switch is on, the battery takes care of running the tag, and the RF signal need only be large enough to be decoded.

For the tag shown in Figure 5.31, the authors report receive sensitivity of -35 dBm, about $15-20$ dB better than a passive tag, and very much what we expect to achieve using diode detectors (Figure 3.35). The ideal line-of-sight range at which the tag can receive data is about 100 meters (ignoring any possible tag antenna gain), but at this range the backscattered signal at the reader will be about -100 dBm — possible but challenging to receive and decode in the presence of the large transmitted signal of the reader. As we noted in chapter 3, a battery-assisted backscatter tag can outrun the ability of the reader to hear it.

The battery switch is sized for a resistance of about 500 ohms in the on-state, in which the remainder of the circuitry requires 60 microamps to operate. (That is, the supply voltage drops about 30 mV across the switch when the latter is on.) The corresponding leakage of the switch in the off-state is around 100 nA (plus the current required by the comparator that senses when the RF power is high enough to turn the switch on). This is much less than the internal leakage of most battery technologies, so the sleep-state power consumption does not limit the endurance of the tag.

The current consumption in the active state provides about 2500 hours (15 weeks) of operation from a button battery, or roughly 80 hours from a printed paper battery. If the tag is never on, except when a reader is interrogating it, this is plenty of endurance for most applications. If the tag is turned on all the time, the battery is likely to be depleted long before the required lifetime of the tag is reached. The effectiveness of the wake-up circuit is critical to achieving commercially relevant lifetimes for a battery-assisted tag, particularly if inexpensive compact batteries are used.

This very simple example illuminates several aspects of the problem of waking up a tag. The effectiveness of the wake-up circuit often determines the battery's useful life. If we make the wake-up circuit very sensitive, it will cause the chip to awaken when there is no reader inquiry pending: there will be many *false positive* indications. If the wake-up circuit is very insensitive, it will miss the reader signal and be no better than a passive tag: there will be many *false negative* indications. Furthermore, if the wake-up circuit just responds to some measure of the average RF power, it has no way of distinguishing between a reader signal and any other RF signal, nor between the reader reading this tag and other readers looking at other tags. If exposure to RF is frequent (from people's cellular handsets, cordless phones, and readers that are interrogating other tag populations), the battery could be depleted prematurely. The more sensitive we make the wake-up circuit, the more likely it is to wake up in error.

One option is make the wake-up circuit smarter. For example, it might be designed to look for a specific pattern in the received signal before activating the battery switch. If the pattern also includes some identifying information, the tag can wake up only when a reader is specifically looking for it. However, as the wake-up circuit becomes smarter, it uses more power, vitiating the advantage of accurate wake-up behavior. Some examples of sophisticated wake-up approaches for battery-assisted tags will be discussed in section 6 of chapter 8.

Another approach is to abandon incident RF as the trigger for tag activity and instead use a timer or alarm. A tag being used as a sensor may simply need to read a value and report it periodically. If the expense of a quartz crystal oscillator is not excessive, a very accurate timer can be provided with only a few microamps of battery current. This approach works best with active tags, as otherwise the reader must transmit continuously to provide backscatter ability to many tags in the field. Once a tag is actively transmitting, it can listen for commands from the reader to change timing, change the data it collects, receive software upgrades, and perform the various other operations needed to support extended life in the field.

An active tag is at least a full-fledged radio transmitter. Transmitters must comply with regulatory requirements to avoid interference, and so will typically contain a quartz-crystal-controlled local oscillator that creates an accurately specified transmission frequency. The local oscillator signal, e.g. at a channel within the 2.4-GHz Industrial, Scientific, and Medical (ISM) band, is mixed with an information-containing signal to form the transmitted data and then amplified and sent to an antenna. The transmitted signal can be a simple beacon essentially saying "I am here," or it can be a protocol-compliant signal containing medium-access, addressing, and data. Beacon tags may only use a transmitter and provide no high-frequency radio receiver (though a low-frequency receiver may be used to trigger the beacons, as described below). A beacon tag can send an identifiable brief transmission upon request, which allows a reader to establish its presence from a considerable distance. If multiple antennas are available, the reader can accurately determine the location of a beacon tag by measuring the time the beacon arrives at the various receiving locations, and/ or by monitoring signal strength or direction of arrival. The ability to locate a tag is very important when long-range tags are used, but it is a continual challenge in realistic radio environments, where diffraction and multiple reflections are common.

A more elaborate active tag may implement a full two-way communications protocol, enabling it to carry out arbitrary exchanges not only with the reader, but also with networked devices reachable from the reader. Some examples of such protocols will be discussed in chapter 8. The active tag will need a receiver to detect and decode the incoming signal; the receiver will include sensitive amplifiers, a mixer to convert the high-frequency received signal (e.g. 2.4 GHz) to some convenient intermediate frequency at which it can be digitized, and elaborate processing firmware to decode the signal and respond appropriately.

An example of this sort of sophisticated active sensor device is depicted in Figure 5.32. The complex radio function, in this case implementing the IEEE 802.11 (Wi-Fi) protocol, is contained in a single integrated circuit — a *system-on-chip* or *SOC*. The SOC's CPU can also manage a simple interface to other chips (SPI), enabling it to talk with various sensors and non-volatile memory (NVM) placed on the same circuit board. Two quartz-crystal-controlled oscillators are available. A 33-kHz oscillator provides a very-low-power clock

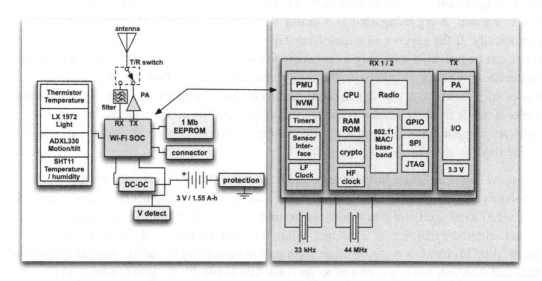

Figure 5.32
An active sensor device, using a system-on-chip radio. *After Folea and Gherciouiu.*

that runs as long as the battery is on, and can be used to awaken the SOC at periodic intervals. The 44-MHz oscillator provides a clock for the CPU and other logic circuitry, and a reference for controlling the local oscillator used to transmit and receive signals in the 2.4 or 5.2 GHz unlicensed bands. The SOC includes the elaborate capabilities needed to encode, modulate, demodulate, and decode standard networking packets (TCP/IP) sent over the 802.11 link. Finally, the CPU is also available to run application-specific firmware, so that the device can be programmed to do what the user needs it to.

Fully capable active sensor nodes like this one are much more power-hungry than a simple semi-passive tag. It is imperative that they be off much of the time to achieve decent battery life, and thus solving the problem of when to awaken is even more important than in the case of a semi-passive solution. When the circuit is asleep, only the LF oscillator and a bit of RAM need to be on, and total current consumption is measured in microamps or even nanoamps. When the processor and radio transmitter are active, current consumption is tens or hundreds of milliamps. In order to conserve even a large battery, the wake-up times must be brief and efficient. In return for this extra complexity, the tag's visibility is greatly increased. For a typical transmit power of 10 dBm, indoor ranges on the order of 50 meters, and unobstructed outdoor ranges of hundreds of meters, can be readily achieved.

For the example shown in Figure 5.32, it takes about 52 ms to awaken the tag's SOC and related systems, at a cost of 0.8 mJ. Data acquisition consumes another 120 ms and 22 mJ, which is highly dependent on the data being acquired. Reception (here ensuring that the channel is clear before transmitting) requires 90 ms and 16 mJ, and transmission of the data 4.7 ms and 10 mJ. The total energy used is 48.5 mJ. With a 1550 mA-hour battery (a typical capacity for a lithium AA cell), about 425,000 wakeups can be managed. That's about 10 months at 1 wakeup per minute, or several years if wakeups are less frequent.

A fully capable active tag can do a lot of neat stuff. The Wi-Fi tag shown supports Ethernet and TCP-IP communications, so it can be integrated into a local network without any special gateways. When it is awake it can listen for commands (potentially from anywhere on the Internet), process and send data and files, receive updates, and generally act like any IP-accessible device. But this flexibility comes at a steep price. A passive tag, using around 20−30 microwatts for the 200 ms duration of a typical inventory, consumes about 4−5 *mJ*. The active tag uses 10,000−12,000 times more energy during a wakeup. This ratio can be reduced with alternative architectures and protocols, to be discussed in more detail in chapter 8, but it will remain large. Transmitters are really useful but energetically expensive.

Active tags can be configured to awaken at timed intervals because they typically include a very-low-power clock that continues to run when the rest of the circuitry is powered down. This works fine for tracking objects that move infrequently and don't need to be found immediately, or for recording sensor data whose time-dependent behavior is fairly predictable, like ambient temperature. It's also easy to configure such a device to be awakened by a digital signal at an alarm input, but then the alarm has to come from somewhere. A motion detector can be used when the tag only really needs to go off if the object it is attached to is moved from one place to another. A popular alternative is to include a localized alarm that goes off when the tag enters a specific spatial region.

A localized alarm can be arranged by including an LF or HF receiver in the tag. A low-frequency transmitter – an exciter – can then be placed in a strategic location to trigger activity in the tag, which may otherwise lie low and save power. An exciter can often be powered externally and thus provide a substantial signal, so that the tag's receiver can be passive and requires little power. Because the exciter and the tag are inductively coupled (see section 6 of chapter 2), range is intrinsically short, so the exciter will only be seen from the tag when the two are in close proximity. Receiving an exciter signal tells the tag where it is without requiring any sophisticated capabilities from the tag. Exciters can be placed at pinch points or exits to trigger transmission when a tag enters or leaves a defined area; tags that track infants in a nursery are often configured this way. When the child passes through the nursery doorway, an exciter mounted near the door triggers beacons, so the system knows that the baby has been moved, and can verify that this is intended (e.g. by detecting the mother's identifying bracelet at the same time). In shipping yards, inexpensive tags can be affixed to trailers (which may spend days in the yard waiting to be loaded or unloaded), and an exciter can be affixed to the yard trucks that move them around. The tag can remain inactive, or send its ID very rarely, when the trailer is stationary, but beacon rapidly when the trailer is moved, making it easy to keep track of all the trailers in the yard. The long range of active tags ensures that the tag will be heard from any location in the yard. An example of this type of arrangement is depicted in Figure 5.33.

The possibilities opened up by local power are numerous; we can't possibly discuss them all here. For example, a powered tag can include a GPS receiver to get global location when satellites are visible and local communications (such as the 802.11 capability discussed above) to exchange location information with tags that can't see the sky. In this fashion, a pile of containers on a ship in transit can collectively be aware of where they are in the world, and even report that information back to a satellite link. The full power of modern integrated circuits and software can be brought to bear upon complex problems of identification, sensing, and location. These benefits are, however, only obtained at a substantial cost in money, size, and maintenance requirements.

5.9 Capsule Summary

Passive tags harvest power from an incident RF signal using diodes arranged to form a charge pump. Multiple stages help boost the output voltage to the level needed to power an integrated circuit, but a practical limit exists to the extent to which circuitry can be used to

Figure 5.33
A low-frequency exciter is affixed to a yard truck, to trigger the active tracking tag when the trailer is moved. *From "A Success Story for RTLS at NYK Logistics", Rick Crawford, RFID Applications Fall 2006, used with permission.*

compensate for low peak voltages received from the antenna. Multiple-stage charge pumps are rather inefficient, converting less than half of the received RF power to useful DC power for the remainder of the tag circuitry. A similar rectifier or charge-pump configuration with a faster response is used to extract the amplitude envelope of the reader signal, wherein can be found the commands and data from the reader.

Tags talk back using backscatter modulation. There is an inevitable tradeoff between the extent to which the scattered power is modulated, and the amount of power retained to run the tag. The best balance between these competing alternatives is obtained by changing not the resistance attached to the tag antenna, but the capacitance. Backscatter efficiency of -3 dB can in principle be obtained with modest effects on the tag power supply (but actual results depend on the provisions for matching tag antenna to IC load).

While the logical processing required by a typical protocol is modest compared to modern processor capabilities, design of tag integrated circuits is unusually challenging due to the intersection of two stringent constraints: very low cost and marginal power. Careful attention to individual aspects of the design, including hand-routing and analog simulation, is required.

A tag is more than just the integrated circuit. Assembly of the tag combines an IC, strap if used, antenna, substrate, and optionally label. Assembly is usually based on forming contact structures (bumps) on the integrated circuit to mate to contacts on the strap or antenna. Printed or etched metal antennas have historically dominated, but antenna fabrication via printed conductive inks shows promise for UHF applications. The IC, strap, antenna and substrate form an inlay, which may be laminated into a printable and thus human-readable label to be used.

The addition of a battery greatly improves the ability to hear tags at long distances and in obstructed environments and increases the things the tag can do. Each improvement in capability is burdened by additional cost, complexity, and maintenance requirements

Further Reading

Passive Tag IC Design

Design and Optimization of Passive UHF RFID Systems, J. Curty, M. Declercq, C. Dehollain, and N. Joehl, Springer, 2006. *This is a well-organized and clearly illustrated book, with an excellent discussion of charge pump operation. However, be aware that in chapter 5, the authors seem to suggest that the radar cross-section of an antenna is 0 when the antenna is matched to the load. This is not correct, invalidating their analysis of PSK and ASK modulations (or the current author has misunderstood Curty et al.'s notation, in which case apologies are offered).*

"Design criteria for the RF section of UHF and microwave passive RFID transponders", G. De Vita and G. Iannaccone, IEEE Transactions on Microwave Theory and Techniques, Volume 53, Issue 9, Date: Sept. 2005, Pages: 2978–2990

"Fully Integrated Passive UHF RFID Transponder IC with 16.7 mW Minimum RF Input Power", U. Karthaus and M. Fischer, IEEE J. Solid-State Circuits, 38 #10 p. 1602 (2003)

"UHF Passive-Tag IC Design", Roger Stewart, IEEE MTT-S, June 2006, Session TSC-110.

"Design of Ultra-Low-Cost UHF RFID Tags for Supply Chain Applications", Rob Glidden, Cameron Bockorick, Scott Cooper, Chris Diorio, David Dressler, Vadim Gutnik, Casey Hagen, Dennis Hara, Terry Hass, Todd Humes, John Hyde, Ron Oliver, Omer Onen, Alberto Pesavento, Kurt Sundstrom, and Mike Thomas, IEEE Communications Magazine, August, 2004, p. 140

"Single-Ended Ultra-Low-Power Multistage Rectifiers for Passive RFID Tags at UHF and Microwave Frequencies", K. Seeman, G. Hofer, F. Cilek, and R. Weigel, IEEE Radio and Wireless Conference (RAWCON) 2006 paper TH2A-1

Chip Assembly Techniques

"Wafer bumping technologies. A comparative analysis of solder deposition processes and assembly considerations", Patterson, D.S.; Elenius, P.; Leal, J.A., Advances in Electronic Packaging 1997. Proceedings of the Pacific Rim/ASME International Intersociety Electronic and Photonic Packaging Conference. INTERpack ASME, 1997. p.337–51, vol.1, Conference: Kohala Coast, HI, USA, 15–19 June 1997

"Manufacturing Multichip Modules", p. 391ff, by Rakesh Agarwal and Michael Pecht, in **Physical Architecture of VLSI Systems**, ed. Robert J. Hannemann, Allan D. Kraus and Michael Pecht; John Wiley & Sons Inc., New York (1994)

"Advanced solder flip chip processes", Rinne, G.; Koopman, N.; Magill, P.; Nangalia, S.; Berry, C.; Mis, D.; Rogers, V.; Adema, G.; Berry, M.; Deane, P. SMI. Surface Mount International. Advanced Electronics Manufacturing Technologies. Proceedings of The Technical Program Edina, MN, USA: Surface Mount Technol. Assoc, 1996. p.282–92 vol.1 of 2 vol. 826 pp. Conference: San Jose, CA, USA, 10–12 Sept 1996

"Multichip Assembly with Flipped Integrated Circuits", Heinen, Schroen, Edgwards, Wilson, Stierman and Lamson, Proc 39th Electronic Component Conference p. 672 1989

"Flip-chip packaging with polymer bumps", Estes, R.H., Semiconductor International (Feb. 1997) vol.20, no.2, p.103

"Reflowable anisotropic conductive adhesives for flipchip packaging", Sea, T.Y.; Tan, T.C.; Peh, E.K., Proceedings of the 1997 1st Electronic Packaging Technology Conference, 1997. p.259 Conference: Singapore, 8–10 Oct 1997

Conductive Inks

"Anisotropic Conductive Adhesive Films for Flip Chip on Flex Packages"
L. Li and T. Fang [Motorola]
4th International Conference onAdhesive Joining and Coating for Electronic Manufacturing, 2000, p. 129
"The Performance of New Conductive Inks for RFID Smart Labels"
Paul Berry [Dow Corning]
Smart Labels USA (IDTechEx), June 2005

SAW Tags

"A Global SAW ID Tag with Large Data Capacity", C. Hartmann, IEEE Ultrasonics Symposium 2000, p. 65

"Anti-Collision Methods for Global SAW RFID Tag Systems", C. Hartmann, P. Hartmann, P. Brown, J. Bellamy, L. Claiborne and W. Bonner, IEEE Ultrasonics Symposium 2004, p. 805

Organic ICs

"Organic Semiconductor RFID Transponders", P. Baude, D. Ender, T. Kelley, M. Haase, D. Muyres, and S. Theiss [3M], IEDM 2003 paper 8.1.1 (03-191)

"Progress Toward Development of All-Printed RFID Tags: Materials, Processes, and Devices", V. Subramanian, J. Frechet, P. Chang, D. Huang, J. Lee, S. Molesa, A. Murphy, D. Redinger, and S. Volkman, Proc IEEE v 93 #7 p. 1330 (2005)

Battery Tags and Active Sensors

"Analysis, design, and implementation of a semi-passive Gen2 Tag", W. Che, Y. Yang, C. Xu, N. Yan, X. Tan, Q. Li, H. Min, and J. Tan, IEEE International Conference on RFID 2009, p. 15

"Ultra-low-power wi-fi tag for wireless sensing", S. Folea and M. Ghercioiu, IEEE International Conference on Automation, Quality and Testing, Robotics (AQTR) 2008, p. 247

"Interaction in pervasive computing settings using Bluetooth-enabled active tags and passive RFID technology together with mobile phones", F. Siegemund and C. Florkmeier, Proceedings of the first IEEE International Conference on Pervasive Computing and Communications (PerCom 2003).

"Sensor measurements for wi-fi location with emphasis on time-of-arrival ranging", S. Golden and S. Bateman, IEEE Transactions on Mobile Computing v. 6 #10, p. 1185 (2007)

"Revisiting RFID link budgets for technology scaling: range maximization of RFID tags", R. Chakraborty, S. Roy, and V. Jandhyala, IEEE Transactions on Microwave Theory and Techniques, v. 29 #2, p. 296 (2011)

Exercises

Rectifiers:

1. We treated a diode as a very simple object that has no current flow until the voltage exceeds $+ V_{ON}$, and unlimited current flow with no additional voltage thereafter. Real diodes are somewhat more accurately represented by equation (5.1), repeated here for convenience:

$$I = I_0 \left(e^{\frac{qV}{kT}} - 1 \right)$$

where the product q/kT is about 38.5 at room temperature. Let I0 = 10−15 A. What voltage across the diode will produce a diode current I = 1 microamp? What voltage is needed to achieve a diode current I = 10 mA? What error in voltage have we made by using an on-voltage of 0.5 V?

V(1 μA): _____   Error vs. V_{ON}: _____

V(10 mA): _____   Error vs. V_{ON}: _____

2. We examined the use of voltage doublers to increase the output voltage of the rectifier stages. An alternative approach is to employ a *full-wave rectifier* circuit. A possible equivalent circuit for a full-wave rectifier attached to an antenna and load is shown below. Using the idealized diode model of Figure 5.3, find the output voltage for a given input voltage. Compare it to the output voltage of a doubler (equation (5.6)). Backscatter modulation:

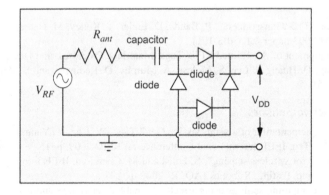

3. An alternative way of imposing an modulation on a tag antenna is shown below. Assume that the radiation resistance and load resistance are both 100 ohms, and that the circuit is operated at 915 MHz. Let the modulation capacitor value be 1 pF. Find the complex impedance of the circuit in the modulated state (when the capacitor is *not* shorted out). Appendix 3 may be helpful. What is the magnitude of the current that

flows through the circuit when the capacitor is present compared to that when the capacitor is shorted? What is the relative phase?

_____ magnitude   _____ phase

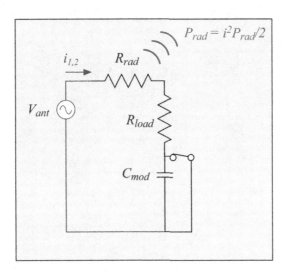

Find the power absorbed in the load in the modulated state as a fraction of the available power P_{av}.

_____ normalized power, modulated state

Find the average load power, normalized to the available power, assuming the modulation occupies the two possible states with equal probability.

_____ normalized power, average

Subtract the complex value of the current through the radiation resistance when the capacitor is present from the current amplitude when the capacitor is shorted to find the change in current due to modulation. Find the absolute magnitude of this current, and square it and multiply by $R_{rad}/2$ to find the peak radiated power.

_____ peak radiated power

(The average radiated power of an OOK signal is half this value.)

Divide by the available power to find the modulation efficiency:

_____ modulation efficiency

CHAPTER 6

Reader Antennas

6.1 Not Just for Insects Anymore

In Chapter 3, we learned that an antenna is a device to produce a distribution of currents and charges that do not cancel when observed from far away. We also introduced several properties of antennas that bear on their utility in an RFID system:

- **Gain and radiation pattern:** the extent to which the power radiated from an antenna is concentrated in some directions in preference to others;
- **Effective aperture:** the equivalent area from which a receiving antenna collects energy;
- **Polarization:** the orientation of the electric field radiated by the antenna;

There are three other key parameters that we got to ignore when thinking about link budgets, but which become very important when we need to hook a reader to an antenna:

- **Impedance:** how much voltage is required to cause a given current to flow in the antenna?
- **Bandwidth:** over what frequency range does the impedance remain reasonably constant?
- **Size and cost:** what we give away to get small and cheap;

In this chapter we'll look at how we get the gain and polarization we want, and how antenna impedance and bandwidth must be traded against antenna size. Armed with these fundamentals we can examine the requirements for different reader applications and see how they map to preferred antenna types. We will touch upon how antennas affect implementation: how to ensure that the beam covers the tags, what polarizations and orientations to use, and something about the specialized connectors and cables used at UHF and microwave frequencies.

6.2 Current Events: Fundamentals of Antenna Operation

Let us first briefly review the discussion of chapter 3, and tie down a few definitions more precisely. Antennas radiate different amounts of power in different directions relative to the antenna structure. The ratio of the power density, which we'll call U, in any given direction to that averaged over all directions is the ***directive gain*** of the antenna in that direction. The directive gain in the direction in which the power density is largest is called the ***maximum directive gain*** or ***directivity.*** The ***efficiency*** is the percentage of the power delivered to the antenna that actually gets radiated as opposed to being absorbed or reflected; for most reader antennas, this quantity is close to 100%, so there is little distinction between the directivity and the product of directivity and efficiency, the ***power gain*** or just ***gain*** of the antenna. Formally:

$$D_{max} = \frac{U(\theta_{max}, \phi_{max})}{\iint_{\theta,\phi} U \sin(\theta) d\theta d\phi}; \quad G = \varepsilon D_{max} \approx D_{max} \qquad (6.1)$$

For antennas where a well-defined beam exists, we can estimate the solid angle of the beam as:

$$\Omega_{beam}(\text{steradians}) \approx \frac{4\pi}{D_{max}} \qquad (6.2)$$

and the beam width of a fairly symmetrical beam as:

$$\theta_{beam}(\text{radians}) \approx \sqrt{\frac{4\pi}{D_{max}}} \qquad (6.3)$$

(Recall that a radian is $180/\pi \approx 57°$.) The archetypal simple antenna, the ***dipole***, is a pair of wires connected at the center to a signal source.

The ***effective aperture*** of the antenna, the area over which it collects power from an incoming signal, is also proportional to the antenna gain. The received power is proportional to the power density at the receiving antenna multiplied by the effective aperture, so a high-gain transmitting antenna is also a good receiver. In RFID, the forward-link-limited range (the distance at which the tag receives enough power to operate) is proportional to the square root of the reader antenna gain, as is the reverse-link-limited range.

The effective isotropic radiated power, EIRP, is the power that would need to be transmitted equally in all directions to provide the same power density U as a real antenna does in the direction of maximum gain. Thus:

$$EIRP = P_{TX}(dBm) + G_{TX}(dBi) \qquad (6.4)$$

Antennas radiate electric fields, which point in a specific direction at each location in space and each moment in time. The behavior of this direction defines the ***polarization*** of the radiated wave. The electric field of a linearly-polarized wave always points in one direction (vertically, for example) at all times and places. The electric field due to a circularly-polarized wave rotates around the axis of propagation each RF cycle at any point along the wave. Intermediate cases (elliptically-polarized waves) are also possible.

In the remainder of this discussion we'll take a somewhat closer look at how antenna configuration influences gain and polarization, and also estimate the impedance and bandwidth of an antenna, which determine over what frequency range it can be used.

6.2.1 Got Gain?

A dipole induces a voltage on another antenna located perpendicular to its axis, but no voltage along the axis. What about intermediate cases? It can be shown that the voltage depends on the sine of θ, the angle between the receiving antenna and the axis of the transmitting antenna. It should also be apparent that the dipole transmitter is symmetric around its axis, so the radiation must also be symmetric around the axis. Thus the dependence of the received signal on the relative location of the receiving antenna – the *radiation pattern* – of a dipole can be depicted as in Figure 6.1.

The maximum directivity of this pattern – the ratio of the power density in the best direction to the average over all directions – varies from about 1.5 for very short dipoles to about 1.7 for a dipole half a wavelength long (16 cm at 900 MHz). Recall from chapter 3 that the received power at the tag is directly proportional to the gain of the reader antenna, and that the received power at the reader is proportional to the square of the reader antenna

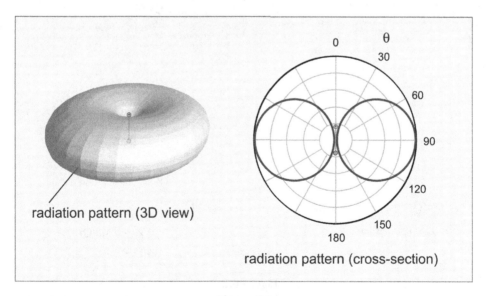

Figure 6.1
Radiation pattern of ideal dipole.

gain: to get longer range, it is helpful to have more gain than a dipole can provide. How do we make the antenna pattern more directive? One possible approach is to use more antennas. For example, let's consider the scheme shown in Figure 6.2: two ideal dipole antennas side by side. In the real world we need to worry about these antennas affecting each other, but for simplicity here we're going to just assume that the antennas are just the same together as they were alone, so each one has an identical sinusoidal current inducing a vector and scalar potential; the total voltage on a receiving antenna is due to the sum of the voltages from the two antennas. What happens?

Recall that the two individual dipoles transmit the same intensity to all directions in the plane of the paper (Figure 6.1). However, the pair of dipoles does not, because the signals must be added together, and as the angle of view changes the distance from the receiving antenna to each dipole changes. This difference means that the signals from the two transmitting antennas arrive with different time delays at the receiver, or equivalently that they are not at the same phase.

To find the effect of this phase difference we need to write an expression for the voltage. We can write the voltage due to a wave traveling away from a single antenna as:

$$V(r) = \frac{V_0}{r} \cos\left(\omega t - \frac{2\pi}{\lambda} r\right) \tag{6.5}$$

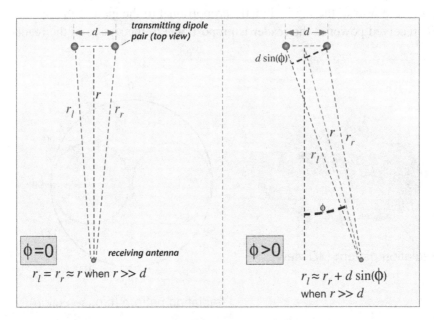

Figure 6.2

Dipole pair transmitting to a distant receiving antenna. Left: receiver perpendicular to plane of array. Right: receiver offset with respect to plane of array.

where λ is the wavelength and $\omega = 2\pi f$ is the angular frequency. (We can verify that this represents a traveling wave: when the time increases by some small increment δ, the first term in the cosine increases by $2\pi f\delta$. If at the same time the distance increases by $f\delta\lambda$, the argument of the cosine will stay the same. That is, the wave is moving at a velocity = distance/time = $f\lambda$, but since the wavelength is the speed of light divided by the frequency, $\lambda = c/f$, the velocity of the traveling wave is just the velocity of light, c.) When two antennas are present, we can write the total voltage as the sum of the two voltages. For large distances r, the difference in distance to the two antennas is insignificant and the effect of the $1/r$ term can be absorbed into the amplitude, as can the effects of absolute distance and time, so that we are only concerned with the difference between the signals from the left and right antennas. Taking into account the change in phase resulting from the different path length traveled from the two antennas (Figure 6.2), we obtain:

$$V(r) = V_{left} + V_{right}$$

$$= V_{right}\left(1 + \cos\left(\frac{2\pi d}{\lambda}\sin(\phi)\right)\right) \tag{6.6}$$

When the angle ϕ is 0, the voltages just add to give twice the voltage of one dipole (and thus four times the power, since power goes as the square of the voltage). However, when the argument of the cosine is equal to π radians, the second term becomes -1, and the voltage induced is 0:

$$for \quad \frac{2\pi d}{\lambda}\sin(\phi) = \pi:$$

$$V(r) = V_{right}(1 + \cos(\pi)) = V_{right}(1 - 1) = 0 \tag{6.7}$$

Since the sine is at most equal to 1, a zero value can only happen if $(d/\lambda) > 1/2$. When two antennas are present, the induced voltage on receiver − the radiation pattern in the plane − is dependent on angle instead of being uniform, but the exact result depends strongly on the spacing between the two antennas, as shown in Figure 6.3.

When spacing is small compared to a half-wavelength, the radiation is not much different from that of an isolated dipole (left side of the figure). When the spacing is a half-wavelength, radiation is directed into a relatively narrow slice in the plane (though the radiation pattern is vertically broad). Further increase in spacing gives a very narrow beam in the forward and reverse directions, but also creates another beam along the axis of the array. (The peak value of gain in this configuration actually occurs at a separation of about $3\lambda/4$, and is about 6.5 dBi.)

Gain or directivity do not by themselves tell us everything about an antenna radiation pattern. The gain of both patterns in the right side of Figure 6.3 is about the same (roughly 5 dBi), but the shape of the patterns is very different. One cannot use gain alone to compare antennas unless they are of a similar type with generally similar patterns.

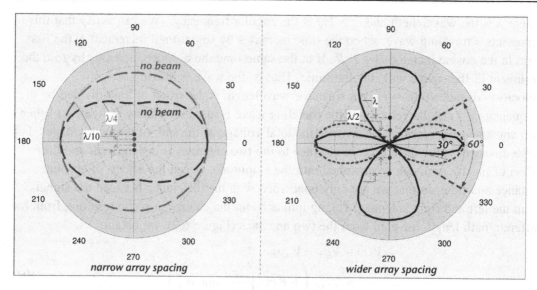

Figure 6.3

Radiation pattern for a pair of ideal dipoles separated by various distances. Left: $\lambda/10$ and $\lambda/4$ spacings. Right: $\lambda/2$ and λ spacings. Positions of the array elements are shown schematically; note that each pattern corresponds to a pair of radiating elements, with two possible spacings shown on the same image for brevity.

The origin of directive behavior in this simple two-element antenna array is characteristic of all antennas, though the details vary greatly depending on the design: the difference in power radiated at two neighboring directions depends on the *change in relative phase* of the paths from radiating current elements to the receiver as the angle of observation changes (Figure 6.2). This change in relative phase arises because of the difference in position (perpendicular to the axis of observation) of the different radiating parts of the antenna. Size is a necessary, though not sufficient requirement for a directional antenna. A large antenna may or may not have high gain, but a small − compared to a half-wavelength − antenna *can't*. Furthermore, because gain is proportional to size and beamwidth is inversely proportional to gain (chapter 3), we can infer that an asymmetric antenna − one that is, for example, much taller than it is long − will produce an asymmetric pattern, in this case wide horizontally and narrow vertically.

The patterns that result from a simple two-element array provide substantially higher directivity than we can obtain from a single dipole − but they are still rather inconveniently symmetrical. For example, let us imagine that we construct a simple array from two dipoles spaced a half-wavelength apart and excited in phase, as shown by the dotted line in Figure 6.3. If we orient this array so that one dipole is to our right and the other to our left, and we read a tag, we can be confident that the tag is either in front of us or behind us − but we don't know which (unless we are standing in the way of one of the beams!).

The two-element array produces a narrow but symmetric pair of beams. In practical applications we'd often like to produce a single beam that helps us address only the region of interest and avoid reading tags in other places. How can we do this?

One simple approach is to put a nice big fat piece of metal behind the array as a ***reflector***. This will work best if the spacing between the array and the reflector is correctly chosen. If the array is 1/4 of a wave from the reflector, the incident waves from the array will strike the reflector (where they suffer an inversion of phase upon reflection) and return to the location of the array delayed by half a cycle, and phase-inverted by half a cycle: the net result is that the reflected wave is in phase with that from the array and adds to it, at least along the direction of the beam. In the other direction, the reflector (if it is big enough) substantially blocks the beam. We obtain roughly a 3-dB increase in gain from removing the backwards-directed beam, and an antenna where most of the transmitted power is concentrated in a single beam, helping to clearly locate the read zone. A much larger benefit can be had if a parabolic reflector is employed instead of a flat plate, though this is obviously more complex.

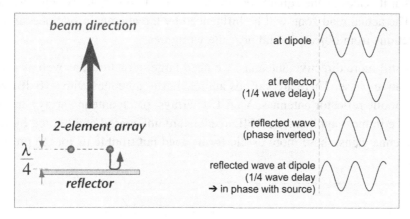

Figure 6.4
Two-element array with a reflector spaced 1/4 wavelength away; in the forward beam direction, the direct and reflected waves add in phase.

We can generally conclude a plausible approach to constructing a $5-10$ dBi single-beam antenna requires a radiating element on the order of 1/2 wavelength in each direction, and a reflector or other element in the direction of desired propagation, to select one beam and reject others. The most common large reader antenna, the patch or panel antenna, generally follows this prescription, though the reflecting ground plane is placed much closer than a quarter-wavelength to the metal patch.

Recall that in the case where the antenna radiates mostly in a single well-defined beam, one can define a useful ***beam width*** θ that is simply if approximately related to the gain G (equation (6.3)). For example, a gain of 6 dBi (a factor of 4) produces a beam about 1.8

radians or 100 degrees; the similar figure for a 10 dBi antenna is about 65 degrees. Real antennas don't have sharp-edged beams, and one has to decide where to measure the beam width; a common choice is the angle at which the power density has decreased by a factor of 2 (3 dB) from the maximum value in the center of the beam.

As we discussed in chapter 3, tags are often forward-link-limited. The region around an antenna in which tags can be read — the *read zone* — is the region in which the incident power exceeds the minimum needed to operate the tag IC. An antenna with a single, more or less symmetrical beam will have a generally ellipsoidal read zone: the range is longest along the center of the beam and falls off towards the edges, and objects displaced by the same distance from the beam will subtend an increasing angle as the distance to the antenna is reduced. A rough estimate of the shape of the read zone can be obtained using the beam widths corresponding to (for example) 3 dB and 6 dB decreases in power; these beam widths may be obtained from the antenna pattern, and might also be documented in the data sheet. The procedure simply assumes that the read range is forward-link-limited and power-limited, so that it scales as the square root of the received power; an example is shown in Figure 6.5. The actual read zone will be influenced by the antenna's sidelobes if present, and by reflections from objects in and near the work area.

To construct still more directive antennas, we need larger structures, as well as very precise control of relative phase. There are various approaches to antennas with >10 dBi gain, including parabolic reflector antennas, Yagi-Uda arrays, patch antenna arrays, and lens antennas. Since in most applications RFID readers are unlicensed and limited by regulation to modest antenna gains, these more exotic forms need not trouble us for the present.

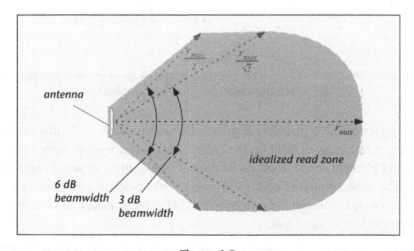

Figure 6.5
Idealized read zone estimate.

6.2.2 Polarization

Currents create vector potentials in the same direction as the current flow. The resulting electric field is constructed from the part of the vector potential that is perpendicular to the direction of propagation. A simple antenna like a dipole, which has current flowing only in one direction, creates a linearly-polarized radiated wave with the polarization direction being the long axis of the dipole. A similar dipole oriented along that field will receive the maximum possible signal; a simple dipole oriented perpendicular to the field will receive no signal at all.

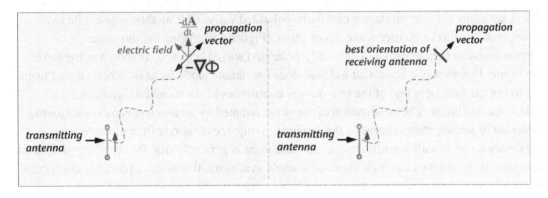

Figure 6.6
The electric field is proportional to the part of the vector potential that is perpendicular to the direction of propagation.

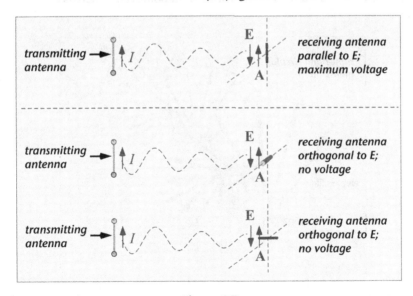

Figure 6.7
No voltage is induced on a receiving antenna orthogonal to the electric field.

As we have noted previously, many tag antennas are long and thin and behave rather like a dipole; this means that a tag oriented perpendicular to the electric field of a linearly-polarized reader antenna will receive no voltage and will not be read. One possible solution is to vary the direction of the electric field with time. If this is done on the same time scale as the RF cycle we obtain a circularly-polarized wave, in which the electric field rotates as a function of time (figure 3.32, reproduced for convenience below as Figure 6.8). As long as the tag antenna is aligned perpendicular to the direction of propagation, the electric field and the tag antenna will be parallel at two times during the RF cycle.

There are many ways to produce a circularly-polarized wave. One method is to excite two orthogonal antennas a quarter-wave out of phase (Figure 6.9). Along the direction perpendicular to both antennas, a circularly-polarized wave is created, as shown in the top of the figure. However, we know that a dipole does not radiate along its axis. When viewed from the top or the side, only one of the two dipoles contributes to the received signal, and the polarization is linear. The polarization of the wave radiated by an antenna can *depend on the direction of propagation relative to the antenna*. While the detailed effect of angle on polarization varies with antenna design, this statement is generally true for circularly-polarized antennas. If the antenna has high gain and a single symmetrical beam, we probably don't care too much: at high angles where the polarization is strongly affected, there isn't much signal power left. However, in some applications it is useful to employ antennas with a narrow beam in one direction and a wide beam in the orthogonal direction, sometimes known as *fan beam* antennas; in this case we must attend to the variation in polarization across the beam.

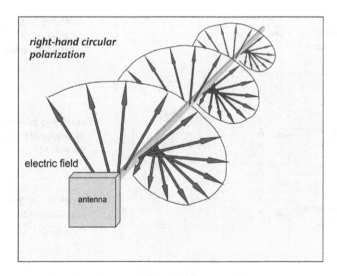

Figure 6.8
In a circularly-polarized radiated wave the direction of the electric field rotates one full turn in each RF cycle.

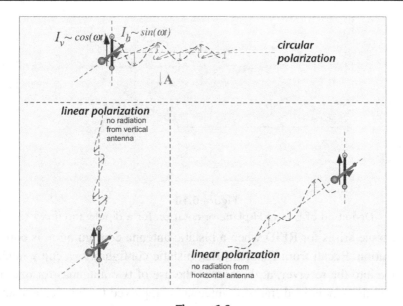

Figure 6.9
Crossed dipoles excited in quadrature (90 degrees out of phase) produce circular polarization in the mutually-orthogonal direction but linear polarization along either axis. (This physical arrangement is sometimes known as a *turnstile* antenna.)

We should note that it is also possible to achieve the same effects, from the point of view of an RFID reader, by varying the polarization on a much longer time scale: we simply switch between a horizontally-polarized and a second collocated vertically-polarized antenna. To be useful in most RFID protocols, one would need to make the switch in the pause between inventory operations – a time period that could be as short as a couple of milliseconds or as long as hundreds of milliseconds, depending on the protocol, setup, and tag responses. The disadvantage of sporadic coverage that results from this alternating-polarization approach must be weighed against the benefit of placing all the instantaneous power in one polarization at any given time, which provides superior read range for single-dipole tags.

A bit of notational convention is also convenient to introduce at this point. The radiation pattern of an antenna is often depicted on planar cross-sections. The planes are sometimes named relative to the viewer – elevation and azimuth, or vertical and horizontal – but it is also common practice to call the plane in which the electric field lies the ***E-plane***, and the orthogonal plane the ***H-plane***. A dipole pattern in the E-plane is a donut like that of Figure 6.1; the H-plane pattern is a circle, since the radiated power is symmetric about the axis. In cases where the orientation of the radiated electric field is readily determined, such as a dipole, no ambiguity is created. With more complex antenna structures it can be harder to see where the E-plane ought to be, and circularly-polarized antennas don't have a unique electric field plane.

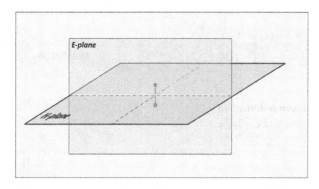

Figure 6.10
Definition of E- and H-plane orientation for a dipole antenna.

An interesting issue arises for RFID when a bistatic antenna configuration is combined with circular polarization. Recall from chapter 4 that bistatic configurations enjoy much reduced transmit leakage into the receiver, at the cost of the use of two antennas for one read location. If a circularly polarized transmit antenna is employed to illuminate a single-dipole antenna, the induced currents in the tag will flow along the axis of the tag antenna, and the backscattered signal will be linearly polarized. Any linear orientation will in general be received by a circularly-polarized receiving antenna. However, some tags (known as ***dual-dipole*** tags − see chapter 7, section 3.6) have two orthogonally-oriented antennas. The current in these antennas will be excited 90 degrees out of phase by an incident circularly polarized wave, and thus the backscattered wave will be circularly polarized but of ***opposite sense***. (The electric field is still rotating in the same direction as the transmitted wave − clockwise or counterclockwise, as the case may be − but the direction of propagation is reversed. Thus, from the point of view of an observer looking along the direction the radiation is traveling, the sense of rotation has reversed.) The receiving antenna should then be of the opposite sense from the transmitting antenna to get the best signal. That is, if a right-hand-circular transmitter is used, a left-hand-circular receiving antenna should be chosen. A monostatic circularly-polarized antenna may have reduced reverse-link sensitivity for dual-dipole tags.

6.2.3 Impedance and Bandwidth

An antenna simultaneously stores charge (***capacitance***), opposes changes in current (***inductance***), and radiates power into the wide world (***resistance***). From the electrical point of view, an antenna looks like an R-L-C circuit. The configuration of the circuit depends on the antenna type. In Figure 6.11 we show two different antennas with their equivalent circuits. The top image is a dipole with a bit of feed line. A dipole that is short compared to the wavelength looks like a series combination of an inductor and capacitor with some resistance. The inductance and capacitance are both roughly proportional to the length of

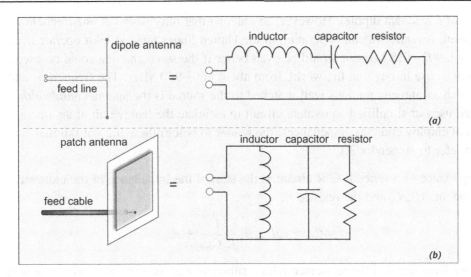

Figure 6.11
Antenna equivalent circuits. (a) dipole antenna; (b) patch antenna.

the dipole; at very low frequencies, the inductance has little effect and the capacitance stores little charge, so the dipole looks like an open circuit. As frequency increases, the inductive reactance increases and the capacitive reactance decreases; at some ***resonant*** frequency they are equal in magnitude and opposite in sign, and neutralize each other, so that the antenna looks electrically like a pure resistance.

The bottom image depicts a ***patch antenna*** — a structure about which we shall have more to say in a moment — as an example of an antenna that requires a parallel rather than series resonant circuit. In this case, the parallel combination of an inductor and capacitor draws no net current at resonance, so again the antenna looks like a pure resistance.

It is important to note that these simple equivalent circuits with fixed component values are only valid over narrow frequency ranges. In particular, the resistance value is due to losses to radiating waves. For dipoles short compared to a wavelength the radiation from the dipole is roughly proportional to the square of the dipole length. Short antennas have very small radiation resistances. At resonance (which turns out to correspond to a length just less than half a wavelength), a dipole radiation resistance is around 60–70 ohms (slightly dependent on the wire diameter). In contrast, a dipole that is only 1/10 of a wavelength long will have a radiation resistance of about 2 ohms.

To get good power transfer between a resistive electrical source and a resistive load, the source and load resistances should be equal: the source and load impedances should be ***matched***. The feed line is often a 50-ohm cable; in this case we'd like the antenna to look like a 50-ohm resistor. A dipole at resonance isn't a bad approximation: a pure resistance of about 65 ohms. No special measures are needed to match the source and load (one of the benefits of a resonant dipole). However, an antenna that only works at one frequency isn't very useful. A reader antenna operating in the United States must at least operate over the ISM band, 902–928 MHz; it would be even better if the same antenna could be used for all the bands in use throughout the world, from about 860–960 MHz. The frequency range over which an antenna remains well-matched to the source is the antenna **bandwidth**. We shall use our simplified equivalent circuit to estimate the bandwidth of an antenna. (We shall employ complex impedances; the reader to whom these are not familiar may wish to refer to Appendix 3.)

The impedance of a series L-C-R circuit is the sum of the impedance of the inductor, $j\omega L$, the capacitor, $1/j\omega C$, and the resistor R:

$$Z_{ant} = j\omega L_{ant} + \frac{1}{j\omega C_{ant}} + R_{rad} \tag{6.8}$$

where we have denoted the resistance with a subscript *rad*, on the assumption that it arises from radiation. At resonance the two reactances cancel exactly, but away from resonance they don't. The change in impedance per Hz near resonance is just proportional to the inductive reactance:

$$Z_{ant} \approx 2j(\omega - \omega_{res})L_{ant} + R_{rad} \tag{6.9}$$

The magnitude of the current flowing for a fixed source voltage is the voltage divided by the magnitude of the impedance:

$$|I_{ant}| = \frac{V}{|Z_{ant}|} \approx \frac{V}{|2j(\omega - \omega_{res})L_{ant} + R_{rad}|}$$
$$= \frac{V}{|4(\omega - \omega_{res})^2 L_{ant}^2 + R_{rad}^2|} \tag{6.10}$$

When the first term (in ωL) is small compared to the second (in R), the current is about what it would have been at resonance; when the first term is large compared to the second, the current is much smaller than at resonance. The edges of the band can be sensibly defined as those frequency displacements where the two terms are equal:

$$4(\omega_{edge} - \omega_{res})^2 L_{ant}^2 = R_{rad}^2 \rightarrow |(\omega_{edge} - \omega_{res})| = \frac{R_{rad}}{2L_{ant}} \tag{6.11}$$

and the bandwidth is twice as large:

$$\frac{2[\omega_{edge} - \omega_{res}]}{\omega_{res}} = \frac{R_{rad}}{\omega_{res}L_{ant}} = \frac{1}{Q}; \quad BW(GHz) = \frac{R_{rad}}{2\pi L_{ant}} \tag{6.12}$$

where the quality factor Q was defined in chapter 4, and the bandwidth is in GHz if the inductance is in nH and the resistance in ohms. We can see that to get a large bandwidth, we want a large value of radiation resistance and a small inductance. Note that the calculation for the parallel resonant circuit (part (b) of Figure 6.11) proceeds in exactly the same fashion, except that admittances are substituted for impedances ($Y = 1/Z$). In this case it becomes convenient to use the capacitance instead of the inductance, and we obtain:

$$BW(\text{GHz}) = \frac{1}{2\pi C_{ant} R_{rad}} \tag{6.13}$$

where the bandwidth is in GHz if the capacitance is measured in pF and the resistance in kΩ.

What are reasonable values for a dipole? An approximate equivalent circuit for a resonant dipole at around 900 MHz is shown in the top of Figure 6.12. Putting the appropriate values into equation (6.12), we obtain:

$$BW(\text{GHz}) \approx \frac{65}{2\pi(60)} \approx 0.17 = 170\ \text{MHz} \tag{6.14}$$

which is an ample bandwidth to cover the whole region of interest for international use (860–960 MHz).

The bottom of the figure provides the corresponding equivalent circuit for a much shorter dipole, the sort of antenna one might be tempted to use for a handheld reader. The inductance and capacitance have been scaled by a factor of 4, but note that the radiation resistance has fallen by much more than that! This is because the radiating length has been

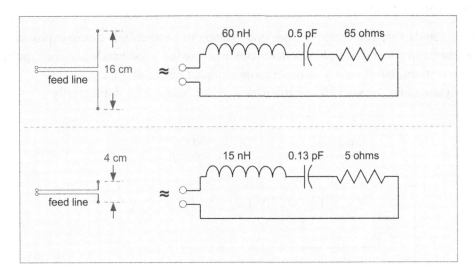

Figure 6.12
Equivalent circuits for dipoles. Top: resonant dipole ($\lambda/2$) at 900 MHz; Bottom: small dipole ($\lambda/8$).

scaled by a factor of 4, so the radiated potential and electric field are reduced by 4 times, but the power (which goes as the square of the field) is reduced by a factor of 16. This short dipole is also no longer resonant at 915 MHz: at this frequency the antenna series inductance has little effect, and the antenna looks capacitive. To match it to a resistive source, we need to add more inductance – that is, we use a ***matching element***. The circuit ends up looking like Figure 6.13. The combined circuit looks like a pure resistance (in this case, 5 ohms). The bandwidth is:

$$BW(\text{GHz}) \approx \frac{5}{2\pi(227 + 15)} \approx 0.003 \equiv 3 \text{ MHz} \tag{6.15}$$

This is seriously inadequate for the US ISM band (26 MHz wide)! The narrow 3 MHz bandwidth is sufficient for operation in e.g. the European RFID band at 865–868 MHz, but as you can imagine, any tiny change in the antenna properties that shifts the resonant frequency will result in poor antenna performance.

A (reader + cable) doesn't behave like a voltage source, but instead looks more a source with a finite output impedance (typically close to that of the cable, say 50 ohms). In Figure 6.14 we show an estimate of the radiated power vs. frequency, assuming that the reader acts like a fixed voltage (chosen to provided 1 Watt to a matched resistive load) and a source impedance equal to the radiation resistance. (We have here ignored the additional problem of transforming the source impedance from 50 to 5 ohms, which represents still another opportunity to narrow the bandwidth of the overall system.) Both dipoles radiate 1 Watt when matched, but the resonant dipole is much more tolerant of variations in frequency than the small dipole!

While the details vary considerably from one antenna to another, this example provides good guidance in general: simple resonant antennas provide good bandwidth for typical RFID applications, but if we try to make more compact antennas for e.g. portable applications, bandwidth and robustness suffer, sometimes dramatically.

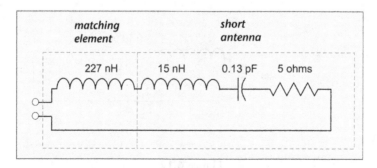

Figure 6.13
A short dipole requires a matching element.

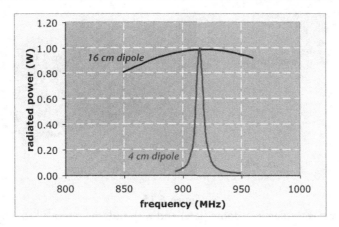

Figure 6.14
Radiated power vs. frequency for resonant dipole and matched short dipole.

When one wishes to actually characterize an antenna, it's quite a bit easier to measure the power that is reflected from the antenna, rather than the power that is transmitted by it. For most reader antennas, we can assume that any power that didn't get reflected was transmitted to the wide world, so if the reflections are small, power is being efficiently radiated. The magnitude of the reflected signal relative to the incident signal is known as the **return loss**, and should be less than −10 dB. This quantity is readily measured with a calibrated **network analyzer**, a commonly-available if specialized microwave instrument. Good antennas maintain −15 to −20 dB return loss across the band of interest. It is difficult to achieve return loss better than −25 dB consistently: the reflected power is so small that it becomes quite sensitive to reflections from objects near the antenna and other non-ideal conditions. The reflected power is also sometimes reported in terms of the **voltage standing-wave ratio** (VSWR), which is the ratio of the largest power to the smallest power when measured by a probe moving along a transmission line carrying power to the antenna. (The basis for standing-wave measurements is historical: in the early history of microwave technology, before network analyzers were readily available, probes inserted in slotted waveguides were used to measure reflected power.) When no reflected wave is present, the same power is measured everywhere along the line, and VSWR = 1. When a reflected wave is present, the transmitted and reflected waves interfere, creating maxima and minima of power separated by a quarter of a wavelength. A return loss of − 10 dB corresponds to a VSWR of about 2:1, and −20 dB to about 1.2:1.

6.2.4 The Patch Antenna

A very popular antenna for RFID reader applications is the **patch** or **panel** antenna. A patch antenna gains its name from the fact that it basically consists of a metal patch suspended over a ground plane. The assembly is usually contained in a plastic radome, which protects

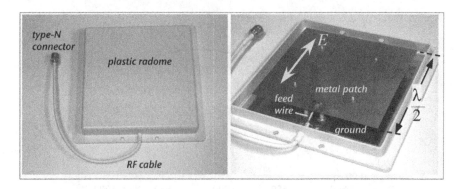

Figure 6.15
Patch antenna, shown with plastic radome (left) and with radome removed to expose patch (right).

the structure from damage (as well as concealing its essential simplicity). Patch antennas are simple to fabricate and easy to modify and customize.

The simplest patch antenna uses a half-wavelength-long patch and a larger ground plane. (Large ground planes give better performance but of course make the antenna bigger. It isn't uncommon for the ground plane to be only modestly larger than the active patch.) The current flow is along the direction of the feed wire, so the vector potential and thus the electric field follow the current, as shown by the arrow in the figure labeled E. A simple patch antenna of this type radiates a linearly polarized wave.

The gain of a patch antenna can be very roughly estimated as follows. Since the length of the patch, half a wavelength, is about the same as the length of a resonant dipole, we get about 2 dB of gain from the directivity relative to the vertical axis of the patch. If the patch is square, the pattern in the horizontal plane will be directional, somewhat as if the patch were a pair of dipoles separated by a half-wave, as in Figure 6.3: this counts for about another 2−3 dB. Finally, the addition of the ground plane cuts off most or all radiation behind the antenna, reducing the average power over all angles by a factor of 2 (and thus increasing the gain by 3 dB). Adding this all up, we get about 7−8 dB for a square patch, in good agreement with more sophisticated approaches (see Balanis, p. 841, in Further Reading for more details). A typical radiation pattern for a linearly-polarized patch antenna is shown in Figure 6.16. The figure shows a cross-section in a horizontal plane; the pattern in the vertical plane is similar though not identical. The scale is logarithmic, so (for example) the power radiated at 180° (90° to the left of the beam center) is about 15 dB less than the power in the center of the beam. The beam width is about 65° and the gain is about 9 dBi. (Note that in the United States, the maximum legal power for a reader using this antenna is 1/2 watt, to remain within legal limits for radiated power density.)

The beam width is rather close to that we obtained for the horizontal plane with a simple two-dipole array (see Figure 6.3): the directivity in this plane arises essentially from the fact that the

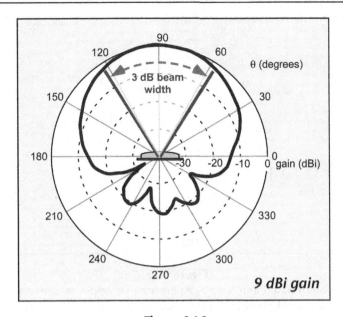

Figure 6.16
Radiation pattern in the horizontal (H-) plane for a typical commercial patch antenna specified for 890–960 MHz operation. Antenna orientation is shown in the inset.

roughly-square patch is about half a wavelength wide. An infinitely-large ground plane would prevent any radiation towards the back of the antenna (angles from 180 to 360°), but the real antenna has a fairly small ground plane, and the power in the backwards direction is only about 20 dB down from that in the main beam. This means the forward-link-limited read range will be about 10 times less backwards than forwards: if you can read a tag 5 meters from the front of the antenna, you will also read a tag 50 cm behind it – so don't put them there!

What bandwidth can we expect from a patch antenna, and what influences bandwidth and impedance? To address this question, we need to consider the currents and charges in the antenna. Currents flowing on the patch induce currents in the opposite direction in the ground plane (Figure 6.17). (To be precise, the currents on the ground plane are equivalent to an image of the patch displaced below the ground plane by the same distance the physical patch is above it.) The radiation from these oppositely-directed currents nearly cancels; what radiation does result occurs only because of the slight difference in time delay (equivalent to phase) from the two plates. To compensate for this near-cancellation, we use a resonant (half-wave) patch, which supports very large peak currents in the patch from only a small current flow in the feed wire.

The equivalent circuit of a resonant patch antenna is like that of the folded dipole in Figure 6.11: a parallel inductor, capacitor, and resistor. In order to say something about the bandwidth performance of this sort of antenna, we need to put some values on the equivalent

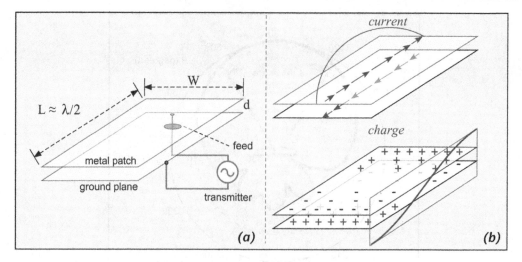

Figure 6.17
a) Schematic depiction of patch antenna; b) Approximately opposite currents flow on the patch and ground plane, inducing opposing charges on the plates.

circuit elements. The radiation resistance can be estimated from the radiated power, and is about 130−160 ohms for a typical square patch (half a wave in both directions). The radiation resistance seen in the center of the patch is actually quite small (1−2 ohms for reasonable geometries, which would be very difficult to work with); this small value is transformed by the patch acting as a transmission line to provide a convenient resistance when the patch is fed at or near its edge. Since the radiation resistance isn't terribly close to 50 ohms, a bit of matching is needed; this is often accomplished by displacing the feed point of the antenna away from the edge a bit, as shown in the figure.

To estimate the bandwidth of a patch antenna we need some idea of the equivalent capacitance to plug into equation (6.13). A wild guess using the charge distribution shown in Figure 6.17(b) would suggest that a feed near the end of the patch sees about 1/3 of the parallel-plate capacitance; we can justify this estimate a bit more formally by equating the impedance of the equivalent circuit to the impedance of the transmission line formed by the patch:

$$\sqrt{\frac{L_{ant}}{C_{ant}}} = Z_{line} \approx Z_0 \frac{d}{W} \tag{6.16}$$

where the value for the line impedance is an approximation value when $d << W$, as is usually the case for a patch antenna, and Z_0 is the impedance of free space, about 377 ohms. We can find the capacitance using the expression for the resonant frequency:

$$\omega_{res} = \frac{1}{\sqrt{L_{ant}C_{ant}}} \tag{6.17}$$

This gives us two equations for the two unknowns (L and C), so we can solve, obtaining:

$$C_{ant} = \frac{W}{\omega_{res}Z_0 d} \tag{6.18}$$

We can rewrite this expression using the fact that the length of the plate is half a wavelength:

$$C_{ant} = \frac{W}{\frac{\pi c}{L}Z_0 d} = \frac{WL}{\pi \mu_0 c^2 d}; \quad \frac{1}{\mu_0 c^2} = \varepsilon_0$$

$$\to C_{ant} = \varepsilon_0 \frac{A}{d}\frac{1}{\pi} = \frac{1}{\pi}C_{parallel\ plate} \tag{6.19}$$

which is to say, the capacitance seen by the feed is about 1/3 of the parallel-plate capacitance of the line, as we guessed at the beginning. A reasonable value for a square patch 8 mm high is about 9 pF. The resulting equivalent circuit of the antenna is shown in Figure 6.18.

Using this value, we can estimate the bandwidth:

$$\delta f = \frac{1}{2\pi \frac{W}{\omega_{res}Z_0 d}R_{rad}} \approx f_{res}\frac{Z_0}{R_{rad}}\frac{d}{W} \tag{6.20}$$

or in terms of the fractional bandwidth:

$$\frac{\delta f}{f_{res}} = \frac{Z_0}{R_{rad}}\frac{d}{W} \tag{6.21}$$

The fractional bandwidth of a patch antenna is linear in the height of the antenna. We need room for a reasonably thick structure to get good bandwidth.

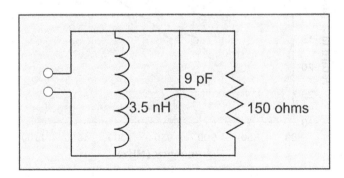

Figure 6.18
Approximate equivalent circuit for a resonant square patch at 915 MHz.

How thick does the antenna need to be? The impedance of free space is 377 ohms, so for the typical radiation resistance of about 150 ohms, we get the further simplification:

$$\frac{\delta f}{f_{res}} = 2.5\left(\frac{d}{W}\right) \tag{6.22}$$

For a square patch at 900 MHz, W will be around 16 cm. A height d of 1.6 cm will provide a fractional bandwidth of around 2.5(1.6/16) ≈ 25%, or about 230 MHz. This turns out to be optimistic by about a factor of 1.5 when we incorporate a more accurate calculation of capacitance and the practical issue of matching to the antenna, so that a 16-mm-high antenna can provide a bandwidth of around 150 MHz − sufficient to cover the whole international operation region, at the cost of a rather thick structure.

Our estimates have also assumed that the patch is suspended in air. A patch printed onto a dielectric board is often more convenient to fabricate and is a bit smaller, but the capacitance of the antenna is increased, so the bandwidth decreases. The calculations we have made also assume a very large ground plane. Real patch antennas often use ground planes only modestly larger than the patch, which also reduces performance. The details of the feed structure affect bandwidth as well.

Return loss for a pair of representative commercial patch antennas is shown in Figure 6.19; both antennas deliver substantially more than the 26 MHz needed to cover the ISM band. Antenna B uses a 16-mm patch height above ground, and the measured bandwidth of about 150 MHz at 10 dB return loss is rather close to that estimated above. However, this antenna also uses a very large (30 × 30 cm) ground plane. Antenna A delivers similar bandwidth but somewhat poorer performance in band, but at about 20 × 20 cm is considerably smaller and

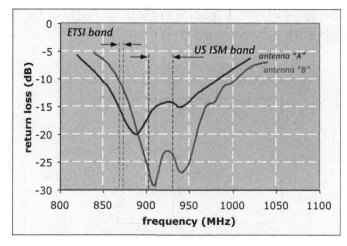

Figure 6.19
Return loss vs. frequency for two commercial ISM-band patch antennas.

more convenient to mount and position. Commercial antennas vary widely in performance, often due to poor centering of the band even when theoretical bandwidth is achieved, and can show return loss as poor as -8 to -9 dB at the edges of the ISM band; such antennas may be wholly unsuitable for global use. Bad return loss will degrade monostatic reader operation due to transmit leakage and should be avoided! Make sure that your antennas achieve -15 dB over the band of interest.

Rectangular (non-square) patches can be used when it is desired to produce a ***fan beam***: a radiated wave whose vertical and horizontal beamwidths are substantially different. Circular patches can be used instead of square patches; fabrication is straightforward though calculating the current distribution is more involved!

It is also possible to fabricate patch antennas that radiate circularly-polarized waves. There are a couple of different ways to do this (Figure 6.20). One approach is to excite a single square patch using two feeds, with one feed delayed by 90° with respect to the other. In this fashion, when (say) the vertical current flow is maximized, the horizontal current flow will be zero, so the radiated electric field will be vertical; one quarter-cycle later, the situation will have reversed and the field will be horizontal. The radiated field will thus rotate in time, producing a circularly-polarized wave. An alternative is to use a single feed but introduce some sort of asymmetric slot or other feature on the patch, causing the current distribution to be displaced. Note that, while circular patches can be used for these techniques, a circular patch does not necessarily radiate circularly-polarized waves! A symmetric circular patch with a single feed point will create linearly-polarized radiation. Finally, a nearly-square patch can be driven at the corner; if the length is just a bit less than resonant and the height a bit more (or vice versa) a circularly-polarized wave will result.

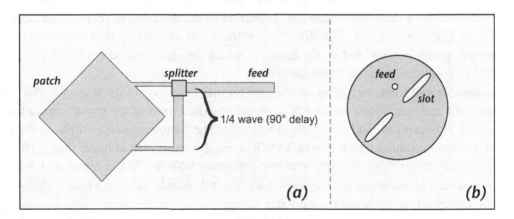

Figure 6.20
Approaches for creating circularly-polarized patch antennas. a) square patch with dual feeds, 90° phase offset; b) round patch with asymmetric slot. Note that square or round patches can be used in both cases.

Recall that readers may use a bistatic antenna configuration, and that if circular polarization is employed, dual-dipole tags will invert the sense of polarization on reflection. Readers that use a bistatic antenna configuration often provide two circularly-polarized antennas of opposite sense packaged in a single housing; this ensures that the antennas are properly oriented when mounted, though the resulting structure is large (about a meter long) and somewhat ungainly. Some designs use a 'bathtub' configuration, with a ground plane substantially larger than the patches, and a recessed region about 5 cm deep in which the patches sit. This type of design provides excellent isolation between the two antennas (around 40−45 dB), and also return loss of −20 to −25 dB over the ISM band.

6.2.5 It's All on the Datasheet (Except the Price!)

Antenna manufacturers characterize the performance of their antennas, typically using an anechoic chamber (a room with non-reflecting walls), and summarize the results on a datasheet. The datasheet typically describes:

- *Gain*: the measured gain, the actual measured peak power divided by the average power density at the corresponding distance, is usually about equal to the directivity for large antennas − that is, all the power that goes into the antenna is radiated. Gain values in the vicinity of interest for most RFID applications − from 5 to 10 dBi − require antenna sizes on the order of half a wavelength, about 16 cm. Gain is usually reported relative to an ideal isotropic antenna (dBi), but sometimes relative to a standard half-wave dipole (dBd).
- *Radiation pattern*: the gain does not determine the radiation pattern. The pattern is usually given in two orthogonal planes. The pattern can be given in terms of the electric field intensity or the power density as a function of the direction of propagation, and may be expressed linearly or in dB. The datasheet may also report the beamwidth in degrees, typically identified by the angles at which the electric field has fallen by either 3 dB or 10 dB from that in the beam center.
- *Bandwidth and frequency of operation*: the normal frequency range is usually that over which the antenna is matched to a 50-ohm cable so that most of the power sent to the antenna is transmitted and very little is reflected. The frequency range will be quoted for a given maximum return loss or VSWR; typically return loss is better than −10 dB (sometimes better than −15 dB), with corresponding VSWR < 2:1 or less than 1.5:1, respectively. Unfortunately, datasheets usually don't include measured data for return loss, but merely quote a worst-case value.
- *Polarization*: the polarization is provided in those cases where it isn't obvious. In the case of circular polarization, some measure of the extent to which polarization is maintained across the pattern may also be given, or patterns may be provided for co- and cross-polarized receiving antennas.

- *Mechanical parameters*: the physical size and weight of the antenna may be critically important in some applications. Antennas intended for outdoor use may report wind resistance and weatherizing.

6.3 Antennas for Fixed Readers

6.3.1 Doors and Portals

Passive RFID readers are often used to monitor the passage of cartons or objects through a door, portal, gateway, or other localized chokepoint between different areas. Antennas for this purpose will be fixed in place during most or all of their useful lifetime. In a typical case, where it is desired to monitor the loading dock of a shipment facility (a warehouse, distribution center, or receiving area), the door is from 3–5 meters across, and the height of the load can vary from less than a meter to 3 meters. The reader should be able to read tags located anywhere within the entry region, but should not read tags located outside this region, in order to avoid false positive reads from cartons in staging areas, objects in storage, or tags in the pockets of folks passing by. An antenna with a single beam and reasonably high gain is required to achieve both the desired read range (at least half the door width) and specificity to the region of interest. It is also desirable to view the portal zone from both sides, to detect tags on either side of a pallet load without requiring the beam to pass through possibly-opaque carton contents. In addition, the beam must also cover the whole region of interest. These requirements are not easily met by a single antenna; it is typical to use four antennas placed on the opposite faces of the doorway (Figure 6.21). The antennas are *multiplexed*: connected in succession to the reader. If a bistatic configuration is employed, four pairs of antennas are used. In most cases the tags to be read are on pallets or cartons carried by forklifts or pallet jacks, and require several seconds to move through the doorway. Read speed is not critical, but fast handling of multiple tags is. When multiple adjacent doors are present, the interior readers are connected to as to view two adjacent doors (that is, a single reader's antennas cover the left

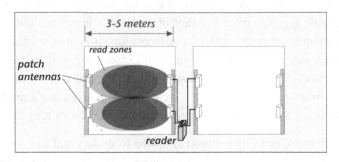

Figure 6.21
Typical portal antenna configuration.

side of the door to the right of the reader, and the right side of the door to the left) to avoid running RF cables over or under the doors.

The characteristics that are important for antennas in fixed applications are:

- *gain*: we clearly want an antenna with a single reasonably-well-defined beam rather than an omnidirectional antenna like a dipole. The largest gain value we can use at full radiated power in the United States (without a license) is 6 dBi; we must reduce the reader power if we use more gain. There's usually not much point in having gain higher than about 9 or 10 dBi.
- *bandwidth*: the antenna needs to be well-matched to the reader or cable over the band we wish to use. In the United States that will be the 902−928 MHz ISM band in most cases; in Europe, 865−868 MHz is normally allocated for RFID, and in Asia, slices may be available in the European and US regions, or at other frequencies extending from around 860 to 960 MHz.
- *beam shape*: the single beam may not necessarily be symmetrical. We may want a fan beam (e.g. tall and thin), so that we can limit the read zone to the extent of the doorway and avoid seeing tags in a truck, or read only tags that are outside the door. (A fan beam is also useful for a reader monitoring cartons moving on a conveyor, where it isn't really necessary to look very high or low, but it is useful to maximize the width of the read zone along the conveyor in order to ensure that tagged cartons remain in the read zone for a long time and thus have the best chance of being read; we'll discuss conveyors in the next section.)
- *polarization*: the choice of polarization is dominated by our choice of tag antenna type and orientation. Single-dipole tag antennas can only be read when copolarized (that is, when the long axis of the tag is along the direction of the electric field from the reader antenna). If the tags are always oriented vertically, we must use a vertically polarized antenna or a circularly polarized antenna to read them; if the tags are horizontal, a horizontally polarized or circularly polarized antenna is needed. When the tag orientation is not controlled, a circularly polarized antenna must be used. (Recall that this results in a 3 dB reduction in power to the tag, and thus a modest reduction in read range.) Dual-dipole tags are more tolerant of variations in orientation, and impose minimal requirements on reader antenna polarization. It is also worth noting that when linearly polarized antennas are used in an open indoor area, it is usually best to employ vertical polarization if possible. The reason is that reflections from the floor are eliminated for a wide range of angles of incidence near Brewster's angle (Figure 3.36), producing somewhat more consistent read results as the position of tags varies.

We can use the tools introduced in section (2) above to understand in a semi-quantitative way what antenna configurations make sense. For example, in the portal configuration of Figure 6.21, how many antennas are needed, and where should they be placed? We can make a first estimate by applying the procedure of Figure 6.5 to roughly sketch out the read zone, using a 6 dB beam width of about 120° obtained from the typical patch antenna pattern of Figure 6.16. In Figure 6.22 we show such an estimate overlaid on a 3-meter-square doorway, assuming the maximum read range in beam center is a conservative 3 meters. It's apparent that a single pair of antennas, either both on one side or both at one elevation, will not cover the whole area of the doorway, but a pair of antennas on each side does a good job.

Naturally, the real world will not be this simple, due to reflections from the doorframe, support posts, forklift and goods when present, and other objects in the neighborhood. If low-lying loads must be monitored, floor reflection is likely to be important. Note that since the beam width decreases as a tag nears the antenna, a tag near the floor or the door sill, and displaced close to the door frame, may not be read effectively.

6.3.2 Interference and Collocation

The reflected signal from a tag is small and easily swamped by the signal from a reader. An example is shown in Figure 6.23; the received power from the interfering reader is found readily using the Friis equation (3.20), assuming both the transmitting and receiving antennas have 6 dBi gain. The power received from an interfering reader 20 meters away, roughly 0.1 mW, is about 40−50 dB (that is, a factor of 10,000 to 100,000) larger than the reflected signal from a tag a few meters away. If the interfering reader on the same frequency as the wanted tag, it can be very difficult to see the small signal against the

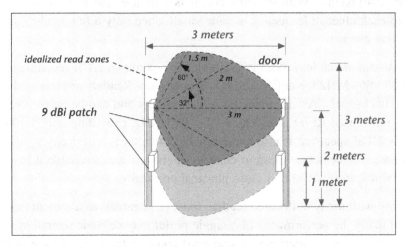

Figure 6.22
Idealized read zone for patch antennas at a doorway, assuming beamwidth parameters from a commercial 9 dBi antenna.

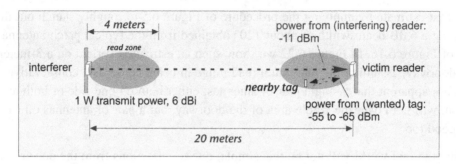

Figure 6.23
Power from a distant interferer can be much larger than a reflected tag signal.

larger one. (The ISO18000-6C standard does make special provisions for minimizing reader-tag interference; we will examine these in more detail in chapter 8.)

The importance of interference depends on the amount of spectrum available, and the physical arrangement of reader antennas. An interfering reader whose signal is many MHz away from the wanted signal may be readily filtered out in the radio's baseband chain, although as we noted in chapter 4, third-order distortion in the RF section of a radio can lead to in-band mixing products when more than one interferer is present. Readers operating on the same channel as the victim reader, or an adjacent channel, are most likely to be a problem. In the United States, the ISM band provides 26 MHz for unlicensed use. Readers configured for frequency-hopping operation (the most common approach) are required to change channels no less often than every 0.4 second, and use all channels with equal probability. Typically 500 kHz channels are used, so there are 50 channels available (allowing for guard bands). With so many channels available, the likelihood of two readers being close to each other in frequency is quite small when only a few readers are in proximity to one another.

In Europe and Asia, much less room is available. For example, ETSI recommendations allow only 865−868 MHz for unlicensed RFID operation. Readers are required to use one of four 200 kHz channels. Asian jurisdictions vary widely and can be worse; for example, in Singapore, unlicensed RFID is allocated 923−925 MHz, providing only 4 500-kHz channels. (The ETSI spectrum is also available, but the total is still pretty small.) Under these conditions, co-channel or adjacent-channel interferers are inevitable if many readers are operating simultaneously and in close physical proximity.

A single facility with many collocated readers must be regarded as a system design problem: optimizing the performance of a single portal may degrade overall system performance. In addition to interference, one must consider the ratio of false negative reads (when a reader fails to read a tag in the read zone) to false positive reads (when a tag in some other area outside the intended read zone of a given read point, is read). Power, duty cycle, and antenna gain and configuration must be arranged to maximize joint performance rather than the performance of a single read point.

The simplest and most powerful lever for reducing interference is to turn the interfering transmitter off when it is not needed. In the context of RFID, this means that a reader should not be on when there are no tags of interest in the field. Portal installations often include presence or motion sensors that activate the reader only when there is something within the portal worth reading; operator intervention may also be required in some applications. Turning the reader on only when needed reduces the duty cycle; when a reader isn't on, it doesn't interfere. Multiple sensors can also provide information on the direction of travel of goods, and good procedures can minimize unneeded reader activation.

Antenna configuration can also be used to reduce interference. In some cases it may be possible or desirable to direct antennas outwards towards a loading dock or other facility access, reducing interference as well as providing some additional locating information. A glance at our representative antenna pattern (Figure 6.16) suggests that one might hope to realize 6−12 dB/antenna (12−24 dB for the link, remembering that reduced transmit radiation indicates reduced receive effective aperture). One may also synchronize the multiplexing configuration of multiple readers, so that only right-looking or only left-looking reader antennas are activated at any given moment. The signal from a left-looking transmitter into a left-looking receiver should be reduced by about the front-to-back ratio of the antenna, typically around 20 dB for a small patch antenna. If circularly-polarized antennas are used in a bistatic configuration, the right-looking receive antenna should be of opposite sense to the right-looking transmit antenna in order to receive backscattered signals from dual-dipole tags; if the same sense is used for all transmitting antennas, the left-looking receiving antennas will be cross-polarized for right-looking transmitters (and vice-versa), reducing any interfering signals.

Physical isolation can be achieved through the use of screens, walls, or tunnels; if the openings are small compared to a quarter-wavelength (about 8 cm here), a conductive screen will effectively block radiation. Common interior construction materials (wood and gypsum 'dry wall') are modestly absorbing and reflecting. Concrete is a modest reflector at 900 MHz: the refractive index is around 3, so the reflection coefficient for normal incidence is around $(3 − 1)/(3 + 1) = 0.5$. Concrete absorption varies considerably depending on the cure state and water content of the concrete. A thick concrete wall may provide only 5−10 dB of attenuation, but multiple partitions of concrete (or even dry wall) will attenuate signals fairly rapidly. Note that to be effective isolators, screens or walls must subtend a large angle relative to the transmitter; radiation can readily diffract around obstacles that are only a few wavelengths across. Rooms containing a substantial load of RF-absorbing materials (moist porous materials, fresh produce, concrete corridors, etc.) will lead to relatively rapid attenuation of distant signals, but common warehouse arrangements with large open areas near the entries are conducive to long-distance propagation.

In the United States, FCC regulations do not permit the coordination of frequencies for unlicensed transmitters, but at the time of this writing coordination of reader frequencies is under consideration for revised European regulations. Frequency planning, where permitted, can help reduce the impact of interference by ensuring that readers on the same or adjacent channels are relatively distant from one another. The resulting increase in distance is proportional to the number of available channels, n; if the area is relatively open so that signal power falls as the inverse second power of the distance, the expected reduction in interference power is $20 \log_{10}(n)$. Recall that these benefits might not be realized if third-order distortion is important (chapter 4): pairs of equally-spaced channels can generate intermodulation products onto the wanted frequency in the mixer and any RF amplification that is used, producing unfilterable interferers even with frequency planning. Recall that third-order distortion also doubles the spectral width due to modulation, making it more likely that a modulated reader signal will overlie the wanted tag signal in frequency.

The ISO 18000-6C (EPCglobal Class 1 Generation 2) standard provides a special operating mode, **Dense Interrogator** operation (often known as **dense reader mode** or DRM), in which the spectra of the backscattered signals are displaced from the carrier frequencies, and the reader transmissions are kept within a narrow band around the carrier. We will discuss dense reader operation in more detail in chapter 8. Dense reader operation makes it possible in principle for the radio's baseband filter to select tag responses even in the presence of adjacent-channel or co-channel interferers. However, it does so at the cost of relatively low data rates for both the tag and reader; for intermediate numbers of readers and tags, it may be better to use high data rates and turn the readers off when all tags have been read.

6.3.3 Conveyor Antenna Configuration

Conveyors are widely used for moving and sorting goods packaged in cartons, boxes, or shrink-wrapped containers. There are various types of conveyors. **Accumulation** conveyors bring loads into a conveyorized system; some are equipped with sensors to detect the size of a given object, and may be able to rotate the object into a desired orientation. **Transport** conveyors move goods long distances at speeds as high as 3 meters/second. Motion can be continuous or intermittent. **Sortation** systems are used to direct cases, totes, or pieces to differing locations. The destination can be a truck dock, warehouse shelf, or a staging/ assembly area. Carousel sorters are used for manual piece sorting, and support rates up to about 70 items per minute per sort zone. A-frame sorters are used to pick or dispense individual items to fill orders, and are frequently used for pharmaceuticals. Tilt tray sorters are used to sort parcels and luggage, and can reach 175 items per minute; cross-belt sorters achieve even higher rates by redirecting items to other conveyor paths as they move along a main conveyor. All these systems require accurate identification of an object to support high-speed automation.

Conveyor antennas are often mounted on a gantry near or surrounding the conveyor (Figure 6.24). Ideally the location and orientation of the antennas is chosen so that all possible tag locations and orientations will be read. If pallets or tall loads are likely to be encountered, it may be necessary to stack two antennas at different heights. The orientation of an object may or may not be controlled on the conveyor. When orientation is uncontrolled, a tag may lie on any face of the item or carton and point in any direction. Simple single-dipole tags are sensitive only to radiation polarized along their axis; in this case a circularly-polarized antenna should be used. The reduction in read range is usually not critical in a conveyorized environment. Dual-dipole tags have reduced dependence on orientation, but are larger and slightly more expensive than single-dipole tags.

A rough estimate of suitable locations may be obtained using the idealized read zone approach depicted in Figure 6.5; an example is shown schematically in Figure 6.25. However, the real read zone is likely to be complex, with patches of good and bad reads, due to reflections from the conveyor components as well as the goods being transported, and the use of far-field antenna patterns may be misleading for tags close to the antenna, as may occur with displaced loads on a conveyor. Empirical confirmation is necessary for these simple estimates.

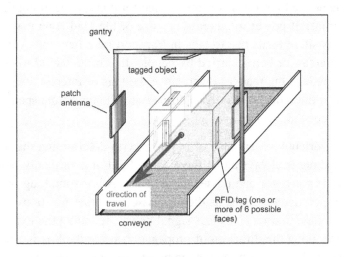

Figure 6.24
Conveyor with antenna gantry showing possible antenna positions and tag positions and orientations.

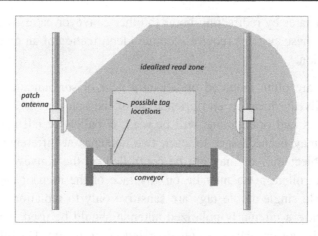

Figure 6.25
Example of antenna positioning to cover tag locations using idealized read zone.

If tags may be located on the bottom of an item (which can be an issue with uncontrolled object types, as in airport baggage transport), it may be necessary to place a reader antenna below the conveyor to illuminate them. Metal rollers will filter the portion of linearly-polarized illumination that is along their long axes, leaving only the perpendicular part to reach the tags; if circular polarization is required, the rollers above the antenna should be replaced with plastic rollers if load requirements permit.

The tradeoff between false negative reads (missed tags) and false positive reads (tags read that are outside the desired region) is of importance for conveyorized applications. Increasing reader transmit power to increase the size of the read zone may also increase the frequency of false positive reads due to tags in transport on other conveyors, on nearby forklifts, in staging areas, or being carried by people. It is often useful to proceed in the other direction and reduce the reader power, since the tags of interest are relatively close to the reader. However, the reduced read zone size means the total time spent in the read zone of each antenna is reduced.

Polarization and tag orientation are also of importance in determining the real read zone. A vertically-oriented tag is always along the electric field of a vertically-polarized antenna as it travels down the conveyor and passes the antenna. A horizontal tag is only lined up with a horizontally-polarized antenna as the tag passes; in other positions, the tag is at an angle to the electric field and receives less signal power, reducing the extent of the read zone. A tag oriented along the direction of propagation when the box is adjacent to the antenna is invisible in that position, but as the box moves along the conveyor, the angle between the propagation vector and the tag will increase, leading to a bimodal read zone with the best reads to the left and right of the antenna.

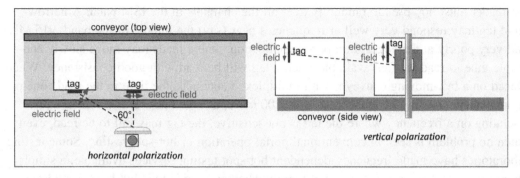

Figure 6.26
Horizontally-oriented tags are increasingly misoriented at long distances from a horizontally polarized antenna; vertically oriented tags are always aligned along the field from a vertically polarized antenna.

Multiplexing antennas also means that the total time to read a tag may be rather short on a conveyor. The read zone for a conveyor is the intersection of the conveyor travel path with the normal ellipsoidal read zone (ignoring the not-necessarily−negligible effects of reflections from the conveyor rollers and walls). If the read zone is a couple of meters long, and the conveyor travels at 3 meters/second, we have a total of about 660 milliseconds to read a tag. With four-fold multiplexing, this is reduced to about 215 msec. While this is ample time to perform several read attempts from the point of view of the native tag protocols (which typically require around 1−5 ms to read a tag), it is much less so if the reader duty cycle is reduced by slow host communications. Modern readers can read tags at a rate of hundreds per second during an inventory, but many readers can only launch about 10−20 inventories per second. Protocol behavior is also important; for example, a Class 0 tag cannot enter an inventory in the middle, but must receive certain synchronization inputs and the tree-walk-start signal; in this case, it can be very useful to place presence sensors on the conveyor to initiate reads when a tag is likely to be present. It is also important to ensure that the software is configured for rapid reading of a small number of tags, rather than exhaustive inventory of a large number of tags; for example, in ISO18000-6C operation, the initial value of the Q parameter must be small. We will discuss protocols in detail in chapter 8. One must also take statistics into account. In preliminary experiments, the author has observed that the distribution of reads on a conveyor is roughly Poisson-distributed; in order to ensure that tags are read at least once with $> 99\%$ probability in a single pass, an average tag should capable of being read 6−7 times during its passage through the read zone.

Tags placed in air or on an empty box are usually rather frequency-insensitive, but tags placed close to a metal object may experience resonant behavior and become much more frequency-selective, as well as more directional. This is a particular issue for conveyorized operation, because of the frequency-hopping behavior of a reader. In US operation, the reader must hop pseudo-randomly over all the channels in the ISM band. A narrow-band tag may respond very well at frequencies near (say) the middle of the band, 915 MHz, and very poorly at the edges. In a portal application, where loads may move slowly and ample time to read is often available, such a tag will be read with good consistency. When placed on a fast-moving conveyor with multiplexed antennas, the time in the read zone of an active antenna may be reduced to 100–200 msec, as noted above. If the antenna is operating on a frequency where the tag is not sensitive, the tag may fail to be read, even when no problem is seen in conventional portal operation or hot-spot testing. Some testing laboratories have made frequency-dependent hot spot testing services available; a simpler alternative is to note in static tests not merely whether a tag is read but how many times it is read per second, averaged over several seconds. Since many jurisdictions provide different and much narrower bandwidths than the United States, tags and tag placements that work in one country may work poorly in another when frequency selectivity exists.

6.4 Antennas for Handheld or Portable Readers

Antennas for handheld or portable applications face stringent constraints on size and weight not encountered for stationary readers.

Antennas for handheld applications need to be small and light, tolerant of hands and other RF-active objects near the antenna, robust against mechanical damage from bumps and drops, and frontside-directed. The last requirement is in accord with the intuitive expectation of a person that when they point a handheld reader in some direction, most tag reads will come from tags generally located in that direction.

Recall from section 2 that a two-element array needs to be larger than about a quarter-wave across to create much of a beam. It's hard to get more than about 4 dBi of antenna gain from a quarter-wave dimension (8 cm). Half-wave structures can provide 6–8 dBi of gain but you have to find room for a 16-cm antenna. This is a difficult tradeoff, because there are substantial benefits to a high-gain antenna for a handheld device. Range is increased for the same transmitted RF power, or equivalently one can decrease the transmitted power to achieve the same range. The latter option is perhaps more valuable as it implies lower DC power consumption by the transmitter and thus longer battery life. In many handheld applications, very long range (>2 meters) may be unnecessary or even undesirable, whereas a well-defined read zone may be very helpful. In stationary applications a narrow antenna beam may unduly constrain the read zone and cause tags to be missed, but in a handheld device the user gains information about the location of a tag from the behavior of tag reads as they redirect the antenna.

Polarization flexibility may not be compatible with the requirements of a handheld device. Recall that a circularly-polarized antenna is convenient when single-dipole tags are to be read without orientation control. However, as we will discuss below, some antenna designs that are very convenient in other respects do not lend themselves to circular polarization.

Handheld antennas are likely to be close to people's hands. A human body is basically a sack of water from the RF point of view. Water has a dielectric constant of around 80 at room temperature and is also strongly absorptive around 900 MHz; radio waves are reflected by body parts, and capacitance of metal plates will be affected when a hand is close to them. To avoid disturbances from nearby dielectrics, it is best to employ a *balanced* antenna, in which two symmetric halves are driven by opposed currents, and no piece of the antenna is used as a ground reference; a simple example is a dipole antenna (Figure 6.27). A balanced antenna may be larger than a *single-ended* antenna, and requires a balanced-unbalanced transformer (*balun*) or other special provisions to be connected to a standard coaxial cable or grounded circuit board.

A dipole antenna may be balanced, but a half-wave dipole is 16 cm long at 915 MHz and thus impractically large for many handheld applications, particularly if configured to radiate in the direction the user is looking. Dipole antennas also have modest directive gain. A dipole can be made compact by bending it; this trick is widely used in designing tag antennas, as we'll see in chapter 7, and can also be employed for reader antennas. An example of such a compact dipole design is shown in Figure 6.28. The width of the dipole has been greatly reduced relative to a half-wave structure, in this case to around 6 cm.

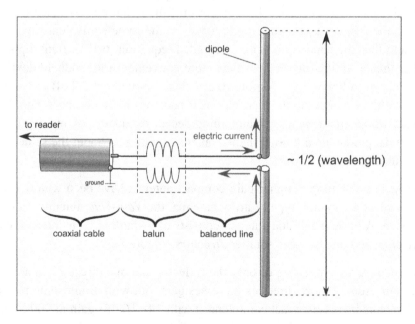

Figure 6.27
Dipole antenna with balun transition to coaxial cable.

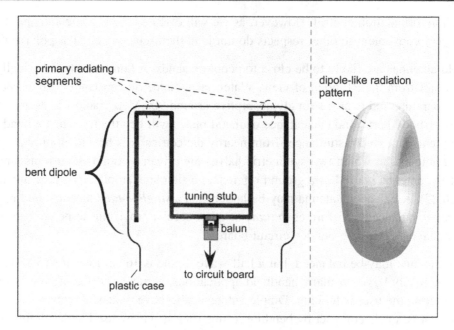

Figure 6.28
A compact forward-radiating bent-dipole antenna; after US patent 7,183,994.

However, as a consequence, the total length of those portions of the antenna that contribute to the radiated wave is also reduced. Because of this, the radiation resistance is much less than would be the case for a straight dipole, for about the same equivalent capacitance and inductance, and thus the bandwidth is also degraded (equation (6.12)). Bent dipoles are much smaller than a straight dipole, and thus more convenient for handheld devices, but gain is generally even lower than a dipole (recall that's only about 2.2 dB to begin with), and the bandwidth of a bent dipole decreases as it becomes more compact. Dipoles can be made compact while simultaneously maintaining decent bandwidth by making the wires fatter at the ends, producing a ***bowtie*** shaped antenna, with gain about the same as a conventional dipole.

Two approaches to obtaining a higher-gain compact antenna have been widely used. The first is to adapt a common type of array antenna, the ***Yagi-Uda*** antenna, to use for a handheld device. A basic Yagi-Uda antenna consists of a single driven antenna, the ***exciter***, a larger ***reflector***, and one or more smaller ***directors*** (Figure 6.29).

A small Yagi-Uda array consisting of only the reflector and one director can achieve about 5−8 dBi of gain. Adding more directors increases gain, but with diminishing returns as the number of directors increases; practical antennas with 14−16 dBi gain at 900 MHz are available but are large and ungainly even for a stationary reader, and quite impractical for a portable or handheld unit. However, a small array using bent dipole elements to reduce the

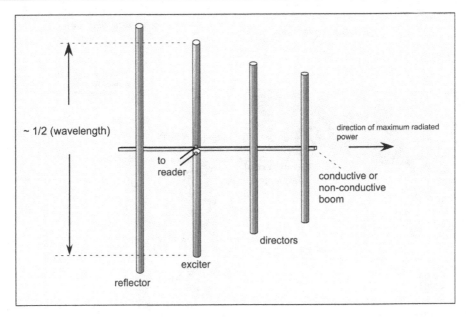

Figure 6.29
Typical Yagi-Uda antenna.

width, and flattened elements to improve bandwidth, can be fit very nicely into a unit that is comfortable to hold, provides about 6 dBi of gain, and whose radiation is directed forward. Because the antenna is balanced, it is relatively insensitive to the presence of the user's hand. The scheme is depicted schematically in Figure 6.30, along with the outline of an actual commercial implementation of this scheme, which is cut out of a thin sheet of copper. (The image shown is a cross-section looking down on the structure; the actual structure is bent to constrain its width, as shown in the schematic depiction.) This structure combines three of the approaches we've discussed above: it uses bent dipoles, a balanced antenna with an integral balun structure, and a Yagi-Uda array. In principle one can make circularly-polarized Yagi-Uda structures, but in this context such a structure would take too much room.

The second possible approach to making a compact, high-gain antenna is to construct a patch antenna that is reduced in size relative to the larger units use for a stationary antenna. A compact patch antenna can be conveniently mounted on the front of a portable reader (Figure 6.31), and this approach has been used in various commercial units. Small patch antennas are also convenient for other portable applications, such as forklift-mounted readers. Circular polarization can be achieved if desired.

One approach that produces an immediate factor-of-two decrease in one dimension of the antenna is to substitute a quarter-wave patch for the standard half-wave patch.

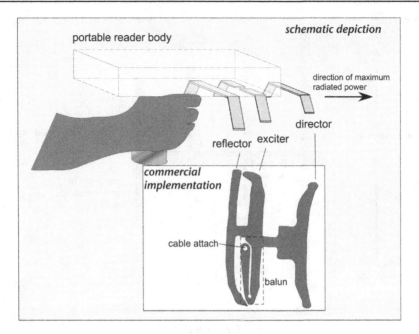

Figure 6.30
A small 3-element Yagi-Uda can be mounted beneath a handheld body. The inset shows a 2-D sketch of a commercial handheld antenna; in use, the ends of each element are bent out of the plane into a shallow U-shape.

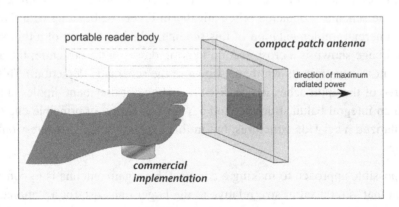

Figure 6.31
Schematic depiction of handheld reader with compact patch antenna.

The gain is somewhat reduced, as the antenna is nearly isotropic in the vertical direction. The bandwidth is roughly the same as that of a half-wave structure. A quarter-wave patch is not symmetrical and will normally be used only for linearly-polarized radiation.

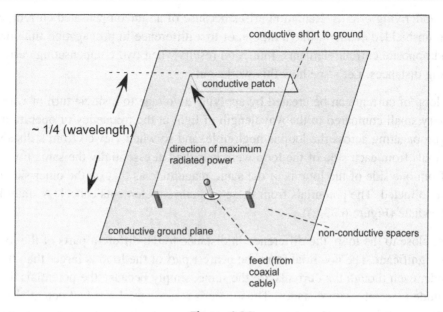

conductive short to ground

conductive patch

~ 1/4 (wavelength)

direction of maximum
radiated power

conductive ground plane

non-conductive spacers

feed (from
coaxial
cable)

Figure 6.32
A patch antenna can be made smaller by using a shorted quarter-wave patch in place of a half-wave structure.

Another approach to shrinking a patch antenna is to place a dielectric material between the patch and ground. Since light propagates more slowly in a dielectric, the resonant length is reduced and the antenna gets smaller. However, the size reduction is in effect achieved by increasing the capacitance of the antenna per unit area using the dielectric material, while the radiation resistance is decreased due to the smaller radiating area. Thus the bandwidth is degraded (see equation (6.13)). Gain is also reduced, since the patch is smaller (refer to section 2.1 above). In order to minimize antenna size, it is further useful to reduce the size of the ground plane to only slightly larger than that of the patch, but this leads to considerable sensitivity of the patch resonant frequency to changes in the total capacitance, which can result from people and objects close to the ground plane. If a small patch is to be used, it should be enclosed in a thick radome to keep people and objects away from the antenna surface.

6.5 Near-Field Antennas

The antennas we have discussed so far in this chapter launch a radiated wave whose power density decreases rather slowly (as the inverse of the square of the distance) as they propagate. However, this is not the only way to induce a voltage on a wire. Tags can also inductively couple to readers.

Inductive and radiative coupling both arise from the same underlying phenomenon: the potentials launched by an electrical current and associated charge. However, they have different underlying origins. Radiation arises because of an uncompensated current, either due to an unshielded element, like a dipole, or to a difference in propagation time between equal and opposite current elements. Induction results when two compensating currents are at differing distances. Let's see how this works out.

A small loop of current can be created by applying a voltage to a single turn of wire. If the loop is very small compared to the wavelength of light at the frequency of operation, the delay in propagating across the loop is negligible, and so when viewed from a distance, the potentials from each side of the loop were launched at essentially the same time. The current from one side of the loop is of the same magnitude as that on the other side but oppositely directed. The potentials from these two current elements cancel. A small loop does not radiate (Figure 6.33(a)).

However, close to the loop, the difference in distance to the different parts of the loop becomes significant. The potential from the nearest part of the loop is larger than that from the far side, even though the currents are the same, simply because the potentials decrease in magnitude inversely with distance. Thus there is a substantial residual potential even though the currents are equal and opposite on opposite sides of the loop; this potential gives rise to an electric field and can be used to create voltage by placing a wire close to the loop. Inductive coupling is obtainable close to the loop even when no radiation is present.

Very small loop antennas (1−3 cm in diameter) can be used in exactly this fashion to inductively couple to tags, even at UHF frequencies. When the reader and tag antennas are

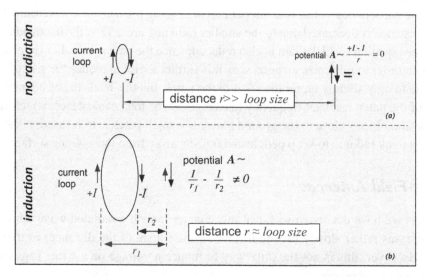

Figure 6.33
Coupling to a non-radiating current loop. (a) Cancellation of potential at long distances; (b) Finite potential at distances comparable to the loop size.

both small loops, only inductive coupling is present: the reader is operating as a ***near-field*** device. Even for small loops, it is necessary to add some dissipation, usually in the form of a large resistance in parallel with the loop, because without radiation there is very little loss in the loop and thus the bandwidth is extremely narrow.

The problem with using a very small loop is apparent from the discussion above: inductive coupling is only significant for distances on the order of the loop size. A loop antenna 1 cm in diameter will have a read range of 1 or 2 cm. Even for near-field operation, many applications require longer read ranges than that. The obvious response is to make the reader antenna bigger.

Two problems crop up when we follow this path. The first is that the inductance of the loop, which is proportional to the perimeter of the loop and thus the radius, increases, making it very difficult to match the loop to a 50-ohm coaxial line.

The second problem is more subtle. The electric field along the center of the wire, which is the joint result of the current flow and net charge, must be zero. Even at low frequencies, where the loop is very small compared to a wavelength, some charge is accumulated on the loop in order to compensate the electric field due to the current flow. Just as water accumulates in a channel when the flow rate decreases, in order to create a net charge, the electric current density must change along the wire. The amount of charge that accumulates is proportional to the time available for it to pile up, and thus inversely proportional to frequency: at low frequencies (where the wavelength is long, and the loop looks small) only a tiny variation in the current along the loop is needed to provide lots of charge, but as the frequency increases a larger and larger change in the current is needed to produce sufficient charge. In addition, the inductive electric field that must be compensated also increases linearly with frequency. The net result is that as the size of the loop becomes significant compared to a wavelength, the current is no longer constant along the loop (Figure 6.34). When the circumference is about 0.4 of a wavelength, the loop experiences a parallel resonance, and exhibits a very large input resistance. By the time the loop reaches about 1 wavelength in circumference, the charge required is so large that the current falls to zero and reverses direction about ¼-way around on both sides of the loop.

Because the current is no longer circulating, but instead accumulating at the top and bottom of the loop, like a dipole configuration, inductive coupling along the axis of the loop becomes small: equivalently, the magnetic field along the axis goes to zero. This occurs at a diameter of about 10 cm at 900 MHz. Thus we have arrived at the following conundrum: with a near-field coupled reader and tag, we can't get long range unless we make the reader antenna large, but when the reader antenna grows large it no longer works as an inductive antenna.

One solution to both problems is to divide the antenna into short segments, and interpose a capacitor between each pair of segments; an example is shown in Figure 6.35. If the

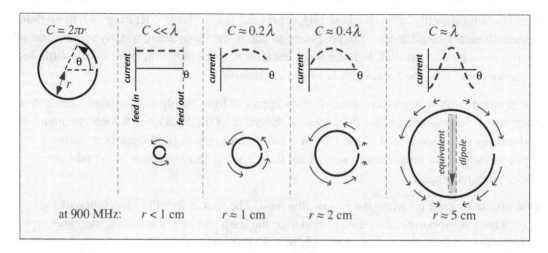

Figure 6.34

Behavior of current around a loop antenna as the size relative to a wavelength is increased.

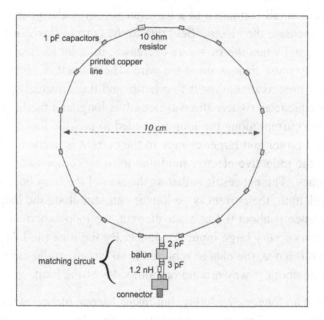

Figure 6.35

Example of a segmented near-field antenna; about 8 cm read range is achieved on the axis (perpendicular to the page). Design details courtesy of Steven Weigand and Nathan Iyer.

capacitor value is chosen so that it resonates with the inductance of the segment, the input impedance of each segment is small and real, so that matching the antenna is straightforward. The capacitors also take care of the charge storage problem, so that the current can be constant in magnitude and (for resonant segments) at the same phase.

A segmented antenna can still couple inductively to a tag along its axis even when the diameter is comparable to a wavelength. Using this approach, read ranges up to 8–10 cm can be achieved.

A segmented loop continues to couple inductively even at large loop diameters, but it also radiates, even though it does so mainly in the plane (like a small loop) rather than along its axis, unlike a conventional loop of the same size. This radiation is inevitable, as it arises from the fact that an observer sees the potential from the different parts of the loop at different times (see Figure 3.2 in chapter 3). Since the radiation is in the plane, it is possible to shield or absorb some of it, but it is important to be aware that practical large antennas constructed to couple to near-field tags will also radiate, and thus the attached reader may read conventional far-field tags if they are close to the antenna.

A different approach to creating a near-field antenna, more appropriate for applications like tracking objects on shelves, uses an unshielded transmission line. A typical example is a *microstrip*: a relatively narrow metal line suspended over a wide ground plane (Figure 6.36). Microstrip is easy to fabricate on a printed-circuit board and is widely used in high-frequency circuitry for moving signals from one place to another with well-controlled impedance and transit time. We've seen a configuration like this previously in examining the patch antenna (section 2.4). We got patch antennas to be good radiators by using them near a resonance, so that the very high local currents could be multiplied by the tiny time difference between top and bottom and still result in a substantial radiated field. When a similar configuration is used without resonance, so that the currents corresponding to a given input power are much smaller, essentially no power is radiated. The currents flowing in the metal strip induce currents in the ground plane that flow in the opposite direction. When viewed from far away, except for the tiny difference in timing due to the spacing between the strip and ground (typically a few millimeters at UHF frequencies), the radiated waves from these currents cancel, and there is very little radiated power.

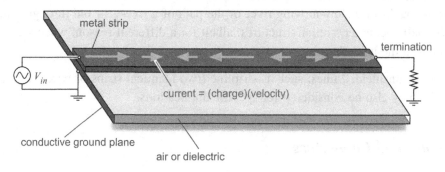

Figure 6.36
Microstrip transmission line.

However, there are still local electric fields between the narrow line and the ground plane, and a tag close enough to the line can extract power from these fields and turn on. "Close enough" here is defined, like in the inductive case, as distances where the $(1/r)$ part of the propagation equation matters, so that for example the field from the current in the strip is larger than that from the countercurrent in the ground plane, simply because the strip is closer. As in the inductive case, the near-field read range of such a structure scales with the size of the features. A microstrip has a spacing or dielectric thickness and a line width. (The line thickness and ground plane size also have more subtle effects.) To get a range of a few tens of centimeters, appropriate to a shelf-monitoring application, the feature size needs to be a few centimeters. If a dielectric is used, it should have a small dielectric constant, to allow the fields to spread outside the dielectric. For example, Medeiros and coworkers report that a microstrip antenna using a 5-cm-wide strip suspended 1 cm above the ground plane in styrofoam (which is mostly air) provided a read range of 7 cm above the antenna.

An ideal transmission is infinitely long and without losses. Real transmission lines have two ends. For example, a shelf antenna is likely to be as long as the shelf is. If the line is simply cut off at the end, an incident wave will be reflected back and will interfere with the incident wave to creating stationary patterns of high and low current: *standing waves*. Standing waves are undesirable in RFID applications, because the variations in current and voltage along the line will lead to places where no tags are read, even though they are close to the line. Therefore, it is usual practice to connect the end of the line to ground using a termination resistor, as shown in the figure. If the value of the resistor is chosen to match the ratio between the voltage and current of a wave traveling along the line, there will ideally be no reflection at the interface. In realistic cases, reflections are greatly attenuated, so that the currents and voltages along the line vary by acceptable amounts.

Coverage of a rectangular region can also be improved by having the strip veer left and right as it travels across the ground plane, rather than moving in the predictably dull straight line depicted in Figure 6.36. Such a structure is known as a meandered antenna, recalling the banks of a slow-moving river or the path of a five-year-old through a toy store. We shall encounter similar structures, albeit for a different reason, when we examine tag antennas in the next chapter.

Other types of unshielded lines, such as stripline (two identical strips or wires separated by a dielectric) may also be considered for specific applications.

6.6 Cables and Connectors

The best antenna isn't very useful if signals can't be delivered to it or extracted from it. Some readers are equipped with an integral antenna, so that no connectors or cables are needed. However, most stationary reader applications require multiple antennas per reader, and mandate flexibility in the placement of those antennas. External antennas require cables to bridge between the reader radio and the antenna. While it is possible to permanently attach the cable at each end by soldering, such an arrangement is very inflexible! It is more common and much more convenient to employ connectors so that cables, antennas, and radios can be readily swapped. Cables and connectors play an important role in real radios.

On the receiver side, losses in cables and connectors generally precede any amplification. As a consequence loss (dB) in cables and connectors adds directly to the noise figure of a radio receiver (equation (4.17)). Similarly, cable and connector losses subtract directly from the transmitted power, so we must either accept reduced read range, or crank up the transmitter output, which costs money both for a bigger power amp and more AC or DC power to supply to it. As we discussed in chapter 4, the noise in a monostatic RFID receiver is often dominated by the phase and amplitude noise in the carrier leaking through to the receiver; in this case, cable loss plays less of a role in determining receiver noise performance, but long cable do increase conversion of phase noise to amplitude noise (section 4.3 of chapter 4). Bistatic readers can achieve thermal-noise-limited sensitivity, and in this case avoiding cable loss becomes of importance, particularly when semi-passive tags are being used. Cable loss also reduces the transmitted power, and thus read range. It is perfectly legal to compensate for such losses by increasing the output power of the transmitter (so long as the power at the antenna remains within legal limits), but providing for additional transmit power requires larger, more expensive power amplifier stages and more DC power consumption.

Minimizing loss often militates directly against the most convenient and inexpensive methods of antenna mounting. Short cables have less loss, but detract from the freedom of the system designer to place the antenna and radio where each can most easily be used and accessed. Low-loss cabling is thick, inflexible, and expensive, increasingly so at higher frequencies. Connector failures can be frustrating to diagnose, and when working on the experimental side it often seems that half of one's time is spent finding the right connector and the other half adapting it to the proper gender! Installers and implementers need to be familiar with cables and connectors.

Almost all remakeable connections employ coaxial cabling. Coaxial cables have a center conductor completely surrounded by a conductive ground shield (Figure 6.37). Signal currents traveling along the center conductor are compensated by counter-currents in the shield, so radiation from the cable is very small.

The space between the center conductor and the shield is generally filled with a low-loss dielectric, although in some cable designs spacers are distributed periodically along the cable with most of the interstitial space filled with air. The ratio of the current in the cable to the voltage between the center conductor and the shield is a real-valued constant: the *characteristic impedance* of the cable. The characteristic impedance is determined by the

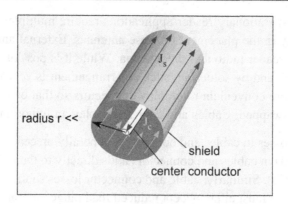

Figure 6.37
Schematic depiction of a typical coaxial cable. Equal and opposite total currents flow in the center conductor and the surrounding shield.

shield radius b, the center conductor radius a, and the relative dielectric constant ε of the fill material:

$$Z_c = \frac{138}{\sqrt{\varepsilon}} \log_{10}\left(\frac{b}{a}\right) \tag{6.23}$$

Common values of Z_c are 50 or 75 ohms. Assuming a typical dielectric constant of around 2, a 50 Ω line requires a ratio $(b/a) \approx 3{:}1$. For any given outer cable diameter, the target impedance constrains the size of the center conductor.

Unlike hollow waveguides, coaxial cables work fine with signals down to DC. The highest frequency that can be used is that at which additional propagating modes, such as hollow-waveguide modes become possible within the cable. The *cutoff frequency* is approximately:

$$f_c = \frac{c}{\pi\sqrt{\varepsilon}(a + b)} \tag{6.24}$$

where c is the velocity of light.

Cable loss results from the finite conductivity of the center conductor and shield, as well as losses within the dielectric material if present. The current flows in a narrow region near the surface of the metal, the *skin depth* (equation (2.2)). Typical skin depths in metals are a few microns at GHz frequencies: losses are much larger than they would be if the current flowed through the whole thickness of the metal. The total area for current is roughly the perimeter of the wires multiplied by the skin depth, so loss increases with decreasing cable diameter. The skin depth decreases as the square root of frequency, so cable loss increases with frequency. Cable loss due to conductor resistance is approximately:

$$\alpha_c = (1.5 \times 10^{-4})\sqrt{\rho f}\left(\frac{1}{a} + \frac{1}{b}\right)\frac{1}{Z_0} \tag{6.25}$$

where ρ is the resistivity of the metal and Z_0 the impedance of free space, 377 ohms. The loss is expressed in ***Nepers/meter***, where a Neper is $(1/e)$ of loss. To convert to dB of loss per meter one multiplies by 8.68. For a 1-cm-diameter, 50 Ω line, the ohmic loss would be about 5 dB/30 meters at 900 MHz. Flexible coaxial cables generally employ some sort of multistrand ribbon material for the shield and are subject to higher losses than would have been the case for a continuous sheath.

A bewildering variety of coaxial cables are commercially available, differing in characteristic impedance, loss, frequency range, flexibility, and cost. Semi-rigid coaxial cables provide the best performance in most respects but at the cost of — surprise! — inflexibility. Semi-rigid cables have a solid copper shield and solid inner conductor, separated by a Teflon variant or other low-loss insulator. The solid shield ensures that essentially no signal leaks out of the cable if it is solidly grounded at the ends; the high conductivity of the solid copper provides low loss. Semi-rigid cables and cable stock are available in a range of diameters and impedances. Semi-rigid cable can be bent by hand or with tube bending tools, and is sufficiently flexible to allow many cm of elastic adjustment over a 10- or 20-cm length of cable, but it is best suited for applications where the connection geometry is essentially fixed.

Flexible coaxial cables come in a wide variety of sizes and construction. They are usually fabricated with a copper or copper-clad steel inner conductor and copper or aluminum ribbon or braid for the outer conductor. Polyethylene and Teflon are common insulators. Foamed insulation may also be used to reduce dielectric constant. From equation [5.53] we see that large diameters are preferred for low loss, but large-diameter cables are relatively inflexible, awkward, and expensive. Small cables are easier to work with, particularly in tight spaces, if the excess loss can be tolerated.

Popular cable nomenclature is based on US military standards. ***Radio Guide*** (RG) designations dating from the 1940's are still widely employed. M17 names (from military standard MIL-C-17, promulgated in the 1970's) are also used. The nomenclature is arranged in a quite arbitrary fashion; there is no simple correlation between the RG or M17 number and cable size, loss, or frequency rating. Performance specifications required in these standards are based on very old technology, and modern cables can easily outperform the RG specs within the same physical dimensions. The nomenclature is often employed in a generic fashion, to denote cables that are similar to but not necessarily in compliance with a given specification. Some commonly-used cables and their nominal diameter are shown in Table 6.1.

Measured loss for some common RG-designated cables is depicted versus outer diameter in Figure 6.38. (Note that the diameter here is the outer diameter of the cable as provided, including the insulating cover; the diameter of the outer conductor b is significantly smaller.) For reasonable cable sizes, a few meters is not much of a concern, but long runs (greater than 10 meters) using small cables (less than 7−8 mm in diameter) should be avoided.

Table 6.1: Common RG Cable Designations

Cable Name	Outer Diameter (cm)	Cable Name	Outer Diameter (cm)
RG8X	0.61	RG58C	0.50
RG122	0.41	RG141	0.48
RG142	0.50	RG188A	0.28
RG196A	0.20	RG213	1.03
RG214	1.08	RG217	1.38
RG218	2.21	RG219	2.40
RG223	0.55	RG225	1.09
RG303	0.43	RG316	0.26
RG393	0.99	RG400	0.50
RG401	0.64	RG402	0.36
RG403	0.22		

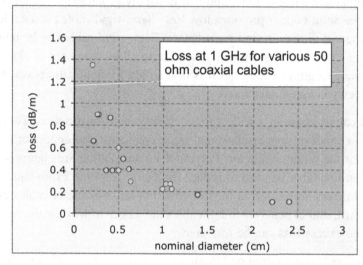

Figure 6.38

Reported loss for various commercial cables plotted vs. actual outer diameter of cable (including insulating jacket).

Connectors must satisfy a number of stringent constraints. The size of the largest open area within the connector sets the highest frequency of use; internal resonances show up as frequency-dependent reflection and loss for frequencies higher than the maximum rated frequency. The ground-to-ground and center-to-center contact resistance between the male and female mating connectors must be reproducibly low even after multiple make-and-break cycles. Spring-loaded female connectors simplify alignment of the center conductor, but the springs can undergo cold working and degrade with repeated usage cycles. Low electrical resistance materials must be employed, but good mechanical properties are also required. Gold plating is often employed to ensure excellent surface properties and corrosion resistance, but of course adds cost to the resulting connector. In some applications, such as connecting a PC-card reader or integral reader module to an antenna, small diameter cables must be used, so very small connectors are needed.

A wide variety of connector types (though fewer than there are of cables) is available to fill these needs. We shall survey a few common types here, proceeding generally from large to small. Cross-sectional sketches of the most common types are shown in Figure 6.39.

In the United States, the Federal Communications Commission requires that unlicensed radio devices be tested for compliance with regulatory restrictions before they can be offered for sale. Until very recently, the FCC regulations sought to ensure that radios in the field used the exact configuration that was tested, unless a "professional installer" was doing the reconfiguration. In order to prevent ordinary folks from messing with the equipment, the Commission requested that antenna connectors be avoided, or, if that was not reasonable, that they be "non-standard" types. Therefore, it is not uncommon to encounter variations of the standard connectors used on RFID radios, whose purpose is nominally to prevent you from making any changes to the manufacturer-provided configuration; *reverse-polarity* connectors (in which the center pin is placed in the nominally female half and the spring in the nominally-male half) are popular. In practice,

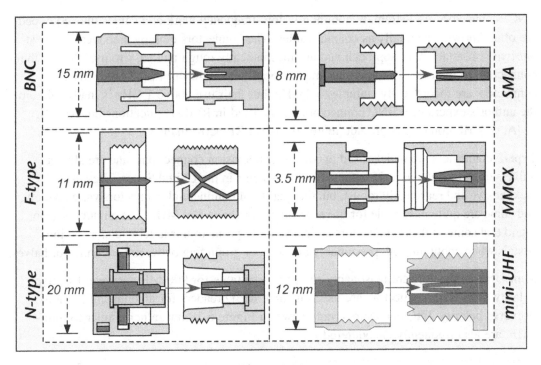

Figure 6.39
Cross-sectional drawings of some common connector types. Note that the scale is NOT consistent between sketches, but is as indicated by the diameter dimension line provided for each type.

with the tremendous popularity of wireless local-area-network equipment, "non-standard" connectors have become widely available in the last few years, and their use no longer presents a significant impediment to reconfiguration.

The highest-performance connectors, such as APC-7 or APC-3.5 types (not shown in the figure), rely on a face-to-face mating of center conductors produced by precision alignment of high-quality outer shells, avoiding the use of springs or clip arrangements. These connectors are employed in instrumentation but are too large, expensive, and slow to be used in most other applications.

Type-N connectors are a very common large connector. N-connectors are versatile and can be used for high power applications. They are physically fairly large and often used to terminate large-diameter low-loss cabling. N-connectors must be aligned and screwed together, using noticeable force. They are mechanically robust and well-suited for outdoor use (with rain protection). They can be used to 10 GHz. Note that both 50 Ω and 75 Ω types exist. The two are visually almost identical except for the center conductor diameter, and will mate with each other (to the extent of being screwed together), but a 75 Ω male will make only intermittent contact with a 50 Ω female, and a 50 Ω male will bend the springs of a 75 Ω female. The resulting connection problems are frustrating to debug: you have been warned!

BNC[1] connectors are extremely common in low-frequency work and cables thus equipped are often known generically as coaxial cables. The connectors employ a convenient twist-on/ twist-off attachment approach and are mechanically tough and easy to use, though the connection to the cable is sometimes made sloppily and will fail in the field. BNC connectors are theoretically suitable to 4 GHz, but their use above 1 GHz is inadvisable in the author's experience. BNC connectors can be used in RFID applications but type-N, SMA, or UHF connectors provide more consistent RF performance.

Type-F connectors are widely used in consumer television connections, and are almost universally encountered in 75-ohm cable TV systems in the United States. They are inexpensive and reasonably robust, but the quality of commercial connectors varies widely, and some are quite unsuitable for use at frequencies above 500 MHz. The author has also found that some of these connectors are particularly prone to introducing distortion into signals when contaminated, presumably due to poor grounding between the connector halves.

Mini-UHF connectors are reasonably robust and suitable for 900-MHz service. Reverse-polarity mini-UHF connectors are used on some RFID readers for antenna connections. (Reverse-polarity TNC connectors, not shown in Figure 6.38 but rather similar to mini-UHF, are also frequently encountered.)

---

[1] It had been the author's understanding that BNC is an acronym for bayonet Naval connector, but posts in discussions groups suggest alternative interpretations. The author is not aware of an authoritative documented origin for the term.

SMA connectors are very commonly used with semi-rigid cabling as well as flexible coax. These screw-on connectors are usable to about 18 GHz. SMA's are reliable and provide low loss and reflection, but they are not intended for an unlimited number of disconnects and are mechanically delicate compared to e.g. N-connectors. In some cases the male pin of an SMA connector is simply the copper center conductor of the attached line, which does not offer the endurance and reliability of custom-made pins. SMA connectors should be kept clean and snugged carefully; use of a torque wrench is not indispensable but will improve the reliability of a connection. The very similar '2.4 mm' connector allows for operation to somewhat higher frequencies.

The MCX and MMCX connectors are more recent additions to the stable; MMCX is the smaller version, widely used for PC-card connections. The small size and press-fit convenience of these connectors make them popular for mounting directly onto printed circuit boards, providing local interconnections within a laptop or other portable system. These connectors are rated for use to 4 GHz. MCX and MMCX connectors are recommended for use with small cables such as RG174 (M17/119) or RG188A (M17/138). These small cables can have very high losses even if the connector works fine. Cables such as these are only suitable for use in lengths of a few tens of centimeters to a meter or so. Other less-common very small connectors include the GPO, U.FL, and AMC types.

6.7 Capsule Summary

Antennas are characterized by their ability to radiate in specific directions − directive gain − and the frequency over which they do it − the bandwidth. High gain means narrow beam widths for antennas that focus their power into a beam. Bigger antennas offer more gain and more bandwidth, if you have the room and the money.

Vector potentials follow currents, and electric fields follow potentials, so the orientation of the radiated electric field − the polarization − is determined by the currents flowing in the antenna. If those currents change orientation as well as direction with time, the polarization may be circular rather than linear − but an antenna that radiates circularly polarized waves in one direction may produce linear polarization in another.

Antennas come in innumerable forms, but a few simple ones are very popular. The dipole and its variants are the archetypal "omnidirectional" antennas, and the patch antenna is an extremely popular directional antenna. Patch antennas are simple and inexpensive, and the gain available from a single patch is a good fit to the gain allowed by most regulatory regimes. Bandwidth is acceptable if the patch-to-ground spacing is big enough.

Stationary readers usually employ patch antennas. Polarization, tag type, and object orientation go together; with single-dipole tags control orientation or use circular polarization, whereas with dual-dipole tags polarization is no longer critical. Antennas should be positioned so that the read zone covers all the places a tag might be; multiplex multiple antennas where needed. Readers interfere with each other and with other radios in the ISM band, who will return the favor if you don't take care. Mitigating tricks include turning readers off when not needed, screening or blocking the direct paths from interferer to victim, frequency planning where allowed, alternating polarization sense, and exploiting Dense-Interrogator-capable readers and tags. Conveyor antennas face similar problems to those used in portals but get much less time to solve them.

Portable and handheld readers require antennas that are small and well-behaved, not an easy mix of requirements. The need is best met by clever use of the space around the reader and the user, rather than by brute-force shrinkage of the antenna structure.

Near-field antennas are straightforward to construct when only a centimeter or two of read range is needed, but somewhat more subtle when ranges comparable to the wavelength of the radiation are of interest, owing to the resonant properties of large loop antennas. Solutions include segmented loops and unshielded transmission lines.

Antennas are usually connected to readers with cables and connectors. Cables allow convenient site selection for antennas but introduce losses; thinner cables are easier to use but lossier. Connectors are indispensable for flexible and reliable attachment of cabling to readers and antennas, but the variety of connectors in common use is sufficient to ensure that the appropriate gender and type of adaptor is never at hand when you need it.

6.8 *Afterword: An Electron's Eyelash*

An RFID reader puts out around a watt at a frequency close to 1 GHz. This requires a peak electric current of about 10 mA. There are about 10^{22} electrons in one cubic centimeter of a metal wire, each carrying 1.6×10^{-19} coulombs of electric charge. If we imagine all of the electrons on the surface of the wire moving in response to an imposed voltage, and changing their minds a nanosecond later, how far do they get? Just what is a milliamp at a gigahertz?

The electric current is the product of the charge per electron, the number of electrons, and the velocity:

$$I = qnv \tag{6.26}$$

Only the electrons within a skin depth of the surface of the wire (say the first five microns or so) actually move, so for a wire with a perimeter of 1 mm, the total number of electrons involved per centimeter of length is:

$$\begin{aligned}
n &= \pi D \delta \rho \\
&\approx 3.141 \cdot (0.1) \cdot (5 \times 10^{-4})(10^{22}) \\
&\approx 1.5 \times 10^{18}
\end{aligned} \tag{6.27}$$

So to produce 1 mA (0.001 A) we need a velocity of:

$$v = \frac{I}{qn} = \frac{0.001}{1.6 \times 10^{-19} \cdot 1.5 \times 10^{18}} \approx 0.004 \text{ cm}/s \quad (6.28)$$

If the electrons travel with this velocity for half an RF cycle (0.5 nsec), they travel on average a distance of:

$$\begin{aligned} d = vt &= 0.004 \cdot 5 \times 10^{-10} \\ &\approx 2 \times 10^{-12} \text{ cm} \end{aligned} \quad (6.29)$$

A hydrogen atom is about 1 Angstrom (10^{-8} cm) across. That is, to produce the radio signals that power and read tags, thereby making the whole enterprise of some use to humanity, requires that electrons in a reader antenna wiggle by a distance about 1/10,000 the size of an atom. It is astonishing that such displacements, and much smaller ones corresponding to received signals, can launch waves that induce even smaller displacements in distant places, and that those displacements can in turn be reliably detected and exploited to identify distant objects.

Further Reading

General Antenna Theory and Practice

The most commonly encountered books in this field, cited previously (in chapter 3), but now of most particular import, are:

Antenna Theory (3rd Edition), C. Balanis, Wiley 2005

Antenna Theory and Design (2nd Edition), W. Stutzman and G. Thiele, Wiley 1997

Antennas (3rd Edition), J. Kraus and R. Marhefka, McGraw-Hill 2001

All three are excellent. Balanis is the most rigorous, Kraus & Marhefka the most accessible and intuitive, and Stutzman and Thiele is a good mix of both.

Antenna Engineering Handbook, ed. R. Johnson, McGraw-Hill 1993. *Indispensable to the serious practitioner. This handbook covers just about every imaginable type of antenna and every common application.*

Practical Antenna Handbook (2nd Edition), J. Carr, TAB Books 1994. *Much less technical, and mainly oriented towards amateur radio operation, but with some useful tricks.*

Alternative Reader Antenna Configurations

"An intelligent 2.45 GHz beam-scanning array for modern RFID system", P. Salonen, M. Keskilammi, L. Sydanheimo, and M. Kivikoski, Antennas and Propagation Society International Symposium, 2000, p. 190

"A dual-polarized aperture coupled microstrip patch antenna with high isolation for RFID applications", S. Padhi, N. Karmaker, C. Law and S. Aditya, Antennas and Propagation Society International Symposium 2001, vol 2, p. 2

"Imaging RFID System at 24 GHz for Object Localization", M. Kaleja, A. Herb, R. Rasshofer, G. Friedsam, and E. Biebl, IEEE Microwave Theory and Techniques Society International Symposium 1999, paper TH2a-4, p. 1497

"Circularly polarized aperture coupled patch antennas for an RFID system in the 2.4 GHz ISM band", M. Kossel, H. Benedickter and W. Baechtold, IEEE RAWCON 1999 p. 235

"Octafilar helical antenna for handheld UHF RFID reader," S. Zauind-Deen, H. Malhat, and K. Awadalla, 28th National Radio Science Conference (NRSC 2011)

"Helical antenna for handheld UHF RFID reader," P. Nikitin and K. Rao, IEEE International Conference on RFID 2010, p. 166

"Compact square quadrifilar spiral antenna with circular polarization for UHF mobile RFID reader", W. Son, H. Lee, M. Lee, S.Min and J. Yu, Proceedings of the Asia-Pacific Microwave Conference 2010, p. 2271

"A circularly polarized square plate antenna with two inclined slots for UHF-RFID reader," P. Charoenchue, C. Phongcharoenpanich, K. Phaebua, and K. Aunchaleeverapan, International Symposium on Intelligent Signal Processing and Communications Systems (ISPAS 2011)

"Compact Yagi antenna for handheld UHF RFID reader," P. Nikitin and K. Rao, IEEE Antennas and Propagation Society International Symposium (APSURSI 2010)

Near-Field Antennas

"RFID smart shelf with confined detection volume at UHF," C. Medeiros, J. Costa, and C. Fernandes, IEEE Antennas and Wireless Propagation Letters 7 p. 773 (2008)

"A shelf antenna using near-field without dead zones in UHF RFID," J. Hong, J. Choo, J. Ryoo, and C. Choi, 2009 IEEE International Conference on Industrial Technology (ICIT 2009)

"Segmented magnetic antennas for near-field UHF RFID," D. Dobkin, S. Weigand, and N. Iyer, Microwave Journal, volume 50, #6, June 2007

"A segmented loop antenna for uhf near-field RFID," Y. Ong, X. Qing, C. Goh, and Z. Chen, 2010 IEEE Antennas and Propagation Society International Symposium

"Loop antenna for UHF near-field reader," Z. Chen, C. Goh, and X. Qing, Proceedings of the Fourth European Conference on Antennas and Propagation (EuCAP 2010)

"UHF RFID reader antenna for near-field and far-field operations," B. Shrestha, A. Elsherbeni, and L. Ukkonen, IEEE Antennas and Wireless Propagation Letters **10** p. 1274 (2011)

Exercises

Gain and read zone:

1. Consider the two-element array shown below:

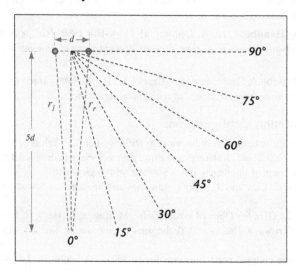

Treating $5d$ as equal to ∞ (and if you can do this without a second thought, you have a future in politics), measure the distances r_l and r_r from the two array elements to the end of the inclined dotted lines shown in the figure and record them, and their difference, below. (Use a ruler on paper, or use the digital version of the figure on the CD.)

Angle	r_l	r_r	$\delta = (r_l - r_r)/d$
0°	_____	_____	_____
15°	_____	_____	_____
30°	_____	_____	_____
45°	_____	_____	_____
60°	_____	_____	_____
75°	_____	_____	_____
90°	_____	_____	_____

Now assume d is two thirds of a wavelength. The phase associated with 2/3 of a wavelength is $(720/3) = 240$ degrees. The difference in phase between the waves received from the left and right array elements is thus $\phi = 240\,\delta$ in degrees. For each case above, calculate this offset angle. Draw two arrows of equal length, one horizontal and one rotated by the angle ϕ. Considering them as edges of a parallelogram, construct the diagonal and measure its length relative to the length of the original arrows. An example is shown below for the case of $\phi = 45°$. This length is the received electric field due to the two array elements, relative to the field from a single element. It is equal to 2 for an angle of 0 degrees, so divide the length by 2 to get the antenna pattern relative to the maximum.

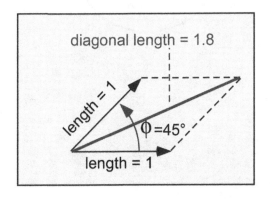

Using your results, plot the electric field pattern of the array on a graph like that shown below. (This is an H-plane slice of the pattern. What does the H-plane slice look like for a single dipole?)

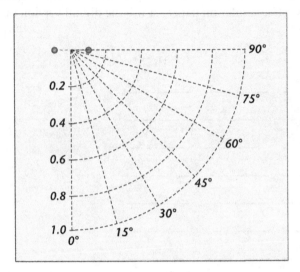

If you're feeling ambitious, also plot the square of the diagonal length, which corresponds to the relative received power.

2. Consider the antenna pattern given below for an ideal patch antenna:

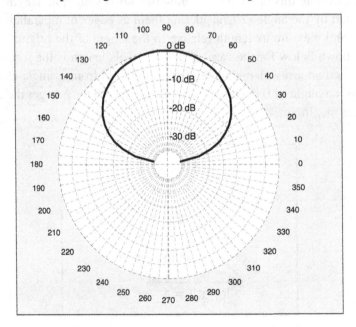

Extract the 3-dB and 6-dB beamwidth. Draw the idealized read zone following the procedure of Figure 6.5, assuming the range in the center of the beam is 4 meters, for a vertically-polarized antenna spaced 1.5 meters from the center of a conveyor, illuminating vertically-oriented tags.

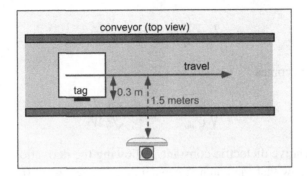

Assume tags are offset towards the antenna by 30 cm relative to the centerline of the conveyor, and that the cartons move along the conveyor at 3 meters/second. How long does a tag spend in the (idealized) read zone?

Now assume horizontal tags and a horizontally-polarized antenna. The voltage at the tag is reduced by the cosine of the angle of incidence θ.

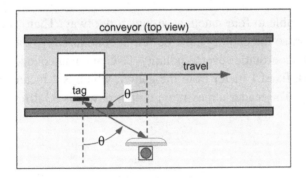

How does the read zone change, assuming that range is proportional to the power (voltage^2)? How long does the tag spend in the read zone?

Time in zone:

_____ s (vertical polarization)

_____ s (horizontal polarization)

Bandwidth:

3. When a dielectric material is placed between the patch and ground plane of a patch antenna, the length and width of the antenna can be reduced as roughly the square root of the dielectric constant, for the same operating frequency:

$$L \rightarrow \frac{L_{old}}{\sqrt{\varepsilon}}; \quad W \rightarrow \frac{W_{old}}{\sqrt{\varepsilon}}$$

Equation (6.16) becomes

$$\sqrt{\frac{L_{ant}}{C_{ant}}} = Z_{line} \approx \frac{Z_0}{\sqrt{\varepsilon}} \frac{d}{W}$$

where ε is the relative dielectric constant. Following the derivation used in the absence of a dielectric, show that the antenna capacitance is unchanged from its value for an air-based patch of similar ratio (W/L).

Assume that the radiation resistance is linear in the ratio of the wavelength in air to the width. Show that the radiation resistance is increased by the square root of ε; demonstrate that the antenna bandwidth is decreased by the square root of the relative dielectric constant.

Cabling:

4. You need to run cable to four antennas around a doorway. There is no time to trench the concrete floor, so the cabling must be run in conduit above the doorway. The reader, with four monostatic reverse-polarity TNC antenna connections, is mounted in a standard rack fixed 1 meter from the gantry, which is 4 meters across around a 3-meter doorway. The reader's maximum output power is 33 dBm. The setup is shown below:

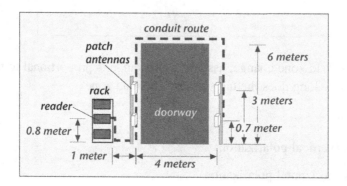

In a cabinet you find two 20-meter and two 6-meter lengths of coaxial cabling terminated in BNC connectors. You measure the outer diameter with a caliper and find it to be 0.2 cm. Estimate the cable loss based on Figure 6.38.

- Should you use this cabling to connect the reader to any or all of the antennas?
- If you order replacement cabling, what should the outer diameter be to ensure that 1 watt is delivered to each antenna?
- Should you remember to check whether the connectors on the antennas are N-male or N-female?
- When you forget to check and purchase cabling terminated in male mini-UHF connectors on both sides, and discover the discrepancy at 11 PM Sunday night for an installation that you promised for 8 AM the following Monday, what do you think the likelihood is that you can find the appropriate adaptors for both sides of each cable in the neighborhood convenience store?

  __ 1/1000

  __ 1/1,000,000

  __ diddly/squat

  __ exactly zero

  __ much higher after consuming the six-pack that you did find there

CHAPTER 7

Tag Antennas

7.1 World to Tag, Tag to World

Tag antennas operate on the same principles as reader antennas, but face some very different practical challenges.

- **Cost**: the total cost, including IC, substrate, antenna, adhesive, die attach, and testing, must be less than US$1 for most applications, and for high-volume supply-chain applications the long-term goal is to reduce total tag cost to less than US$0.05. In contrast, a moderate-quality patch antenna for a reader application has a purchase price around US$150.
- **Size:** in supply chain applications tags must fit onto a 4-inch- (100 mm-) wide label. Since the natural resonant size, half a wavelength, is about 16 cm, many tag antennas must reduce their size. In addition, many applications require total thickness < 1 mm, eliminating many potential structures from consideration.
- **Polarization:** in many applications the orientation of tags, or the objects to which they are attached, cannot be controlled. A tradeoff must be made between the use of circularly polarized reader antennas, sacrificing range, or the use of polarization-diverse tag antennas, adding cost and size to the antenna structure.
- **Matching to the IC load:** recall from chapter 5 that tag IC's consume very little current, and need reasonable input voltage (at least enough to turn the charge pump diodes on) − that is, the IC's have a high (parallel) input impedance. Antennas and associated matching structures need to provide as high an output voltage as possible from a given incident electric field, despite size constraints, and match to the high input impedance for good power transfer.
- **Getting along with the neighbors:** a reader antenna can be placed in a precisely-shaped plastic radome of known composition, and has usually has some hope of a clear field of view in the beam direction. In contrast, tags label objects and ideally should do so irrespective of the dielectric or conductive properties of those objects.

In meeting these constraints, tag designers have created a variety of unusual antenna structures (Figure 7.1), which we shall attempt to classify and explain.

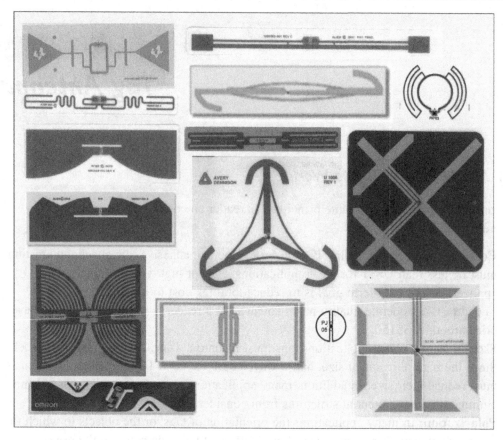

Figure 7.1
An assortment of commercial designs for passive UHF tags (not to scale).

7.2 Impedance Matching and Power Transfer

The first goal of the tag antenna must be to deliver power to the tag IC to turn it on; if that doesn't happen, nothing else matters. We have previously examined how much power an antenna of known gain can receive, arriving at the Friis equation (3.20). Is this actually the power delivered to the tag IC, or is there a chance of something being lost in translation? To figure this out, we must consider more carefully the question of how electrical power is delivered to a load.

Any real source of voltage or current has some associated limitations on how much voltage or current can actually be supplied; in the case of an antenna, this takes the form of a linear source impedance (Figure 7.2). When the source impedance is purely resistive, it is fairly easy to show that the maximum power is delivered to a load resistor of the same value.

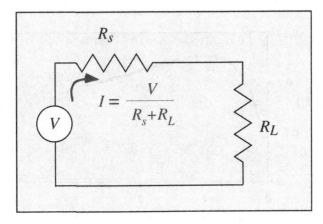

Figure 7.2
Voltage source with source resistance R_s and load resistance R_L.

The current is obtained from Ohm's law:

$$I = \frac{V}{R_s + R_L} \tag{7.1}$$

The power dissipated in each resistor, from equation (3.3), is:

$$P_s = \frac{V_S^2}{2R_s}; \quad P_L = \frac{V_L^2}{2R_L} \tag{7.2}$$

but since the voltage on each resistor is just IR from Ohm's law, and the current is the same through both source and load (it has nowhere else to go), we can also express this as:

$$P_L = \frac{I^2 R_s}{2}; \quad P_L = \frac{I^2 R_L}{2} \tag{7.3}$$

Substituting the current into the expression for power, we find:

$$P_L = \frac{R_L V^2}{2(R_S + R_L)^2} = \frac{1}{R_S} \frac{(R_L/R_S) V^2}{2(1 + (R_L/R_S))^2} \tag{7.4}$$

The second form is plotted as a function of the ratio of the load to source resistance for a fixed source resistance in Figure 7.3. The load power for a fixed source resistance is maximized when the source and load resistances are equal: this is known as a ***matched load***. Since the situation is entirely symmetrical, the same remark is true when we vary the source resistance while keeping the load fixed: the best power transfer always occurs when the source and load resistance are the same.

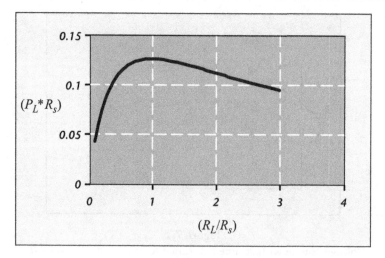

Figure 7.3
Load power multiplied by the source resistance, as a function of the ratio of load to source resistance.

In the more general case where the source and load include a **_reactance_** — that is, a complex part, due to some inductance or capacitance or both — the expression for the current is still of the same form but with a complex part (see Appendix 3):

$$I = \frac{V}{(R_S + jX_S) + (R_L + jX_L)} \tag{7.5}$$

A resistor doesn't know or care what the phase of the current flowing through it is; the dissipated power is only sensitive to the size of the current:

$$P_L = \frac{|I|^2 R_L}{2} = \frac{V^2 R_L}{|(R_S + jX_S) + (R_L + jX_L)|^2} = \frac{V^2 R_L}{2|Z_S + Z_L|^2} \tag{7.6}$$

where | | denotes the modulus of the complex quantity — that is, the length of the vector in the complex plane, and we have introduced the **_impedance_** $Z = R + jX$. The largest power will be obtained when the denominator is small. For fixed values of the real source and load resistances, the denominator is minimized when the imaginary part is 0 (Figure 7.4): that is, when the reactances of the source X_S and load X_L cancel each other out. This cancellation is managed by providing reactances of different sign for the source and load; thus, if the load is capacitive (negative reactance) then the source must be inductive (positive reactance) and of the same magnitude. When the real parts of the source and load impedances are the same, and the complex parts are of the same magnitude but opposite signs, the two impedances are **_complex conjugates_**, differing only in the sign of the

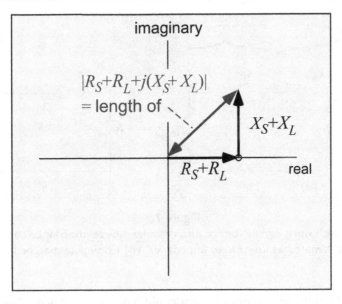

Figure 7.4

The shortest possible length of a complex vector with a fixed real part occurs when the imaginary part ($X_S + X_L$ here) is equal to 0.

imaginary part, so the condition of maximum power transfer is also known as ***conjugate matching***.

Returning to the problem at hand, we can treat the tag antenna electrically as a voltage source, the open-circuit voltage V_{OC}, connected through a complex impedance consisting of the radiation resistance R_{rad} and any inductances or capacitances, absorbed into a reactance X_{ant}; we similarly treat the IC as a linear but complex load impedance (Figure 7.5(a)). (Recall from chapter 5 that this is only an approximation; the diodes that form the input charge pump are non-linear devices, and will result in the generation of some harmonics of the input frequency. In the interests of simplicity we shall treat the IC as a linear load, but the harmonics generated by the IC are not necessarily negligible and can affect regulatory compliance of the tag-reader system.) Maximum power transfer will occur when the source and load reactances are conjugate-matched and cancel, producing the situation in Figure 7.5(b). It is this condition that is normally assumed in calculating the gain of an antenna, and therefore the received power calculated using the Friis equation is only delivered to the tag IC when this conjugate matched condition exists:

$$P_{Av} = \frac{I_{ant}^2 R_{load}}{2} = \frac{V_{oc}^2}{8R_{rad}} \tag{7.7}$$

When perfect matching does not exist, the power delivered to the IC is less than that predicted by the Friis equation. We can use equation (7.6) to find the ratio of the power

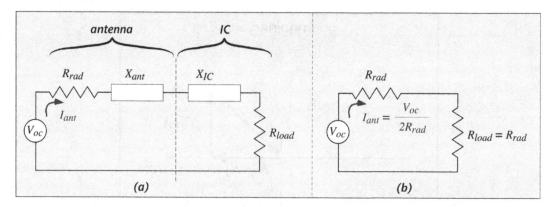

Figure 7.5
(a) Antenna with open-circuit voltage and complex source impedance connected to IC,
approximated as linear load impedance. (b) Conjugate-matched case.

delivered to the IC to the maximum power it could have received according to the Friis
equation, the **power transfer coefficient** τ:

$$\tau = \frac{P_L}{P_{Friis}} = \frac{V_{oc}^2 R_{load}}{2|Z_{ant} + Z_{load}|^2} \frac{8R_{rad}}{V_{oc}^2} = \frac{4R_{load}R_{rad}}{|Z_{ant} + Z_{load}|^2} \qquad (7.8)$$

The power transferred to the tag IC, and thus the tag read range, are maximized when $\tau = 1$
When the power transfer coefficient is less than 1, the read range is proportional to the
square root of τ.

So the problem of designing a tag antenna becomes the problem of constructing an
antenna that provides a good conjugate match to the IC. Recall from chapter 5 that a
tag IC can be approximately represented as the parallel combination of some capacitance
and some resistance. The capacitance is typically around 0.5 pF, and the resistance is
inversely proportional to the power consumption of the IC and the efficiency of the
charge pump. For an IC that requires an 0.5-volt input, with a 25% efficient charge
pump, and consumes 25 microwatts, the current flow will be 100 microamps and the
equivalent input resistance 5000 ohms. However, we should **not** simply set $R_{load} = 5000$
in the equivalent circuit of Figure 7.5, because this load resistance is in parallel with the
capacitance of the IC, not in series. The impedance of a series combination of (say)
1000 ohms and 0.5 pF is in general very different from that of a parallel combination of
the same components (Figure 7.6(a)). Fortunately, at any particular frequency, we can
find new values of resistance and capacitance such that a series resistance and
capacitance present the same impedance as the original parallel circuit (Figure 7.6(b)).
We can insert these equivalent series values into Figure 7.5 and thus establish the
requirements for a matched antenna.

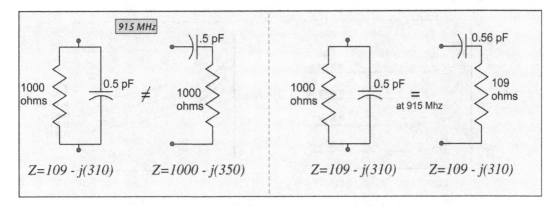

Figure 7.6
(a) A resistor and capacitor may produce very different impedances when connected in series or in parallel; (b) At any given frequency, a parallel circuit can be represented by a series circuit using appropriately modified component values.

The relationships between the series (s) and parallel (p) component values that produce the same impedance at a given frequency $f = \omega/2\pi$ are:

$$R_S = \frac{R_p}{1 + (\omega C_p R_p)^2}; \quad C_S = \frac{1 + (\omega C_p R_p)^2}{\omega^2 C_p R_p^2} \tag{7.9}$$

These relationships are plotted for representative values of the parallel component values in Figure 7.7. The series and parallel capacitances don't differ a lot, but the series resistance is inversely related to the parallel resistance. For reasonable values of around 0.5 pF capacitance and 5000 to 10,000 ohms load resistance, the corresponding series values are 0.5 pF and 12 to 24 ohms. So in summary, the matching problem for the antenna looks like Figure 7.8: for best range, the antenna ought to present a load that looks like a series resistance value of about 18 ohms, and a series inductance that resonates with 0.5 pF. (At 915 MHz, that's about 60 nH.)

A second benefit of producing a good match to the tag impedance is that we can in general obtain some *voltage amplification*. To see how this arises let us first consider the simplified schematic of Figure 7.5(b). In the case where only the two resistors are present and are of equal value, the voltage across the load resistor is simply half of the open-circuit voltage. When one wishes to calculate the current flowing through the circuit, the simplified schematic is quite sufficient if the reactances of the antenna and IC are equal and opposite. However, the voltage on the IC is **not** the voltage across the equivalent series resistor, but rather the voltage across the series combination of the resistor and capacitor. This is an important distinction, because while the series resistor is typically small in value, the reactance of the series capacitance is quite substantial. The voltage across the IC is obtained

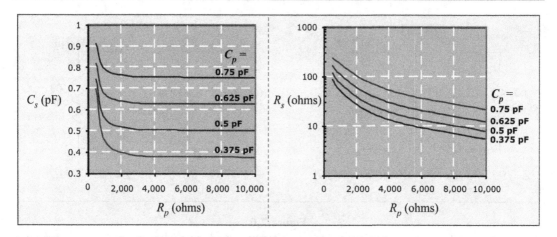

Figure 7.7
Values of series capacitance C_s and resistance R_s that produce the same impedance as parallel capacitance C_p and resistance R_p at 915 MHz.

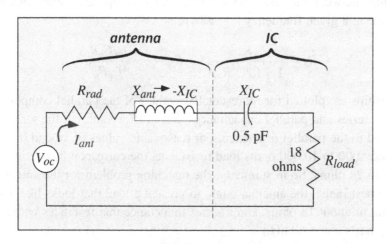

Figure 7.8
Typical load presented to tag antenna by tag *IC*.

by multiplying the current in the circuit by the impedance of the series R-C circuit. When the circuit is conjugate-matched, this is simply:

$$\frac{|V_{IC}|}{V_{oc}} = \left| \frac{R_{load} + \frac{1}{j\omega C_{IC}}}{2R_{load}} \right| \approx \frac{1}{2\omega C_{IC}R_{load}} \tag{7.10}$$

For typical component values this is quite a bit larger than 1. For example, if $C_{IC} = 0.5$ pF, $X_{IC} \approx -j350$ ohms. For a series equivalent load resistance of 18 ohms, we can expect a voltage multiplication of about $(350)/(36) \approx 10$. This amplification is very helpful in achieving good read range, since as the reader will recall from chapter 5, a tag needs a high enough voltage to turn the input diodes on before it can rectify effectively. A tag 5 meters from a reader antenna radiating the maximum allowed power of 4 W EIRP experiences an electric field of about 3 V/meter. If the tag is 9 cm long, the open-circuit voltage is roughly 0.14 V, not nearly enough to drive the input rectifier. However, after matching, about 1.3 V will appear on the IC, more than sufficient to allow the rectifiers to function.

The manner in which we obtain the antenna reactance $X_{an\ t}$ is also important, particularly because it bears upon the bandwidth over which tag performance will be acceptable. The argument is essentially the same as that of section 2.3 of chapter 6. At the frequency where the tag and antenna are perfectly matched, the reactances cancel and the total resistance seen by the voltage source is just $2R_{rad}$. As frequency changes, the reactance will increase in magnitude; when the reactance becomes comparable in magnitude to the resistance, the power transfer coefficient will be reduced by a factor of 2 and the read range by a factor of about 1.4. We can thus conveniently define the tag bandwidth as

$$|X_{ant} + X_{IC}| = 2R_{load} @ \left(|f - f_{ctrr}| = \frac{BW}{2} \right) \tag{7.11}$$

The best bandwidth will be obtained when the antenna reactance is a simple inductor. Obtaining the same reactance by using, for example, the series combination of an inductor and a capacitor will produce a stronger frequency dependence:

$$\underbrace{j\omega_0 L_{simple}}_{\frac{d}{df} = 2\pi L_{simple}} = \underbrace{j\omega_0 (2L_{simple}) - j\frac{1}{\omega_0(C_{res})}}_{\frac{d}{df} = 6\pi L_{simple}} \qquad C_{res} = \frac{1}{\omega^2 L_{simple}} \tag{7.12}$$

In general, the bandwidth is inversely proportional to the Q factor, which is the ratio of the total reactance to the resistance. For a simple series resonant circuit, Q is about twice the voltage amplification factor (equation (7.10)). Thus, voltage amplification must be traded against bandwidth. Antennas with large reactance (that is, large values of inductance and small values of capacitance) and small values of resistance may be adjusted to be matched to the tag at one frequency, with good power transfer and voltage multiplication, but performance will degrade at other frequencies. Antennas with small reactance will provide better performance over frequency.

We are now equipped to evaluate various tag antenna alternatives against these requirements.

7.3 Dipoles and Derivatives

The author would likely be in much better financial condition if the title of this section referred to financial risk mitigation tools rather than the half-wave dipole we briefly encountered in chapter 6, but might have become the subject of protesters' wrath. Nevertheless, the humble dipole forms the basis of a great many approaches to tag antenna design. The native half-wave dipole is shown both geometrically and in electrically equivalent form in Figure 7.9.

It is apparent that an unmodified half-wave dipole is not a terribly good antenna for a typical tag IC. The antenna reactance is very small, so there's nothing to remove the quite substantial reactance of the tag IC (corresponding to an imaginary impedance of about 350 ohms). The power transfer coefficient is on the order of $4(65)(18)/350^2 \approx 4\%$, due to the large uncompensated reactance of the IC capacitance. In consequence, we don't even get far enough to discover that the radiation resistance is poorly matched to the effective series resistance of the tag. Finally, the physical antenna, at 16 cm long, is too big for labels and many other applications. If we are to preserve the essential simplicity of the dipole but still obtain decent performance, we need to make some changes.

7.3.1 Wiggling Wires

The easiest problem to tackle is the mechanical one. If we want to make the dipole smaller, we just put the squeeze on it (Figure 7.10). By bending the wires, we can reduce the linear extent taken up by 16 cm of wire to something less, depending on how much bending we're inclined to do. A dipole that is shortened in this fashion is known as (surprise!) a **bent dipole**. As more bends are added, the dipole starts to look like a river meandering across flat terrain, and is known as a **meandered dipole**. With enough bends, we can make the dipole much shorter for the same length of wire.

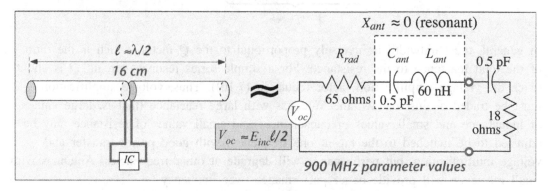

Figure 7.9
Half-wave dipole as a tag antenna.

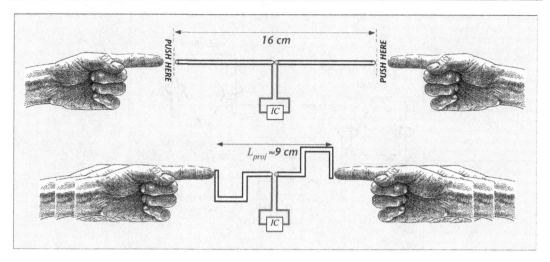

Figure 7.10
A half-wave dipole can be made shorter by bending the wires.

Bending the dipole is not without electrical consequences. Radiation resistance is a consequence of power lost to radiated waves, and radiated waves, as the reader will recall, arise from uncompensated currents. The linear flow of current along a dipole is all in one direction and the radiation from all the current along the dipole adds to create a vector potential in the same direction as the current flow. When the dipole is meandered, the direction of the current flow in neighboring arms of a meander is inverted, so that to a good approximation these currents cancel when viewed from a long distance, and contribute no radiation (Figure 7.11). The more squeezing we do to fit the antenna into a smaller space, the more effective the cancellation is. For a densely meandered antenna, a pretty good guess at the radiation resistance is obtained by considering that only the parts of the antenna that are oriented in the original dipole direction contribute to radiation. The radiated potential is thus reduced by the ratio of the length of radiating current in the meandered dipole to that in the original dipole, and the radiated power (being proportional to the square of the electric field) by the square of that ratio. That is, if the length of the antenna in the original direction is L_{proj}, the radiation resistance is approximately:

$$R_{rad,meander} \approx \left(\frac{L_{proj}}{L_{half-wave}}\right)^2 R_{rad,half-wave} \approx \left(\frac{2L_{proj}}{\lambda}\right)^2 65 \qquad (7.13)$$

In addition, the capacitance and inductance of a meandered structure are not the same as those of a straight structure. The self-inductance of a wire results from the magnetic vector potential **A** along each part the wire, which is created by the current flowing in the same direction on other parts of the wire. All the current in a dipole flows in the same direction, so each current element contributes to the potential at all positions along the wire, albeit

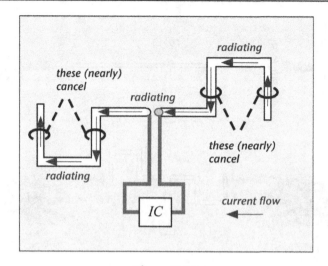

Figure 7.11

The current in neighboring arms of a meandered structure flows in opposite directions and does not radiate.

with a decreasing effect inversely proportional to distance along the wire (Figure 7.12(a)). When the dipole is bent, all parts of the wire are no longer parallel. Current flowing in segments perpendicular to any particular location makes no contribution to the magnetic vector potential at that location on the wire (Figure 7.12(b)). Furthermore, the potential arising from neighboring currents flowing in opposite directions tends to cancel. The result is that a densely-packed meandered structure has significantly less inductance per unit length than a straight dipole. The capacitance per unit wire length of a meandered structure is also reduced, simply because the charge on the wire is packed closer and closer, increasing the voltage on the wire for the same total charge.

Since the resonant frequency of the antenna is inversely proportional to the produce of inductance and capacitance, when these are reduced, the resonant frequency is increased. A meandered dipole has a higher resonant frequency than a straight dipole for the same wire length. To obtain 900 MHz operation from a meandered dipole, we will need a larger total length than would have been the case using a straight dipole.

Some experimental data for bent antennas made from 0.6-mm-diameter wire is summarized in Figure 7.13, confirming the expectation that as the antenna is bent more tightly, the inductance and capacitance fall and the resonant frequency increases for a fixed wire length. Note that the data shown here is for monopoles above a large ground plane. A monopole can be regarded as seeing half the voltage of the corresponding dipole, so the capacitance is doubled and the inductance is halved relative to a dipole. Since the monopole is half as long, the capacitance per unit length is increased by 4-fold relative to a dipole, and the inductance per unit length is unchanged. At the most densely folded configuration, a wire length of about

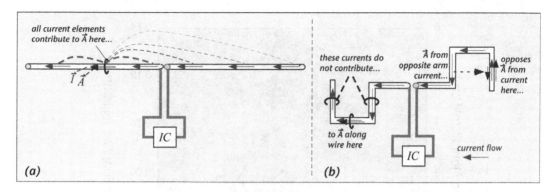

Figure 7.12
Meandered wires have less self-inductance due to orthogonal and opposing current flows.

17 cm for the monopole, or a 34-cm-long dipole, would be needed to deliver a resonant frequency of around 900 MHz. This is more than twice as long as a straight dipole.

Even once the dipole antenna is squeezed down to an acceptable projected length, and extended enough to achieve resonance, we still face the problem of matching. Let us consider as an example a meandered resonant antenna with a projected length of 9 cm (so that it will fit on a label), and a tag IC with equivalent circuit as in Figure 7.8. The radiation resistance will be roughly $65(9/16)^2 \approx 20$ ohms, a good match to the equivalent series resistance of the tag. However, the reactance of the antenna is about 0. The power transfer coefficient from equation (7.8) is about:

$$\tau \approx \frac{4(18)(20)}{\left|20 + \left(18 - j\frac{1}{2\pi(915 \times 10^6)(0.5 \times 10^{-12})}\right)\right|} \approx 1.2\% \tag{7.14}$$

That is, only about 1/100 of the received power makes it into the tag. Since the forward-link-limited range is proportional to the square root of the power, this corresponds to a 10-fold reduction in read range vs. a matched antenna − hardly desirable. How can we improve the match of antenna to IC?

One approach is to make the antenna wire longer than resonance, while keeping the same projected length (that is, more meandering!). Dipole antennas longer than resonance are inductive, and thus have positive imaginary reactance which will tend to cancel out the capacitance of the IC. We need a reactance of about $j350$ ohms or 60 nH at 915 MHz, which at the inductance density of Figure 7.13 would require about 15 cm of wire. While this length will be reduced somewhat using printed lines, which have a modestly higher inductance per unit length, several additional cm of wire will be required. An example of such an approach in a commercial antenna design is shown in Figure 7.14 (see Rao *et al.* in Further Reading for more details on this structure). The total length of the

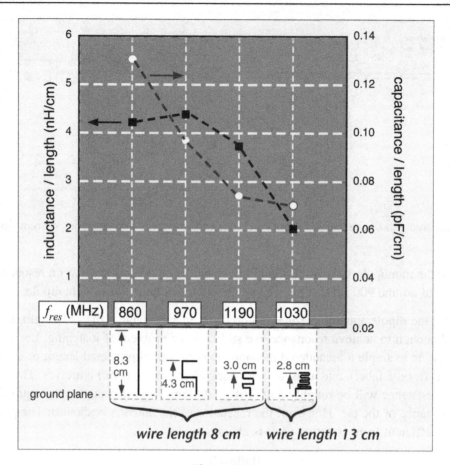

Figure 7.13
Experimental values of inductance and capacitance per unit length, and resonant frequency, for increasingly-dense bent wire monopoles

antenna is about 33 cm, meandered to fit within the 10-cm label constraint. The additional straight bar at the top of the structure acts as a bit of shunt capacitance, helping to trim the radiation resistance to the input resistance of the tag IC.

7.3.2 Match L with L of L

A second approach to adapting a short antenna to the capacitive IC load is to add a matching structure. One possible structure uses a combination of shunt and series inductors; such an approach is shown in Figure 7.15. The shunt and series inductors are realized as lengths of conductive line connecting the antenna to the IC. This is a planar variant of the very old technique of the T-match, in which an antenna is matched by adding a wire in shunt connected along its length.

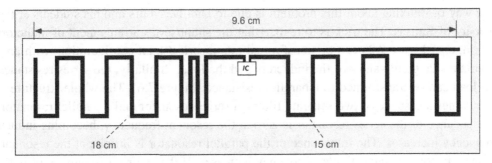

Figure 7.14

Meandered tag antenna, after Rao *et al.*; note that the structure is designed to operate at 830 MHz as shown, but the trace length is readily trimmed to provide good performance at higher frequencies.

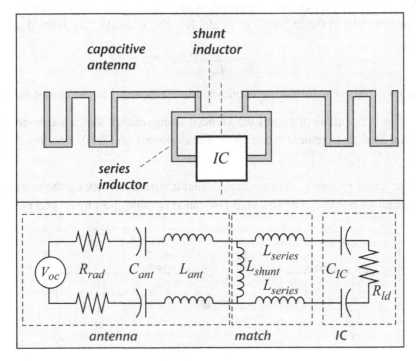

Figure 7.15

Shunt/series inductors match a capacitive antenna to a capacitive load.

The T-match is versatile and widely used, often in combination with the other techniques described in this chapter. The T-match gives excellent power transfer at the design frequency, but many choices can be made for the series and shunt elements. Some work only near the design frequency and others will provide good broadband performance, which is very helpful in a practical antenna design. How does one arrive at an optimal configuration?

A nice way of thinking about this problem is due to Dan Deavours and his students at the University of Kansas. The trick is to realize that the shunt/series arrangement of inductors shown in, e.g., Figure 7.23 can be transformed into a series/shunt configuration, with an appropriate change in values of the inductors and the load. Similarly, the series resistance of the load can be transformed to a parallel resistance (Figure 7.6). The whole structure then becomes a cascade of two separate filters: a series resonator and a parallel resonator. The impedance of the series resonator is zero at the resonant frequency, becoming inductive as frequency increases. The impedance of the parallel resonator is infinite at the resonant frequency, becoming capacitive as frequency increases. If the resonant frequency of both is about equal, the changes roughly cancel as we move away from the resonant frequency, producing good power transfer and good broadband performance.

Let's look at how this plays out, first in general and then with a specific design example. The series/shunt to shunt/series transformation is shown in Figure 7.16 (with the names slightly shortened to keep the expressions readable). The quantity β is defined as:

$$\beta = \frac{L_{sh}}{L_{sh} + L_{ser}} \qquad (7.15)$$

(The derivation of this transformation is outlined in Exercise 6[1] at the end of the chapter.)

So we start with the equivalent circuit we've been using, except that it's convenient here to use the parallel load arrangement (see Figure 7.6 and equation (7.9)). The initial equivalent circuit is shown in Figure 7.17.

Performing the shunt–series to series–shunt transformation, we obtain the modified equivalent circuit of Figure 7.18. The load resistance is multiplied by β^2 and the

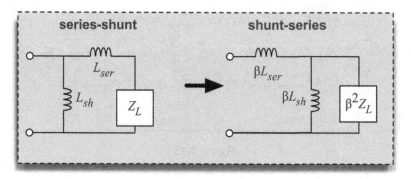

Figure 7.16
Transformation of shunt/series to equivalent series/shunt configuration.

---

[1] Exercise 6 is not very funny. Mathematics is generally not funny. Speaking of which, did you hear about the y who had to set his square equal to the square of x, and then add 1? He was so ticked off he went hyperbolic.

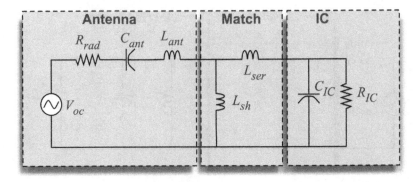

Figure 7.17
Equivalent circuit of T-matched antenna and IC load, using parallel load configuration.

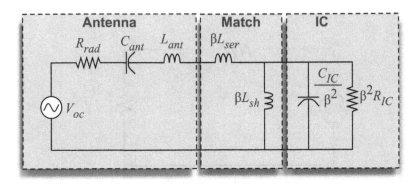

Figure 7.18
Equivalent circuit after transformation.

capacitance is divided by the same quantity. (Recall that the impedance of a capacitor is $1/(\omega C)$, so to multiply the impedance we divide the capacitance.)

If we simply regroup the circuit elements, we can now recognize the matching circuit, combined with the reactive part of the antenna and the IC, as forming a cascade of a series and parallel resonator, as shown in Figure 7.19.

At resonance, the series resonator has zero impedance, and the shunt resonator has infinite impedance, so the circuit collapses to the very simple result shown in Figure 7.20. Optimal power transfer will be obtained when the radiation resistance and the transformed load are equal (Figure 7.3).

We can now outline a procedure for determining a reasonable first-cut T-match design. (When we get to the end, you'll see that a few iterations are likely.) We start with the usage requirement for the overall size of the antenna, which, as noted above, roughly determines the radiation resistance (see equation (7.13)), a target frequency

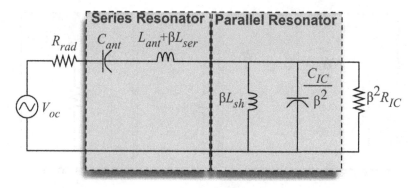

Figure 7.19
The transformed equivalent circuit is a cascaded series and parallel resonator.

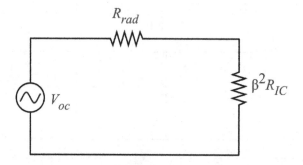

Figure 7.20
At resonance, the series and parallel resonators disappear from the circuit.

(e.g., 915 MHz), and the known tag IC impedance, using the parallel-resistor model. Knowing the radiation resistance allows us to estimate the quantity β that provides good power transfer:

$$\beta^2 = \frac{R_{rad}}{R_{IC}} \tag{7.16}$$

Knowing the value of β, we can specify the transformed value of the IC load capacitance:

$$C_{ICtr} = \frac{C_{IC}}{\beta^2} \tag{7.17}$$

If we then require that the parallel resonator composed of the transformed shunt inductance and the transformed IC capacitance resonate at the target frequency f_{ctr}, we can determine the value of the transformed shunt inductance. Since we already know β, we can then find the value of the original shunt inductor in the T-match. In this derivation, we make use of the fact that the resonant frequency is inversely proportional to the square root of the product (LC), as derived in Appendix 3, section 2.

$$f_{ctr} = \frac{1}{2\pi\sqrt{LC}} = \frac{1}{2\pi\sqrt{\beta L_{sh}(C_{IC}/\beta^2)}} \rightarrow \frac{1}{2\pi\sqrt{(L_{sh}C_{IC}/\beta)}}$$

$$(2\pi f_{ctr})^2 = \frac{\beta}{L_{sh}C_{IC}} \rightarrow L_{sh} = \frac{\beta}{(2\pi f_{ctr})^2 C_{IC}}$$

(7.18)

Now we know both β and L_{sh}, so from the definition of β we can find L_{ser}:

$$\beta = \frac{L_{sh}}{L_{sh} + L_{ser}} \rightarrow L_{sh} = \beta L_{sh} + \beta L_{ser}$$

$$L_{sh}(1 - \beta) = \beta L_{ser}$$

$$L_{ser} = L_{sh}\frac{(1 - \beta)}{\beta}$$

(7.19)

Finally, knowing the series inductance of the T-match lets us to estimate the required reactance of the unmatched antenna, roughly equivalent to setting the native antenna resonant frequency. For example, in the meandered dipoles discussed in the previous section, this requirement roughly sets the length of the meandered part of the antenna.

$$f_{ctr} = \frac{1}{2\pi\sqrt{(L_{ant} + \beta L_{ser})C_{ant}}}$$

$$(2\pi f_{ctr})^2 = \frac{1}{(L_{ant} + \beta L_{ser})C_{ant}}$$

(7.20)

To build the resulting structures, we need a way of translating the requested values into dimensions. The inductance can be very roughly estimated using low-frequency formulas for isolated structures. For a trace of length l and width w, the inductance is roughly

$$L = \frac{\mu_0}{2\pi}\ell\left(\ln\left[\frac{\ell}{w}\right] + \frac{\pi}{2}\right); \quad \mu_0 = 4\pi \times 10^{-7} \ H/m$$

(7.21)

Thus, a bit of trace 1 cm long and 1 mm wide has an inductance of around 8 nH. The reactive elements of a meandered antenna can be estimated from, for example, Figure 7.13. However, the design is inevitably iterative, because these simplistic estimates ignore mutual inductance − the interaction between currents on neighboring pieces of wire − as well as radiation from the matching structures.

Setting the two resonant frequencies equal to the target frequency will give the best power transfer at that frequency, but it is also possible to intentionally offset them to improve bandwidth, at the expense of peak efficiency. A slight advantage is also obtained by allowing the matched IC resistance to be a bit bigger than the antenna radiation resistance. An example broadband antenna is depicted in Figure 7.21.

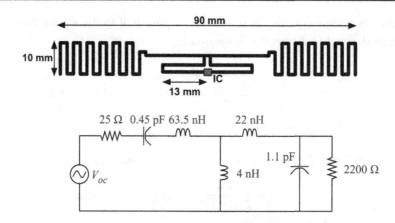

Figure 7.21

Example broadband meandered dipole with T-match, and approximate equivalent circuit.
After Deavours and Dobkin.

The resulting impedance of the matched antenna and IC (that is, looking to the right of the antenna radiation resistance) is depicted in Figure 7.22. The resonant frequencies of the series and parallel equivalent resonators are also shown in the diagram; we see that they are slightly displaced down and up in frequency, respectively. The optimal match is when the rest of the antenna is just equal to the radiation resistance: that is, resistance = 25 ohms and reactance = 0 ohms. This matching condition is not satisfied for any frequency, and thus the power transfer coefficient is always less than 1. However, in return, we obtain pretty good power transfer all the way down to about 860 MHz and all the way up to about 1 GHz. The antenna is designed to be centered a bit above the middle of the various international bands (around 860−960), to allow for a slight reduction in optimal frequency when placed on a dielectric material such as cardboard.

An alternative, and in some ways more general, way of thinking about matching problems is to use a graphical tool that treats impedances in terms of their equivalent reflection coefficients, the Smith Chart. (See Appendix 4 for a description of this useful graphical tool.) An example is shown in Figure 7.23. For simplicity we have depicted matching a single-ended circuit (like a monopole) rather than a balanced dipole, and we have used somewhat unrealistic values of antenna and tag impedance to move the loads a bit farther from the edge of the chart so that they are easier to see. Our goal is to transform the radiation resistance of the antenna to the complex conjugate of the IC impedance. (One can equivalently start from the IC impedance and transform to the radiation resistance − that is, proceed from right to left instead of left to right.) The real radiation resistance of the antenna (point 1) is added to the reactance of the antenna capacitance and inductance; in this case, we have assumed that the antenna is shorter than the resonant condition and so is slightly capacitive (point 2). The shunt inductor moves the antenna impedance along a circle of constant conductance onto the top (inductive) half of the chart (point 3). The series inductor then moves the load along a circle of constant resistance until the total impedance

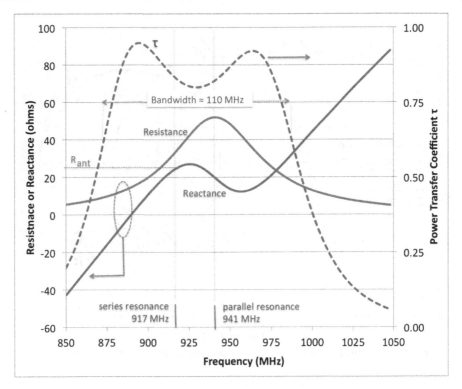

Figure 7.22
Resistance and reactance $(R + jX)$ of the components to the right of the radiation resistance; matched condition is $R = 25$, $X = 0$. Also shown is the power transfer coefficient τ (equation (7.8)).

(point 4) is the conjugate of the load presented by IC. By increasing the conductance of the shunt inductor (which is just a matter of making it shorter) we can move farther around the circle to access smaller series resistances, at the cost of requiring increased series inductance to complete the match. The inductance values required are a few tens of nH and readily realized. A detailed example is provided in the exercises.

One of the ancillary benefits of matching the load is an increase in the voltage applied to the IC, which as the reader will recall is very helpful in turning on the diodes in the IC charge pump. For a pure reactive power match the available power must remain constant, so the voltage is increased by the square root of the ratio of resistances, where the load resistance should be regarded as the shunt value in this context.

7.3.3 Getting Loaded

Another method of making a shorter antenna with suitable impedance is to flare the end of the antenna out to a larger structure; this increases the antenna capacitance and is known as

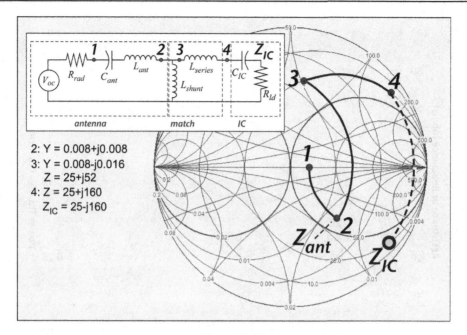

Figure 7.23
Example of shunt-series inductance matching depicted on a Smith chart.

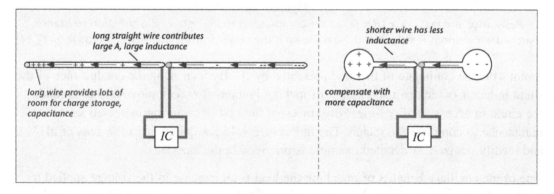

Figure 7.24
A dipole can be shortened without unacceptably large reactance by adding capacitance to the tips.

capacitive tip-loading (Figure 7.24). Since the magnitude of capacitive reactance decreases as the capacitance increases, a tip-loaded dipole looks more inductive than a conventional dipole of the same length, and is easier to match.

Classic tip-loaded dipoles often used metal spheres for loads, with capacitance proportional to the sphere's surface area. Printed tag antennas must use flat structures, and the added capacitance is roughly proportional to the perimeter of the loading shape rather than its area. Examples of measured data for a wire monopole 3.1 cm long loaded by disks of varying size are shown in Figure 7.25. Disks with radii of around 1 cm provide capacitance of about 1 pF, and allow a resonant antenna at 900 MHz that will fit into a 10-cm projected length constraint (although just barely — it helps to flatten out the loading structure, which has little effect on its capacitance).

An example of a commercial antenna structure that employs tip loading and inductive matching is shown in Figure 7.26. This specific structure also used printed conductive ink

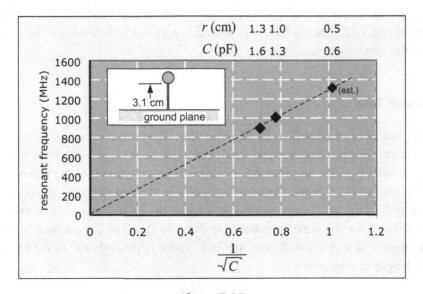

Figure 7.25
Resonant frequency of a 3.1-cm wire monopole loaded with flat disks of varying radius *r*.

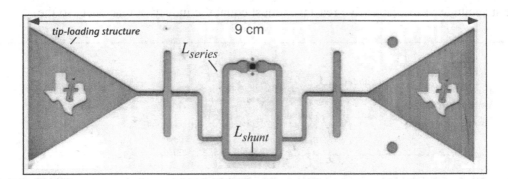

Figure 7.26
Texas Instruments Class 1 Generation 2 inlay (2006), using tip load and inductive matching to fit within 9 cm projected length.

to form the antenna structure. As we noted in section 6 of chapter 5, such inks have a sheet resistance of around 20 milliohms/square. The narrow part of the antenna structure of Figure 7.26 is about 90 squares long, producing an ohmic resistance of 1.8 ohms, tolerable compared to the radiation resistance of around 10–20 ohms. On the other hand, if one constructs a meandered inductive dipole of the type depicted in Figure 7.14, the antenna represents about 330 squares using 1 mm lines. The corresponding ohmic resistance of around 7 ohms is substantial compared to the expected radiation resistance, and will result in noticeably degraded performance. Different material systems may impose different constraints on tag antenna design.

A second commercial antenna design that combines all three techniques we have discussed is shown in Figure 7.27. The tip load here uses line reversal and could also be regarded as a variation of the meandered structure.

7.3.4 Fat and Thin

The antennas we have examined so far are wire-like. They are convenient to manufacture, but have relatively high inductance and small capacitance. As a consequence, the antenna reactance is relatively large, and thus bandwidth is reduced relative to the potential bandwidth of the tag IC. It is well-known that a dipole antenna that uses thicker elements is more broadband; in the simplest view, this is because the capacitance is increased and inductance reduced for the same resonant frequency, so the Q of the antenna is less and bandwidth larger. For a cylindrical wire, the inductance is proportional to the logarithm of the ratio of length to diameter:

$$\ell \propto L \ln\left(\frac{L}{d}\right) \tag{7.22}$$

The inductive reactance $j\omega$ l thus also scales as the log of the aspect ratio. At resonance the magnitude of the capacitive reactance must equal the inductive reactance, so the

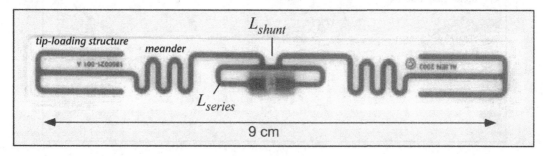

Figure 7.27
Alien Technology Class 1 inlay (2005), using tip load, meandered lines, and inductive match.

capacitance must scale as $1/\ln(L/a)$. On the other hand, the radiation resistance is basically a function only of the length of the wire, since it depends only on the integral of the current (as long as the wire width is small compared to a wavelength). Therefore the quality factor of the antenna, the ratio of reactance to resistance, scales with the logarithm of aspect ratio, and the bandwidth inversely:

$$Q \propto \ln\frac{L}{d} \rightarrow BW \propto \frac{1}{\ln\left(\frac{L}{d}\right)} \tag{7.23}$$

The logarithm changes very slowly, so big variations in the shape of the antenna are needed to produce modest improvements in bandwidth. In Figure 7.28 we show the dependence of relative bandwidth (defined here as the range over which the reactance of the antenna is less than 50 ohms) vs. the inverse logarithm of the aspect ratio for cylindrical wire antennas. To improve bandwidth from (say) 7% to 14% we must decrease the aspect ratio of the antenna from 10,000:1 to about 80:1 — a substantial change!

The flat printed structures used for tag antennas behave in a qualitatively similar fashion, but are different in that for large linewidths, the current or charge near the center of the wire must be accounted for, which does not happen for a cylindrical wire. The result is that scaling is more like:

$$\ell \propto L\left(\ln\left(\frac{L}{w}\right) + 0.2\frac{L}{w} - 0.3\right) \tag{7.24}$$

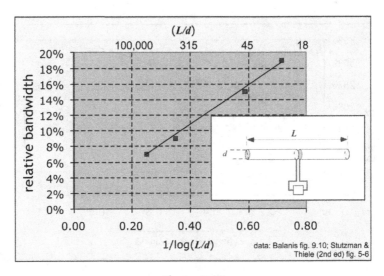

Figure 7.28
Bandwidth vs. aspect ratio for resonant cylindrical wire dipoles, after Balanis and Stutzman & Thiele

(as long as thickness is negligible compared to width) and therefore the benefits of chunky structures are reduced relative to a cylinder. In Figure 7.29 we show bandwidth estimated from measurements on 5-cm-high copper ribbon monopoles at 1 GHz, as the width of the antenna is varied from about 2 mm to 3 cm. We see that ribbons behave similarly to wire monopoles at the narrowest widths, but that the very wide ribbons provide less benefits in Q and bandwidth than would be expected by extrapolation from cylindrical wires.

Nevertheless, flat ribbon-like antennas provide a very considerable improvement in bandwidth relative to meandered wire antennas. A comparison of the measured equivalent circuit parameters for two representative structures of similar projected length is shown in Figure 7.30. A 20-fold reduction in aspect ratio (based on length rather than wire length) results in a roughly 3-fold improvement in Q and thus in bandwidth. (Note that voltage multiplication at the IC will also be reduced by the same factor.)

It is clear that broad thin structures provide significant benefits in operating bandwidth, despite one dimension (thickness) being much less than length or breadth, Of course, once the reactance of the antenna and matching structure are reduced to the smallest possible value, the bandwidth will still be limited by the reactance of the IC input. Very broad structures may also increase manufacturing cost, depending on the techniques used for antenna definition. Nevertheless, fat tag antennas have seen wide commercial deployment. Some typical broadband structures are shown in Figure 7.31. Aspect ratios are roughly 5:1 to 7:1 for these structures.

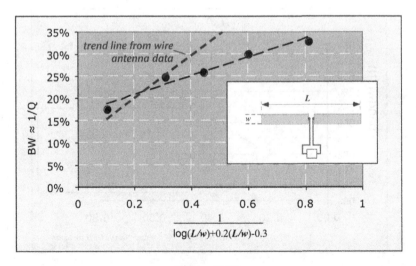

Figure 7.29
Bandwidth estimated from measured Q(1 GHz) for thin ribbon monopoles, 5 cm high, vs. rough estimate of inductance scaling. Trend line is extrapolated from data of Figure 7.28.

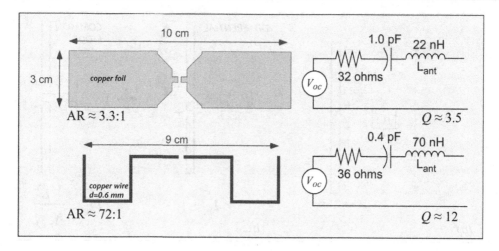

Figure 7.30

Comparison of measured equivalent circuit parameters for two antennas of similar length but varying aspect ratios.

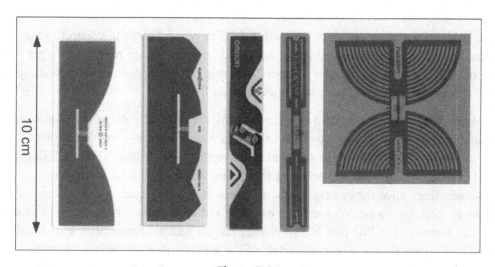

Figure 7.31

Commercial broadband structures, approximately to scale. *From left Alien Technology (2), Omron, Rafsec (2).*

7.3.5 Folding Up

Another dipole-like structure sometimes encountered in tag antennas is the *folded dipole*. A folded dipole is constructed from a conventional dipole by attaching a second length of wire to the ends of the first dipole, where the spacing between the two is small compared to a wavelength (Figure 7.32(a)). Folded wire dipoles are popular antennas whenever

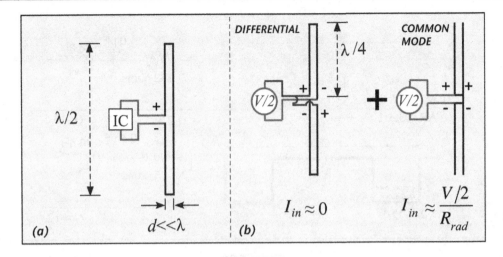

Figure 7.32
(a) Folded dipole; (b) Decomposition of voltage into differential (transmission-line) and common-mode components.

relatively high-impedance line is to be used instead of coaxial cable to connect to the antenna, because a resonant folded dipole has a radiation resistance four times larger than the equivalent conventional dipole – about 280 ohms is typical. A resonant folded dipole is a good match for 300-ohm twin-wire transmission line.

The folded dipole can be analyzed by decomposing the applied voltage into a part that is differential – that is, the two wires receive opposite voltages – and a part that is common, where the left and right segments receive the same voltage (Figure 7.32(b)). The differential voltages propagate along something that looks like a twin-wire transmission line; since the current on the left wire is always equal in magnitude and opposite in direction to that on the right wire, the radiation from these currents cancels and so the transmission-line part of the voltage has no radiation resistance associated with it. In the special case where each half of the antenna is a quarter-wavelength, the transmission line segments transform the short at their end into an open, so that the transmission line voltage draws no current at all.

Since the impedance of a dipole is not a very sensitive function of its cross-section, the pair of wires can be regarded as a regular dipole as far as the common-mode voltage is concerned. Thus the common-mode current at resonance is just what would have resulted from half the applied voltage flowing through the radiation resistance of the ordinary dipole. Half of this current flows on the left wire segments and half on the right; however, the actual input connection only sees the current flowing on the left segment. The net result is that for a given input voltage, we obtain ¼ of the current that we would have observed from a conventional dipole; this means that a resonant dipole has a radiation resistance four times larger than a standard dipole, roughly 260–280 ohms.

In a tag antenna application, a resonant folded dipole without matching is a somewhat better performer than a resonant single dipole. The open-circuit voltage for a given electric field is twice as large as that of a conventional dipole (and thus roughly equal to the produce of antenna length and incident electric field). At resonance, the antenna looks like a pure resistance of just less than 300 ohms, looking at a nearly-pure capacitive reactance $-j(350)$ ohms. Thus a bit more than half the open-circuit voltage appears across the IC load. The power transfer coefficient is better than a standard dipole, but not very good, mainly because of the poor match of the high-impedance antenna to the low-series-impedance load:

$$\tau = \frac{4R_{load}R_{rad}}{|Z_{ant} + Z_{load}|^2} \approx \frac{4(280)(18)}{|280 - j350|^2} \approx 10\% \tag{7.25}$$

And, of course, a resonant folded dipole is too big for most applications.

We can use the same approach to matching a shorter-than-resonant folded dipole we used before, with a shunt and series inductance realized as conductive lines. Folded dipoles are easier to match to the tag IC load than a conventional dipole, and allow operation with lower tag power (and thus higher values of the equivalent resistance of the IC). A folded dipole provides more voltage to the load for the same quality factor Q of the matching network.

Structures with additional wires can be fabricated. Two shorting wires produce a 3-fold-larger open-circuit voltage and a 9-fold increase in radiation resistance vs. a standard dipole, and three wires produce respectively 4× and 16× increases – that is, the radiation resistance is proportional to the square of the total number of wires, and the open-circuit voltage is linear in the number of wires. Since the resistance increases faster than the open-circuit voltage, adding wires doesn't increase the actual voltage delivered to an unmatched IC significantly, but again the larger value of the radiation resistance simplifies matching by moving the antenna load closer to the open-circuit part of the Smith chart.

The author has not encountered commercial tag designs that are classic unmatched folded dipoles. However, variants of the folded dipole have achieved commercial significance. The Alien Technology I-tag designs can be regarded as folded dipoles with shunt/series inductive matching (Figure 7.33). Dual-dipole designs used by Symbol Technologies (now part of Motorola) and Avery Dennison are somewhat similar to three-armed folded dipoles, though they are asymmetric and detailed analysis is somewhat involved.

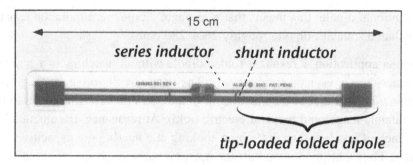

Figure 7.33
Commercial tag using folded-dipole-like antenna with matching structures.

7.3.6 Polarization

Electric fields point in a direction in space; if they are part of a traveling wave, that direction defines the ***polarization*** of the wave. Electric fields induce currents in conductors whose surfaces are parallel to them, and do very little when the conductive surface is orthogonal to the field. Tag antennas are necessarily made of thin conductors and usually will have no sensitivity to radiation perpendicular to their plane. In addition, most of the antennas we've examined are elongated in one direction, and will readily interact with fields polarized along that direction, while having little to do with fields along the short axis. This means that some combinations of polarization and orientation will enable tags to be read, and other combinations will result in tags that are invisible to the reader's illumination (Figure 7.34).

As a consequence, long thin tag antennas must be aligned with the polarization of the reader antenna in order to be read. This can be managed in several ways:

- **Orientation control**: a linearly-polarized reader antenna can be used if the tags are always oriented with their long axes along the direction of polarization, or if the reader antenna can be physically rotated to achieve this alignment.
- **Circular polarization**: a circularly-polarized reader antenna will interact with tags in any orientation in the plane perpendicular to the tag-reader line. Circularly-polarized radiation will also read tags that are aligned along the direction of propagation (and thus invisible when placed directly in front of the antenna), if the objects to which the tags are attached move across the read zone, so that they can be seen from varying angles.
- **Dual-dipole or polarization-diverse tag antennas:** an antenna structure with elongated features in orthogonal, or nearly-orthogonal, directions, will interact with electric fields in any direction in the plane of the antenna (Figure 7.35).

Orientation control is the simplest means of dealing with polarization, but it is often not practical. As we noted in chapter 5, antennas that are circularly-polarized in the center of

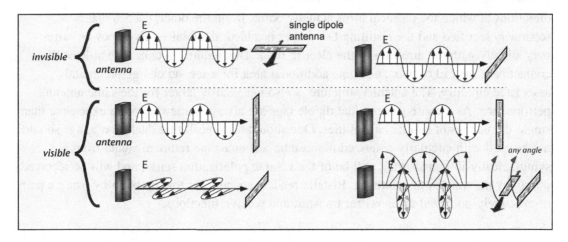

Figure 7.34
Dipole-like tags aligned to reader polarization are read; tags orthogonal to reader polarization are not.

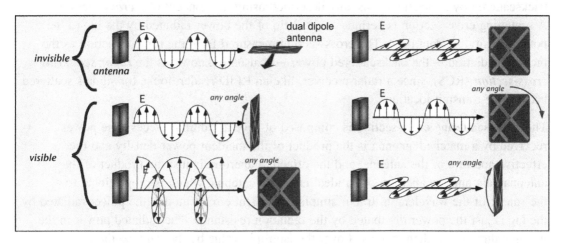

Figure 7.35
Dual-dipole tags are more tolerant of polarization variations.

their beam may be linear or nearly so in directions away from the beam direction; in addition, ellipticity often increases towards the edges of the operating frequency band. Dual-dipole antenna designs provide the most robust means of obtaining tags that can be read irrespective of the direction of polarization. Note, however, that to make use of such a capability requires that the tag IC have two independently-rectified inputs. If one simply combines the received signals from two different antennas and rectifies the sum, the result is a polarization-sensitive antenna, albeit the optimal direction may be in between the directions in which the physical arms actually point. If, on the other, each signal is separately rectified and the resulting DC power is added, the total received power varies only slightly with the direction of the electric field. This requires a chip with at least three contact pads instead of two, and some additional area for a second charge pump and associated circuitry. The antenna structure is also necessarily larger for the same antenna performance. As a consequence, dual dipole tags are always somewhat more expensive than single-dipole tags of similar capabilities. One should also recall that dual-dipole tags should not be used with circularly-polarized monostatic antennas: the returned signal from a symmetrically-illuminated tag will be of the reverse polarization sense, and will be received poorly by the transmitting antenna. Bistatic readers account for this problem by using a pair of oppositely-polarized antennas for transmit and receive functions.

7.3.7 Radar Scattering Cross-Section

As we discussed in some detail in chapter 5, tags communicate with a reader by varying the amount of power they scatter back to the reader antenna. The amount of power backscattered by an antenna is usually described using the concept of a ***cross-section***. A scattering cross-section is defined as the ratio of the power radiated by the tag to the power density incident on it. The cross-section measured from the same direction as the incident radiation − the backscattered power − is usually known as the ***radar scattering cross-section*** (RCS), since a radar receiver, like an RFID reader, looks for signals scattered back to the transmit location.

The radar scattering cross-section is composed of several familiar pieces. The power received by a matched antenna is the product of the incident power density and the effective aperture of the antenna, and the effective aperture is just the product of the antenna gain and the aperture of an ideal isotropic antenna, which is proportional to the square of the wavelength. In our simple equivalent circuit model, the power radiated by the tag is just the power dissipated by the radiation resistance. The radiated power in the incident direction is then increased over the isotropic value by the gain, so the gain multiplies the result twice (going in and going out). The result is:

$$A_{sc} = \frac{\lambda^2}{4\pi} G^2 \left| \frac{2R_{rad}}{Z_{load} + Z_{ant}} \right|^2 \tag{7.26}$$

Thus the radar scattering cross-section is determined by the antenna gain, and the ratio of the radiation resistance to the total load impedance. For a resonant dipole with a matched load, the radar scattering cross-section is about 220 cm^2. When instead a short is presented to the antenna, the current doubles relative to the matched load so the scattered power increases fourfold: the RCS is about 880 cm^2. Note that these cross-sections are greatly in excess of the physical cross-section of the wire (typically around 1 or 2 square centimeters for a thin wire dipole). The identification of the cross-section with the physical extent of the antenna is reasonable for antennas that are several wavelengths in extent, like a parabolic dish, but is fundamentally incorrect for wire-like antennas. In our simplified equivalent circuit, no current flows when an open circuit is present at the load, the radar cross-section falls to 0. In reality, a much smaller current flows along each segment of the antenna, limited by the high capacitive impedance of the segments, and the radar cross-section is finite but small.

The scattering cross-section of a matched antenna depends only on the wavelength and the gain, and for the small antennas used in tags, gain is at most about that of a dipole (2.2 dBi). So in principle, the scattering cross-section of a tag could be nearly independent of the size. However, very small antennas have high reactive impedances and small radiation resistances − that is, high Q values − so the bandwidth over which large RCS is obtained will be reduced relative to a larger antenna. Real antennas will also face reduced radiative efficiency due to ohmic losses, which we have neglected for simplicity in most cases. Ohmic losses are of course particularly significant for antennas fabricated using printed conductive ink, which is less conductive than pure metal layers.

We examined the problem of modulating the backscattered power from the point of view of the IC in chapter 5. Reactive modulation (changing the load reactance rather than the load impedance) produces the best compromise between power delivered to the load and backscatter modulation. The backscattered signal in this case is roughly proportional to the unmodulated signal. Thus, the best reverse-link performance ought to be obtained from tags with large radar cross-sections, and the most robust reverse-link performance when the RCS is large over a broad band. Of course, as we discussed in chapter 3, tags will often be forward-link-limited. In general a tag with a lower radar cross-section is receiving less power, though the absorbed power falls less rapidly than the scattered power as the load resistance increases; tags with small RCS may become reverse-link limited.

Some measured data for commercial tags mounted on paper is summarized in Figure 7.36. The general trends are as we might expect: the largest values are intermediate between those of a matched load and a shorted dipole. Larger tags have larger scattering cross-sections over a broader frequency band. The compact Rafsec and Alien tags ((d) and (e) in the figure) have small RCS values. The very compact, heavily meandered (f) has a measured RCS of only a few square cm; it seems likely that this tag has reduced radiative efficiency due to long thin lines, as well as high Q, and is only well-matched on certain substrates.

We are now (finally!) equipped to return to the question of backscatter modulation efficiency in the context of a realistically matched antenna. Let us consider a matched equipped with a modulation transistor that can short out the IC load to modulate the backscattered signal (at the cost of loss of power to the IC during this time). How is the

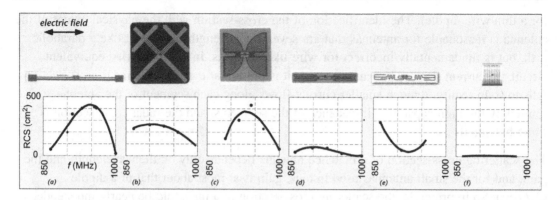

Figure 7.36
Measured unmodulated radar scattering cross-section for several commercial tag designs.

radar scattering cross-section affected? For some plausible parameter values, the two possible states are summarized in Figure 7.37. The radar cross-section of the matched antenna (case (a) of the figure) is given by equation (7.26) with $Z_{ant} + Z_{load} = 2R_{rad}$: that is,

$$A_{sc} = \frac{\lambda^2}{4\pi} G^2 \tag{7.27}$$

The difference between the matched and modulated states is:

$$\Delta A_{mod} = \frac{\lambda^2}{4\pi} G^2 \left| 50\left(\frac{1}{50} - \frac{1}{25 - j35} \right) \right|^2 \approx \frac{\lambda^2}{4\pi} G^2 \cdot 1.02 \tag{7.28}$$

So the modulation amplitude is about equal to the scattered power in the matched state — a similar result to that we obtained without the matching network — though in this case the modulation is basically phase-shift keying rather than amplitude-shift keying. Of course, the IC power falls by 3 dB since the IC is shorted half the time.

What about the case where the reactance of the load is modulated? Let's assume that we subtract $10j$ ohms to the IC load above in the unmodulated state (inducing a slight mismatch) and subtract $10j$ ohms from the load above in the modulated state (a total of $20j$ change). The math is a bit ugly here so let's just quote the results: the backscattered modulated signal power is about 35% of the available power, while the power delivered to the load is decreased by only about 8%.

The complex current flowing through the antenna radiation resistance is shown in Figure 7.38. Note here that the current flowing in the unmodulated case is the distance from the point (0,0) (not shown in the figure for clarity). Thus it's easy to see that the backscattered current between the shorted and matched states (the 915 MHz points in (a)) is about the same magnitude but differs in phase: ASK at the load has been converted to PSK

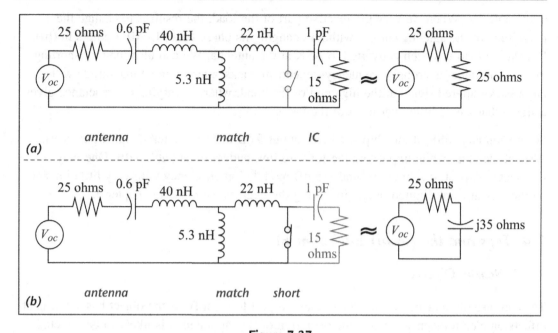

Figure 7.37
Example of load modulation with antenna matching. a) IC matched to antenna. b) IC shorted
out by modulation transistor.

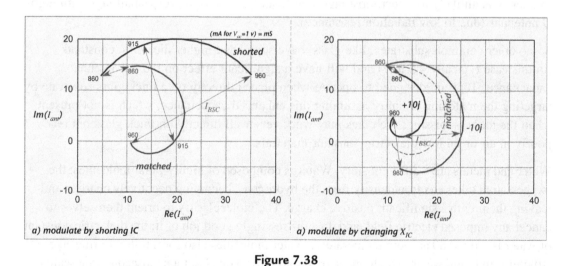

Figure 7.38
Antenna current at Voc = 1 V (= conductances) vs. modulation state, from 860 to 960 MHz.
a) Modulate by shorting IC load; b) modulatte by varying reactive part of IC load.

at the antenna. When we vary the reactive part of the load, the result is to change the magnitude of the antenna current without changing its phase (the 915 MHz points in (b)): PSK at the load has been converted to ASK at the antenna. We can also see that shorting the load produces a strongly frequency-dependent backscattered signal magnitude with a large backscattered signal at the high end of the band, whereas varying the reactance gives a signal that varies little over most of the band.

So in summary, PSK at the chip sacrifices about 5 dB of backscattered signal power vs. ASK at the chip, while improving the forward link budget by about 2.5 dB. One may reasonably expect to achieve around −5 dB modulation efficiency with very little impact on the forward-link-limited range, justifying the guesses we made in chapter 3.

7.4 Tags and the (local) Environment

7.4.1 Nearby Objects

Tags are usually attached to the object they purport to identify. If the object has significant effects on electromagnetic fields, the operation of the tag antenna is likely to be affected.

In supply-chain applications, tags will typically be embedded in a paper or plastic label and placed on a cardboard box. Boxes vary from about 3 mm to 1 cm thick, and are composed of thin sheets of paper and adhesive. Dry paper has a relative dielectric constant of around 3; a thin sheet of paper close to the tag will slightly increase the capacitance of the antenna and have little effect on the inductance. The consequent shift of a few percent in resonant frequency is unlikely to affect most tags significantly, except for very small tags with high-Q antennas (due to low radiation resistance).

Some other common substrates, like glass, have somewhat higher dielectric constants (in this case typically 4.5 to 7) and will have a significant effect on the tag antenna capacitance. Tags are designed to operate when placed directly on a thick glass substrate by targeting the matching circuitry assuming this enhanced capacitance, which is not present when the glass is missing; such tags may work very well directly on thick glass but read poorly in air or on low-dielectric-constant materials.

Water and metals are a different story. Water is composed of highly-polar molecules: the oxygen atom takes electron density from the hydrogens, becoming negatively charged and leaving them with a significant positive charge. The molecules try to orient themselves to cancel any imposed electric field, and do a distressingly good job of it: the dielectric constant of water is around 80 at room temperature. Water molecules also form transient ring-like structures in liquid water, which break apart and reform on about a nanosecond timescale. As a consequence, water is strong absorber as well, with an absorption coefficient of about 5 nepers/meter or about 40 dB/meter, at 900 MHz. Absorption increases rapidly with frequency, and decreases with increasing temperature. Thus water is a strong reflector and absorber of microwave radiation, and substances that contain lots of water (like just about everything we eat, drink, or clean with) are also active at microwave frequencies.

(Just for completeness, we take a moment to debunk a persistent misunderstanding: despite the fact that microwave ovens do work by heating liquid water, liquid water does not have any resonant behavior around 2.45 GHz – that's just an available frequency for industrial use. The molecules in the liquid are too closely coupled to display any sharp resonances at all. Isolated water molecules in air have fundamental rotational frequencies in the hundreds of GHz; there is a transition between two excited rotational states at around 22 GHz.)

Because of the high dielectric constant of water, the electric field inside the water is greatly reduced. The value of the electric field just outside the water and just inside of it must be continuous, so the only way to manage both requirements is to have the electric field also be small just outside the water. We can't arrange the field to be small if there is only an incident wave, but adding a reflected wave does the trick if the reflected electric field is in the opposite direction and nearly as large as the incident wave (Figure 7.39).

As we move away from the water interface, the relative phase of the transmitted and reflected waves changes, because they are moving in opposite directions: *standing waves* result (Figure 7.40). A quarter of a wavelength from the surface, the two fields point in the same direction instead of opposing one another, and they add to create a larger field than was present with no water surface. At a half-wavelength from the surface another null occurs.

A wire-like tag antenna will experience a reduced electric field as it nears an aqueous surface. In equivalent-circuit terms, the open-circuit voltage will fall when the distance to

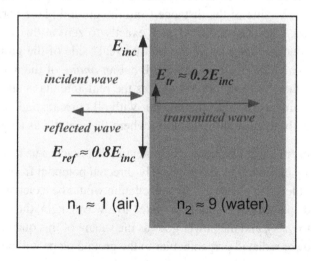

Figure 7.39
Relationships among the incident, transmitted, and reflected waves at an air-water interface.

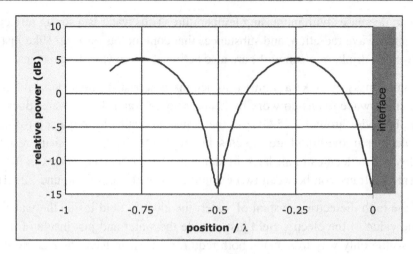

Figure 7.40
A standing wave forms near a reflecting interface, causing variations in field strength and received power with position.

the interface is less than about 1/8 wavelength (4 cm or so). Thus the received power delivered to a fixed load also falls as the tag nears the interface.

Similar but more drastic effects occur when a tag is close to a metal interface. In this case the electric field within the metal must go to exactly zero within a skin depth or two of the surface, so the reflected wave is of the same magnitude as the incident wave. For all common metals, the skin depth is a few microns to perhaps ten microns at 900 MHz, so the metal surface can be regarded as a perfect reflector for practical purposes. The electric field will be proportional to the sine of the distance from the ground plane normalized to a half-wavelength, $\sin(4\pi\, h/\lambda)$, Because the field goes exactly to zero at this very sharp interface, we can view the problem (as long as we are on the "real" side of the ground plane) as equivalent to one in which the metal is removed, but an *image* of the antenna is placed behind the former surface by the same distance as the real antenna is above the surface. The image is exactly the same as the real antenna but with all currents and charges reversed, so that the fields go exactly to zero at the location where the metal was (Figure 7.41).

When we view this antenna from far away, the potential that reaches us from the current on the real antenna is partially cancelled by the oppositely-directed potential from the image. If the two antennas were in the same place ($h = 0$) cancellation would be exact and there would be no radiation. When the displacement is small compared to a wavelength, the net radiated field is proportional to (h/λ), and the power goes as the square of this quantity. For larger displacements, the power radiated perpendicular to the ground plane will be proportional to $\sin^2(4\pi\, h/\lambda)$, and thus very roughly we'd expect the radiation resistance to scale with the square of the sine (ignoring radiation pattern changes). Thus the open-circuit voltage

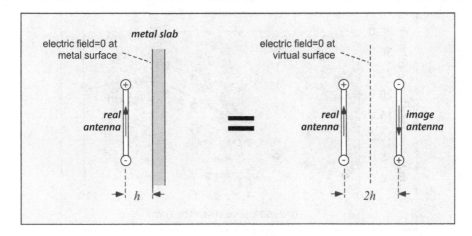

Figure 7.41
When viewed from the real-antenna side, an antenna near a metal surface is equivalent to an antenna and its image with no metal.

(proportional to electric field) and the radiation resistance both should fall rather rapidly to 0 as the tag antenna approaches within about 1/8 of a wavelength (<4 cm) of the metal surface.

Capacitance and inductance of the antenna are also affected by the presence of the image dipole, though the exact behavior is rather dependent on the antenna shape. The inductance and capacitance of a wire scale roughly as the logarithm of the ratio of length to radius, $\ln(L/a)$. The relative effect of a neighboring wire is roughly proportional to the log of the ratio of length to spacing, here $\ln(L/2h)$. Thus as long as $\ln(h/a)$ is large compared to 1, self-inductance and capacitance dominate, but when the separation from the ground plane is only a few times larger than the width or radius of the wires, the inductance falls and the capacitance increases. A non-resonant antenna with a matching network designed for specific values of reactance will deliver less power to the IC as these component values change, even if the resonant frequency of the antenna is relatively unaffected. This is known as *detuning* of the antenna. Detuning is of particular importance when a high-Q, narrow-bandwidth antenna is used.

The measured radiation resistance of some simple copper-ribbon antennas near a ground plane is shown in Figure 7.42. It is generally clear that radiation resistance falls rapidly as the antenna nears the ground plane, though there is some scatter near zero, due to the difficulty of accurately locating the ground plane and accurately measuring very small radiation resistance values.

The inductance and capacitance for the same monopoles near a ground plane are shown in Figure 7.43 and Figure 7.44. The inductance fluctuates rather gradually with distance, with nothing very drastic happening even at close spacings, as one might expect from the fundamentally logarithmic scaling of inductance. The capacitance is roughly constant until

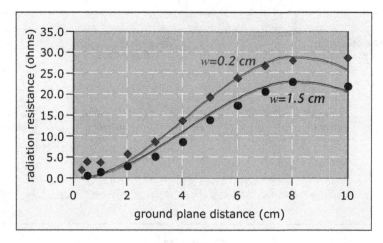

Figure 7.42
Measured radiation resistance at 1 GHz for a 5-cm-high monopole near a ground plane; note equivalent dipole resistance is twice the value shown here. The lines are a sinusoidal model fit to the peak value of the data.

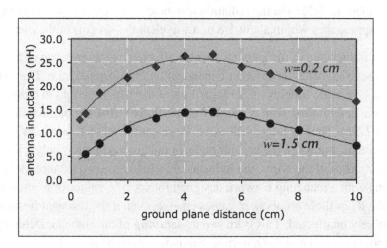

Figure 7.43
Series inductance for copper ribbon monopoles near a ground plane; note equivalent dipole inductance is twice the value shown here. The lines are cubic polynomial fits to the data.

the ground-plane separation approaches about 2–3 times the width of the antenna, and then increases rapidly due to the parallel-plate capacitance between the monopole and its image.

Using the data above, we can infer behavior of a tag matched with the approach discussed in section 7.3.2 above. For example, we can match the 1.5-cm dipole antenna with a shunt inductance of about 9 nH and a series inductance of 38 nH to a plausible IC load (15-j300 ohms). We can then modify the equivalent circuit parameters of the antenna to reflect the

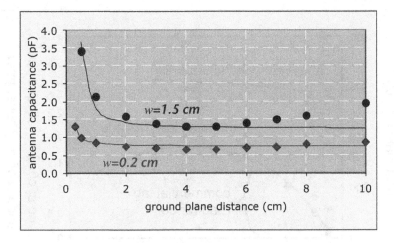

Figure 7.44

Series capacitance for copper ribbon monopoles near a ground plane; note equivalent dipole capacitance is half the value shown here. The lines are fits to a simple end-charge model.

presence of the ground plane, using the data in Figure 7.42 through Figure 7.44, and note the behavior of the voltage delivered to the load. (Since the matching inductors are realized as lines on the antenna, one also ought to allow them to vary, but as we saw the relative inductance variations are modest for narrow lines so we will ignore this effect for simplicity.) The results of such a simulation exercise are summarized in Figure 7.45. The frequency at which the maximum voltage is delivered to the load, denoted the best match frequency, decreases rather slowly until the antenna approaches within 1 cm of the ground plane. Since the bandwidth is more than 50 MHz, the slow drift has essentially no effect on the voltage on the load. Nevertheless, the load voltage falls as the antenna approaches the ground plane, due mainly to the decrease in the open-circuit voltage from the reduced electric field. Some experimental data is shown for tags with antennas that are about 9×2.5 cm (circles) and 9×2 cm (diamonds); the measured read range qualitatively agrees with that predicted from the simple circuit model, although the smaller tag does particularly well at small tag-metal spacings.

Is it possible to construct a tag antenna that will work within a few millimeters of a metal? Fortunately, the answer is yes, and indeed we have already encountered one structure: the patch antenna (Figure 7.46). The patch antenna works by performing an impedance transformation. In the center of the patch, large currents flow; the radiation from these currents is almost perfectly cancelled by the image patch, so for a given current density the radiated electric field is relatively small. A large current and small radiation resistance combine to create an acceptable radiated field. At the edge of the patch, the current falls nearly to zero and a large voltage is maintained between the patch and the ground plane: the small impedance has been transformed to a much larger one, much more appropriate for

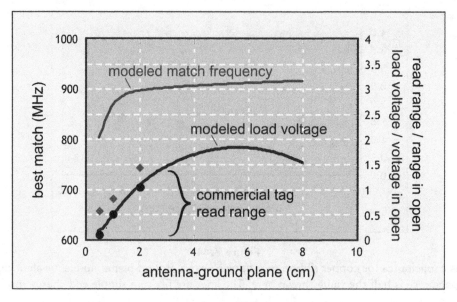

Figure 7.45
Modeled frequency for best matching, and relative voltage to the IC load at 915 MHz, as a 1.5-cm ribbon dipole approaches a ground plane. Also shown are measured relative read ranges for two commercial tags with ribbon-like antennas.

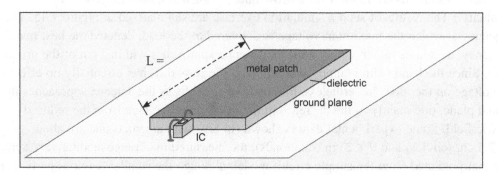

Figure 7.46
Patch antenna configured to drive an IC. Matching structures may also be incorporated using microstrip.

the relatively high-impedance tag IC input. Adding microstrip tuning structures, or recessing the feed point, can be used to match the patch impedance to the tag.

However, recall that a typical reader antenna with robust bandwidth for operation in the 900-MHz region used on the order of a 1 cm patch-ground spacing h. This is hardly practical for a tag! If we make a tag patch antenna that is of a thickness acceptable for most applications — say $h = 0.5$ mm — the bandwidth of the antenna will be only a few MHz, making it unacceptable for use in the US and other countries with wide operating bands available, and unlikely to be robust to small variations in spacing and dimensions even when only a few MHz of bandwidth is available for a reader. Very thin patch antennas are not practical, so patch-antenna-based tags are only a good solution when a relatively thick tag is acceptable.

A single-ended patch antenna like the one shown in Figure 7.46 also requires a connection from the metal patch level to the ground plane, so that we can connect the IC: this can be a *via hole* or a wraparound connection on the side of the dielectric. Both are relatively expensive to fabricate. We can avoid the need for a via hole by instead fabricating a patch dipole, where the IC sits between two half-wave patches, but at the cost of a larger structure.

To make a smaller tag structure, a quarter-wave patch antenna can be employed. The length of the antenna is thus reduced by about a factor of 2, a considerable practical benefit. For example, for 900 MHz operation, a half-wave patch with air dielectric is around 15–16 cm long, not including the IC and associated structures, whereas a quarter-wave patch is only 8 cm long. The tradeoff is that is it necessary to provide a conductive short at the end of the patch: that is, metallization needs to cover the side of the dielectric, increasing manufacturing cost. Plastics with higher dielectric constants can shrink the antenna, but increasing dielectric constant also reduces bandwidth.

To avoid the need for a via hole, while maintaining a reasonably small structure, a loop balun can be employed to invert the patch output. Such a structure is shown in Figure 7.47(a). The loop is simply a printed conductive line, which forms a microstrip transmission line with the ground plane under the tag. If the distance around the loop is about 1/2 of a wavelength in the transmission line, the signal traveling around the loop to the far side of the IC will arrive 180° out of phase with the signal that travels along the short distance to the near side of the IC (Figure 7.47(b)). A virtual ground is created at the IC between the two input pads, so no via is needed. By adjusting the length of the transmission line legs, the phase of the signals can be adjusted to improve the match to the reactive IC load, a task best managed with the aid of the Smith Chart. Instead of a loop balun, an open-circuited length of line about 1/4 wavelength can be used to ground one pad of the IC, since the open circuit at the end of the line is transformed to a short at the IC (see Appendix 3). The quarter-wave dipole can be made even more compact by folding it in half, so that the shortened end is beneath the IC and balun or matching structure (Figure 7.47(c)).

An important advantage of any tag antenna that has a ground plane is improved antenna gain. A patch antenna with a large ground plane typically has a directive gain of 7–9 dBi. Recall from chapter 3 that the antenna's directive gain adds to the forward and reverse link budgets. As long as the tag antenna is facing the reader, this directive gain doubles the read range; an efficient antenna and IC will provide read range in the tens of meters (at least in the frequency range in which the antenna is well-tuned), and an inefficient antenna can still

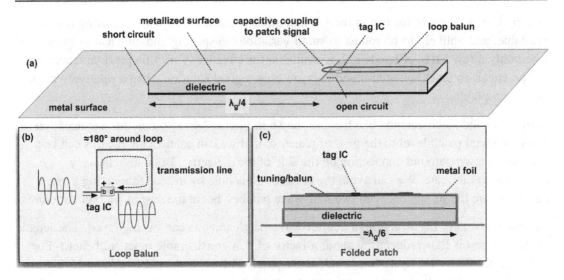

Figure 7.47

(a) Quarter-wave patch antenna with loop balun. The length of the antenna section is 1/4 of the wavelength of the waveguide formed by the metal and dielectric. (b) Operation of a loop balun. (c) Folded quarter-wave patch.

provide acceptable read range even though much of the incident energy is wasted. The tradeoff is that the tag will not be seen if it is not oriented face-on to the reader, but a tag shadowed by a large metal surface would be invisible from the back in any case.

It is also possible to construct a conventional matching network for a tag near a ground plane. For example, we can use the same shunt/ series inductive matching we have used before with small values of radiation resistance, though of course the effects of the image antenna must be considered in estimating the inductance of the short lines used for matching. As the tag nears a metal surface, the radiation resistance falls rapidly, whereas the reactances change rather slowly. Thus, the Q of the antenna goes up, and the bandwidth goes down. The loss in efficiency in the target band is partly compensated by the increase in directive gain (in the good direction).

An example design, due to Deavours and colleagues, is depicted in Figure 7.48. A conventional tip-loaded T-match is employed, but wide lines are used in order to reduce the Q of the antenna (see section 7.3.4 above). The equivalent circuit in air (b) has notably less inductance as a consequence. The transfer coefficient in air is quite good: about 82% in the US ISM band, 902−928 MHz.

The antenna is typically mounted on a thin foam plastic spacer for use on a metal object; this is sometimes known as a **_Foam Attached_** _Tag_ or FAT. When the antenna is

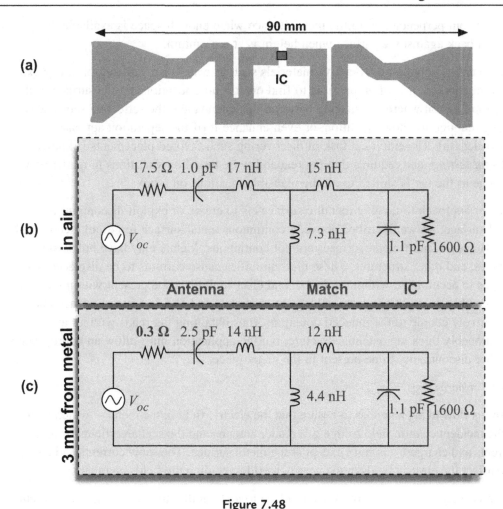

Figure 7.48

(a) Tip-loaded T-match optimized for air or metal use. (b) Equivalent circuit in air. (c) Equivalent circuit within 3 mm of ground plane.

placed close to a ground plane (c), the radiation resistance falls dramatically: from 17.5 to 0.3 ohms. The inductances change modestly, and the antenna's capacitance increases. The transformation constant β (see section 7.3.2 above) falls slightly: it is about 0.33 in air and 0.27 near metal. The transformed IC resistance of about 115 ohms is not a good match, so the power transfer coefficient falls to about 40% in the center of the band, and 15% at the band edges. The degraded match is partially compensated by the increase in antenna gain; the net result is a 40% reduction in read range, acceptable in many applications.

Large metal pieces can also be tagged by hanging the tags by a wire or string so that they are spaced away from the metal. Tags can be mounted on spring-loaded supports so that they pop out perpendicular to the metal surface when enough space is available, but can be pressed back against the surface when height is at a premium.

When cartons containing RF-active materials such as water or metallized-plastic packaging are to be marked, it is often possible to find one or more locations on the surface of the cardboard carton where the spacing between the antenna and the reflective substances is large enough to allow operation, or even enhance it (if the separation approaches ¼ wavelength). The empirical task of discovering such favored placements is known as *hot-spot testing*, and requires that tag readability be tested by variations in reader power or range as the tag is shifted to all the plausible locations on a box.

Another approach to tagging metallic surfaces is to create or exploit discontinuities in the metal surface. As we described above, a continuous metal surface forces all electric fields to 0. However, most metal surfaces are not continuous. Metals may have holes, slots, recesses, and other structures. These discontinuities cause currents to be displaced and charges to accumulate within the metal, and electric fields to be present within the features. It is possible to attach conventional tags to span slots or holes in a metal surface and capacitively couple to the induced potentials, thus obtaining tag reads without an objectionably thick tag antenna structure, but the application must allow an appropriate slot or other discontinuity to be present in the metal piece.

7.4.1.1 Nearby Tags

Currents flow in tag antennas to ensure that the electric field in the metal — which is the sum of the incident electric field from e.g. a reader antenna and the scattered fields from the currents and charges — is zero, except at the metal surface. The same currents and charges also affect the electric fields in the nearby world, including the fields on other tag antennas.

When two tag antennas are placed very close to and parallel to one another, the structure becomes very similar to the folded dipole antenna we studied in section 7.3.5 above, with one important distinction: the ends are generally not connected together. The structure looks like a transmission line with an open-circuited end. If the antennas are a half-wave long or nearly so, the open-circuit load at the end of the transmission line is transformed into a short-circuit at the center of the line by the quarter-wave lenth of transmission line formed by the two wires. This is a big problem: the transmission line (composed of the two antennas) short-circuits any voltage that either individual dipole antenna may develop due to an incident field. The result, that overlapped tag antennas cannot be read, is often somewhat erroneously explained by assuming that the tags compete for the incident power.

Even when tags are seemingly distant from one another, interaction effects can be substantial. An important and particularly simple case to analyze is the case of a plane of tags illuminated in the plane. In this simple geometry, tags effectively cast shadows on the tags behind them. The shadows are not sharp as optical shadows are, but are broad and diffuse, extending laterally and growing deeper as more tags are added. Let us examine an example geometry in which a reader signal is sent to a linearly-polarized receiving antenna (simulating what signal a tag would receive), with varying numbers and types of commercial tags placed in the plane between the transmitting (reader) antenna and the emulated tag (Figure 7.49). Tags are spaced by 5 cm in depth and 20 cm laterally, so that the distance from the receiving antenna to the more distant tags, when many tags are present, is quite large compared to the size of a tag, and the interaction can no longer be regarded as purely local.

Some experimental results for this configuration are shown in Figure 7.50. When no tags are present, the received signal is about −8 dBm (somewhat less than predicted from the Friis equation, due to several effects including antenna mismatch, transmit modulation, and cable loss). A column of six Alien Technology 'Squiggle' tags placed in front of the receiving antenna reduces the received power by 5 dB; a similar column of 'I' tags reduces received power by 10 dB (a factor of 10!). (These two tags are respectively the second right

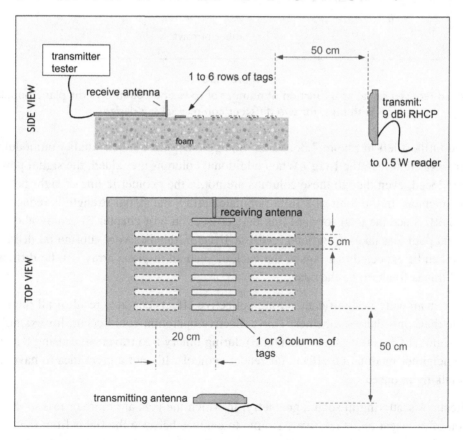

Figure 7.49
Experimental setup for characterizing effects of in-plane tags on received signal.

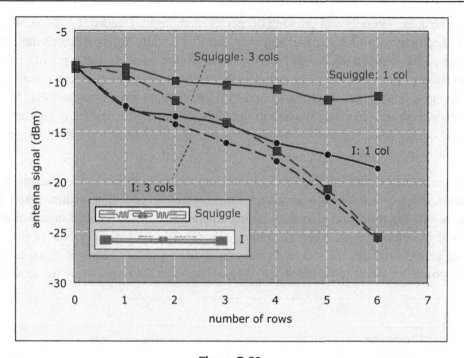

Figure 7.50

Measured received signal as a function of number of rows and columns in the plane populated with tags, for two different commercial tag designs.

right and farthest left in Figure 7.36, so the Squiggle tag has a much smaller unmodulated radar cross-section than the I-tag.) When additional columns are added, the signal power is further reduced, even though these columns are not in the geometric line of sight between the two antennas. In the limit of a fully-populated array, the signal strength is reduced by about 18 dB. Since the total forward link budget (section 6 of chapter 3) is only about 40 to 45 dB, and path loss takes up around 25 dB at 1 meter, this is a very substantial decrease in signal. It can be expected that tags at the back of such an in-plane array will be difficult to read and this is found to be the case.

When tag shadowing is significant, strong collective effects can also result if all the tags modulate their impedances together. This collective operation could occur, for example, with EPCglobal Class 0 tags (see chapter 8) during binary tree traversal, causing tags to mistake neighbor modulation effects for reader symbols. It's not a great idea to have a lot of tags talking at once.

The effects of scattering in such a geometry, in which the tags are more or less in the direction of propagation, is relatively simple to analyze because the fields from the transmitting antenna and from the tag antennas arrive more or less in phase. A tag near the reader antenna receives the reader signal just after it is launched, but the scattered wave from that tag has a long way to travel to reach the receiving antenna (or the tag in the back row). A tag in the back row receives the reader signal after a delay of several nanoseconds, but its scattered signal reaches the receiving antenna almost immediately (Figure 7.51). The total time delay (transmitter ⇒ scattering tag ⇒ receiver) is approximately constant for all the scattering tags, so the scattered fields from all the tags can simply be added up, without much regard for the exact location of the tags doing the scattering.

As a consequence, we can construct a very simple model that captures the essential aspects of tag shadowing behavior. We assume n identical tags of length 1 are separated by a distance d. The current on each tag is the ratio of the local voltage (electric field multiplied by half the tag length) to the impedance of the tag:

$$I_n = \frac{E_n \ell_n}{2Z_n} \quad \text{where } \ell \text{ is the element length} \tag{7.29}$$

We simplify the situation further by assuming that the current is constant along each tag, and that only the electric field due to the vector potential $\mathbf{A}$ is significant; that is, we ignore any capacitive coupling between the tags. Finally, we use a key approximation due to Hill and Cha: we assume that for the purpose of calculating the current in the last tag of the array, the current on all the other tags has the same magnitude, and that the phase differs by exactly the difference in the phase of the incident wave. In forward scattering, this means that all the vector potentials simply add in phase. The electric field is thus:

$$E_n = E_{inc} - j\frac{\omega\mu_0\ell I_{ant}}{4\pi}\sum_n \frac{1}{nd}$$

$$I_{ant} = \frac{E_n \ell_n}{2Z_n} = \frac{E_{inc}\ell - j\dfrac{\omega\mu_0\ell_n^{\,2}I_{ant}}{4\pi}\displaystyle\sum_n \frac{1}{nd}}{2Z_n} \tag{7.30}$$

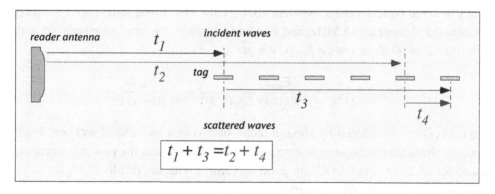

Figure 7.51
Array of tags illuminated in endfire, with example time delays for scattering from first and penultimate tags to the final tag in the array.

We can solve for the current:

$$2I_{ant}Z_n = E_{inc}\ell - j\frac{\omega\mu_0\ell_n^2 I_{ant}}{4\pi}\sum_n \frac{1}{nd}$$

$$I_{ant}\left(2Z_n + j\frac{\omega\mu_0\ell I_n^2}{4\pi}\sum_n \frac{1}{nd}\right) = E_{inc}\ell$$

$$I_{ant} = \frac{E_{inc}\ell}{\left(2Z_n + j\frac{\omega\mu_0\ell I_n^2}{4\pi}\sum_n \frac{1}{nd}\right)}$$

(7.31)

We can rewrite this expression in a more physically appealing form using $\omega = ck$:

$$I_{ant} = \frac{E_{inc}\ell}{\left(2Z_n + j\frac{\mu_0 c}{4\pi}\frac{2\pi}{\lambda}\ell_n^2\sum_n \frac{1}{nd}\right)}$$

$$= \frac{E_{inc}\ell}{\left(2Z_n + j\frac{Z_0}{2}\frac{\ell}{\lambda}\frac{\ell}{d}\sum_n \frac{1}{n}\right)}$$

(7.32)

where Z_o is the impedance of free space, 377 ohms. The current flowing in the last tag in the array is the ratio of the incident voltage to a modified impedance composed of the impedance of the tag and a mutual inductance from the other tags. Since the impedance of the tag antenna Z_n is not necessarily purely real but could be capacitive or inductive (i.e. it could have a positive or negative imaginary part), the total impedance could be larger or smaller than the tag impedance, so that either shadowing or focusing could result from such an endfire array.

Let's look at some typical values. Assume six tags are 14 cm long and spaced 5 cm apart, and illuminated at around 915 MHz, and that an isolated tag is tuned to resonance with an input resistance of 60 ohms (twice R_{rad}). We get:

$$I_{ant} = \frac{E_{inc}\ell}{(120 + j190(0.4)(2.8)(2.3))} = \frac{E_{inc}\ell}{(120 + j510)}$$

(7.33)

The induced current is reduced by about a factor of 5 versus an isolated antenna: shadowing has resulted. Since this is the current into a fixed tag impedance, the power is decreased by about a factor of 25 or about 14 dB, in good agreement with the results of Figure 7.50 given the simplicity of the approximations.

The dependence of the shadow depth on the array also depends on the impedance of the tag. If the individual tag impedance is large and capacitive (as is the case for an antenna that is significantly shorter than resonance and not fully matched), then the mutual impedance of the other tags in the array, which is inductive, will subtract from the local impedance and the array will actually increase the current instead of shadowing it. This is more or less how a Yagi-Uda array works. On the other hand, an inductive tag impedance will add to the mutual inductance of the nearby coupled tags, and shadowing will occur even for small mutual inductance (one or two rows of tags).

The situation is substantially more complicated when the paths of the scattered waves are not in the same direction as the incident waves. An example of a geometry where phase must be accounted for is shown in Figure 7.52. When arrays of tags are placed in planes perpendicular to the direction of propagation, the path length from the reader antenna to a receiving tag is no longer similar to the path length from the reader antenna to a scattering tag and back to the receiving tag. For example, waves scattered from the back plane of tags travel farther to get to the front plane than a wave transmitted from the reader antenna, the difference being roughly twice the interplane gap (once for the reader signal to reach the back plane, and again to return to the front plane). In consequence, as the gap is changed, the scattered signals from the back plane could arrive in phase with signals from the reader, increasing the total signal and making it easier to read the tags in the front, or they could arrive out of phase, making it harder to read the tags in the front plane. For those tags far

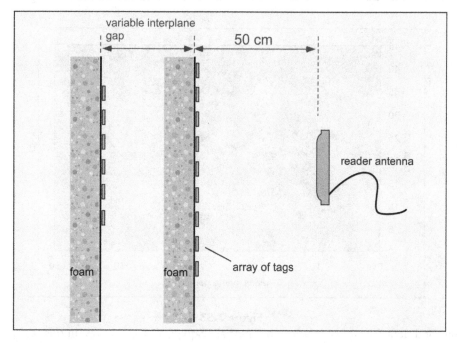

Figure 7.52
Tags arranged in planes perpendicular to the direction of propagation.

from the center of the array, the lateral distance must also be taken into account. Therefore in this geometry we would not expect a simple monotonic change in readability, but an oscillatory behavior as scattered waves move into and out of phase with the incident waves.

Some experimental data in this configuration is depicted in Figure 7.53. This particular experiment employed commercial tags (Alien 'Squiggle' class 1 Generation 1) and a commercial reader (WJ Communications MPR5000) running a class 1 anti-collision algorithm. It is readily apparent that the number of tags read in both planes varies significantly depending on the interplane gap, with a periodicity of about ½ of a wavelength. This is the same periodicity we discussed in connection with reflection from a high-dielectric or conducting surface (Figure 7.40). Configurations of this type can arise when stacks of identical cartons containing RF-transparent materials, identically tagged, are constructed for transport on a pallet. In this case, the read performance in a given geometry will be a sensitive function of not only the tag orientation and tag type, but also the carton size.

What measures can be taken to mitigate tag scattering effects? The simplest approach is to employ tags with reduced scattering cross-sections. For example, the variation in readability as a function of interplane gap seen above is exacerbated when I-tags, with a higher scattering cross-section, are substituted for Squiggle tags; Squiggle tags are better choices for an array of tags. A low scattering cross-section will result if the effective load resistance greatly exceeds the radiation resistance. Since the scattered power decreases roughly as the

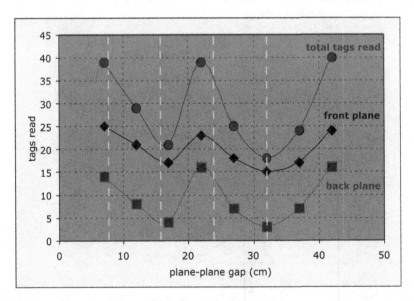

Figure 7.53
Experimental data for tag reads using commercial ('Squiggle' class 1) tags. In this test 27 tags were placed in the front plane and 18 in the back plane, and a 1/2-watt commercial reader was used.

square of the load resistance, but the load power decreases only linearly, substantial reductions in scattering can be achieved with modest effects on forward-link-limited range; see Turner in Further Reading for more details.

A more complex approach is to reduce tag scattering only after a tag has been read. This requires that the tag IC present a switchable load to the antenna, and furthermore that the IC maintain a switched state for some time after power is removed (since little power is received when the tag is in the low-scattering state). This approach has been described by Kruest, and an alternative implementation suited for shunt-matched tags has been advanced by the current author.

7.5 Near-field and Hybrid Tag Antennas

We have previously described near-field UHF operation: configurations in which coupling between the reader and tag antennas is dominated by inductive rather than radiative coupling. In discussing reader antenna design for this application, we found that only very small reader antennas, with consequent short read range, can operate without significant radiation.

Nearfield tags are often directed towards applications where the size of the tag antenna is highly constrained. In this case, a simple loop antenna can be used. The loop antenna's radiation resistance falls as the fourth power of the radius when the loop is small compared to a wavelength (Figure 7.54): that is, small loops do not radiate and by reciprocity don't receive radiation either. A tag loop antenna less than about 2 cm in diameter will couple only inductively even if the reader antenna radiates substantial power.

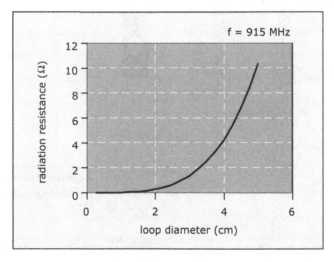

Figure 7.54
Radiation resistance of a loop antenna vs. diameter at 915 MHz.

A simple loop antenna has a parallel resonance at a circumference of about 0.45 wavelength, which at 915 MHz is a diameter of about 5 cm. For diameters much smaller than this, the loop looks like simple inductive load. The inductance of a simple loop of radius R and wire radius a is roughly:

$$\ell \approx \mu_0 R \ln\frac{R}{a}; \quad \text{for } \frac{R}{a} = 20, \quad \ell(\text{nH}) \approx 38 \cdot R(\text{cm}) \tag{7.34}$$

Thus a loop with a radius of 1.6 cm has an inductance of around 60 nH. and will resonate with a typical IC load. Smaller loops may use feed lines to add some inductance.

As we have seen previously (Figure 7.26 and Figure 7.27), shunt/ series inductance matching structures for dipole-like antennas are often convenient to implement as a tapped loop. In many cases the loop is of a reasonable size and configuration to act as a near-field antenna. Tags of this type will couple inductively to an inductive reader antenna, using the center loop, and radiatively to a conventional antenna using the dipole. Such antennas can be regarded as *hybrid* near/far-field antennas. One can use a loop match along with a densely-meandered dipole-like antenna to make a very compact hybrid tag; performance in far-field applications is compromised but acceptable for some uses. Examples of these configurations are presented in Figure 7.55.

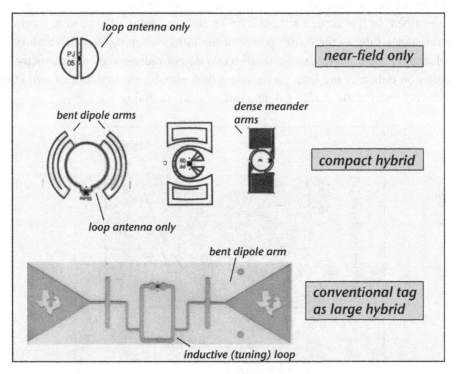

Figure 7.55
Examples of near-field, compact hybrid, and conventionally-sized hybrid tags.

7.6 Capsule Summary

In order to transfer the maximum amount of power from antenna to IC, the antenna and IC should have the same resistance, and the reactance (usually capacitive) of the IC must be matched by some inductance from the antenna. When the IC and antenna are conjugate-matched, the power transfer coefficient is 1 and the received power is described by the Friis equation. Although the most accurate representation of the IC input is as a parallel resistance and capacitance, it is often convenient to transform the load to the equivalent series resistance, usually around 10–20 ohms, and capacitance, about 0.5 to 1 pF. A key goal of antenna design is to match the antenna output to this impedance. A byproduct is some reactive amplification of the antenna open-circuit voltage, highly beneficial in supplying enough voltage to turn on the rectifying diodes in the IC input, but high voltage amplification is accompanied by bandwidth reductions.

Many tag antennas are variations of a dipole. A resonant dipole is too big for most passive tag applications, and doesn't provide a good conjugate match to an IC. Many tricks are used for fitting a big antenna into a small space. The antenna can be bent to reduce its length; an antenna with multiple bends meanders back and forth. Meandered antennas can fit into small spaces but require considerably more wire length than a straight antenna. A meandered antenna longer than resonant length looks inductive and can match a capacitive load. Short antennas can be matched using a shunt inductor fabricated as a length of conductive line between the two halves of the dipole. Adding extra conductor width, in the form of a disk or other flared shape, to the end of a dipole lowers its resonant frequency. The whole antenna can be made fatter; as the aspect ratio of an antenna goes down, so does the quality factor, albeit only at a logarithmic rate. The dipole can be rolled around at the ends to make a folded dipole with a higher source impedance.

Tags reply to readers by backscattering. The amount of power they scatter depends on the modulated radar scattering cross-section. The scattering cross-section is not determined by the physical cross-sectional area of the antenna, but by the wavelength, gain, and impedances of antenna and load.

Tag antennas can be detuned when they are placed on typical dielectrics like glass, though the consequences are reduced if the tag has broad bandwidth. Metal and aqueous fluids cause bigger problems, by driving the electric field (and thus the radiation resistance) to zero or a small value at the interface. A tag antenna that works in air will work poorly very close to a metal surface and vice versa.

Tag antennas will also interact with each other. When an array is dense in the direction of propagation, tags in the rear of the array are shadowed and difficult to read. In more complex arrays, scattering can add to or subtract from the incident wave, so that results may behave in an oscillatory fashion with changes in array spacing.

UHF tags can be configured to couple only inductively by using a small loop antenna instead of a dipole. Tags with both dipole and loop structures in their antennas can couple to a conventional radiated signal or an inductive reader antenna.

Further Reading

"UHF passive RFID tag antennas," D. Deavours and D. Dobkin, in Microstrip and Printed Antennas, ed. D. Guha and Y. Antar, Wiley 2011.

Matching and Antenna Performance

"Analysis and design of wideband passive UHF RFID tags using a circuit model," D. Deavours, IEEE International Conference on RFID 2009, p. 283

"Performance degradation of RFID system due to the distortion in RFID tag antenna," J. Siden, P. Jonsson, T. Olsson, and G. Wang, 11th International Conference on Microwave and Telecommunications Technology (CriMiCo 2001), Sevastopol, Crimea, Ukraine, September 2001, p. 371

"On the read zone analysis of radio frequency identification systems with transponders oriented in arbitrary directions," K. Rao, D. Duan, and H. Heinrich, APAC Microwave Conference 1999, p. 758

"Power reflection coefficient analysis for complex impedances in RFID tag design," P. Nikitin, K. Rao, S. Lam, V. Pillai, R. Martinez, and H. Heinrich, IEEE Transactions on Microwave Theory and Techniques, Volume 53 #9, p. 2721 (2005)

"Antenna design for UHF RFID tags: a review and a practical application," K. Rao, P. Nikitin, and S. Lam, IEEE Transactions on Antennas and Propagation, Volume 53 # 12, p. 3870 (2005)

"Impedance matching concepts in RFID transponder design," K. Rao, P. Nikitin, and S. Lam, Fourth IEEE Workshop on Automatic Identification Advanced Technologies, 2005, pp. 39–42

"Design of UHF small passive tag antennas," Chihyun Cho, Hosung Choo, and I. Park, IEEE Antennas and Propagation Society International Symposium, 2005 p. 349

"RFID tag design, sub fractional performance effects," G. Hassman, Mentor Graphics, November 2010; http://www.mentor.com/electromagnetic-simulation/

Near-Metal Antennas

"An RFID tag capable of free-space and on-metal operation," N. Mohammed, M. Sivakumar, and D. Deavours, IEEE Radio and Wireless Symposium 2009, p. 63

"Improving the near-metal performance of UHF RFID tags," D. Deavours, IEEE International Conference on RFID 2010, p. 187

"Folded dipole antenna near metal plate", P. Raumonen, L. Sydanheimo, L. Ukkonen, L, M. Keskilammi,. And M. Kivikoski, Antennas and Propagation Society International Symposium, 2003. IEEE, vol 1 p. 848–851

"Planar Wire-Type Inverted-F RFID Tag Antenna Mountable on Metallic Objects", L. Ukkonen, D. Engels, L. Sydanheimo, M. Kivikoski, IEEE Antennas & Propagation Symposium, Monterey, CA, USA 2004

Specialized Antennas

"Design and Development of a Miniaturized Embedded UHF RFID Tag for Automotive Tire Applications", S. Basat, K. Lim, I. Kim, M. Tentzeris and J. Laskar, Electronic Components and Technology Conference 2005, volume 1 p. 867

"Multi-standard UHF and UWB antennas for RFID applications," T. Deleruyelle et al., Proceedings of the Fourth European Conference on Antennas and Propagation 2010, p. 1

"Low cost silver ink RFID tag antennas", P. Nikitin, S. Lam, K. Rao, IEEE Antennas and Propagation Society International Symposium, 2005, p. 353

"Investigation of RFID tag antennas printed on flexible substrates using two types of conductive pastes,"
K. Janeczek et al., Third Electronic System Integration Technology Conference, 2010, p. 1
"Rapid prototyping RFID antennas using direct-write," J. Hoey et al., IEEE Transactions on Advanced
Packaging, volume 32 #4 p. 809, 2009
"Reliability of passive RFID of multiple objects using folded microstrip patch-type tag antenna", L. Ukkonen,
D. Engels, A. Sydanheimo, and M. Kivikoski, IEEE Antennas and Propagation Society International
Symposium, 2005 page 341

Antenna Scattering

"UHF RFID and Tag Antenna Scattering: Part 1: Experimental Results", Microwave Journal, May 2006, p. 170,
and "Part 2: Theory", Microwave Journal, June 2006, p. 86, both by D. Dobkin and S. Weigand
"Cloaking circuit for use in a radio frequency identification and method of cloaking RFID tags to increase
interrogation reliability", J. Kruest, US Patent 5,963, 144, granted October, 1999.
"Input impedance arrangement for RF transponder", C. Turner, US Patent 6,870,460, granted March, 2005.
"Mutual coupling of stacked UHF RFID antennas in NFC applications," X. Chen, L. Feng, and T. Ye, Antennas
and Propagation Society International Symposium, 2009, p. 1

Exercises

Matching Antenna and IC:

1. A new tag IC from the Silicon Valley startup company Fundless Networks is reported
    to consume a DC power of 0.3 microwatts at an input voltage of 0.5 V. Treat the IC
    load as a simple parallel resistor and find the resistance value:

    $R_p =$ _____ ohms

    The input capacitance is 0.5 pF. What are the series equivalent input circuit values at
    915 MHz?

    $R_s =$ _____ohms $C_s =$ _____ pF

    What is the voltage amplification factor for a conjugate-matched antenna?

    $|V_{IC}/V_{oc}| =$ _____

    Assume the input capacitance was matched using a simple series inductor, and that the
    antenna looks like a voltage source and radiation resistance matched to the load series
    resistance. What is the bandwidth of the overall antenna-IC circuit?

    $BW =$ _____ MHz

    What is the voltage multiplication factor at the band edge at 902 MHz?

    $|V_{IC}/V_{oc}| =$ _____

2. After sitting through three hours of PowerPointless slides and a lunch whose fat content is
    measured in ounces rather than grams, Bob the lazy RF designer snags a prototype IC from
    Fundless' VP of Sales, Sal E. Closer. Being Bob, he doesn't want to design a matched

antenna and instead simply attaches the IC to the 915-MHz resonant dipole antenna he took from Amy's desk (see Figure 7.9). What is the power transfer coefficient?

$\tau =$ _____ at 900 MHz

Should this be Bob's raise at his next performance review?

3. Consider the matching problem shown below. What shunt inductance is required to move the impedance from point 2 to point 3, assuming a frequency of 915 MHz?

inductance = _____ nH

Assume the inductance of a straight conductive line is

$$\ell(\text{nH}) = 2L\left(\ln\left(\frac{4L}{w}\right) - 1\right)$$

where L is the length of the line in cm and w is the width of the line. If the lines used in the tag antenna are 2 cm wide, how long is the shunt inductor?

length = _____ nH

Repeat the exercise for the series inductor, remembering that the physical implementation will split this into two series inductors with the IC between them:

inductance = _____ nH length = _____ cm

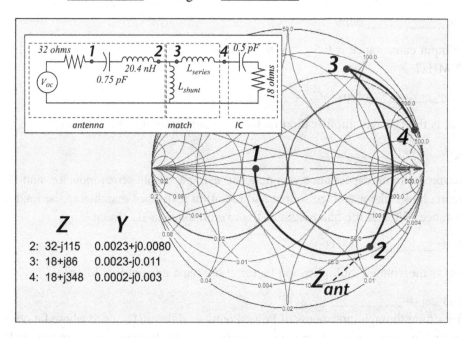

	Z	Y
2:	32-j115	0.0023+j0.0080
3:	18+j86	0.0023-j0.011
4:	18+j348	0.0002-j0.003

Radar Scattering Cross-Section

4. A tag is placed in an anechoic (reflection-free) chamber, 1.25 meters from a linearly-polarized test antenna with a gain of 4 at 915 MHz. The transmitted signal is 10 dBm. The measured reflected signal, corrected for antenna reflections, is -50 dBm. What is the radar cross-section of the tag?

RCS = _____ cm^2

Metal Surfaces

5. The tag of problem (3) is to be used mounted on the bottom of a metal car body using a 5-mm-thick foam spacer. Near the metal surface, the series model of the antenna becomes 1.5 pF and 10 nH in series with 1 ohm. The open-circuit voltage is reduced from its value in the open by $2 \sin(2\pi(0.5/8.2)) = 0.19$. Ignore any change in the matching inductors and calculate the value of the voltage presented to the IC, presuming that 1 V was present for the same antenna illumination with the tag in the open.

V(IC) = _____ V

What is the power transfer coefficient?

$\tau =$ _____

If the tag needs 0.5 V to turn on, how will the read range be affected?

6. Derivation of Series–Parallel Transformation: Imagine we have a load (like our integrated circuit) connected through a T-match that we model as a shunt impedance Z_{sh} followed by a series impedance Z_{ser}. Can we find values Z_{shtr} and Z_{sertr}, and perhaps a transformed load Z_{ldtr}, such that, looking from the left, the series–shunt and shunt–series arrangements present the same impedance, no matter what the load is?

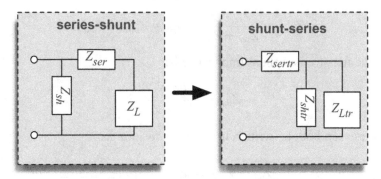

If this can be done, it has to work for the two limiting loads: an open circuit and a short circuit ($Z_L = 0$ and ∞). Using the formulas for series and parallel impedances (Appendix 3), we can obtain expressions for the input impedance for short-circuit and open-circuit loads, as shown in the following figure

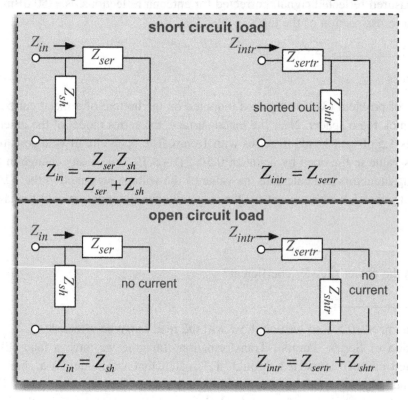

Now we require that the transformed versions are equal to the non-transformed versions in both cases:

$$Z_{sertr} = \frac{Z_{ser}Z_{sh}}{Z_{ser} + Z_{sh}}; \quad Z_{sh} = Z_{sertr} + Z_{shtr}$$

Show that the transformed shunt impedance is:

$$Z_{shtr} = \frac{(Z_{sh})^2}{Z_{ser} + Z_{sh}}$$

Note that these can both be written in a similar fashion as:

$$Z_{shtr} = \beta Z_{sh}; \quad Z_{sertr} = \beta Z_{ser}; \quad \beta \equiv \frac{Z_{sh}}{Z_{ser} + Z_{sh}}$$

Using this fact, we can write the statement that the impedance is the same for ANY load as:

$$\frac{Z_{sh}(Z_{ser} + Z_L)}{Z_{sh}(Z_{ser} + Z_L)} = Z_{sertr} + \frac{Z_{shtr}Z_{Ltr}}{Z_{shtr}Z_{Ltr}}$$

$$\frac{Z_{sh}(Z_{ser} + Z_L)}{Z_{sh}(Z_{ser} + Z_L)} = \beta Z_{ser} + \frac{\beta Z_{sh}Z_{Ltr}}{\beta Z_{sh}Z_{Ltr}}$$

Now comes the hard part: SHOW that this equation is always true if:

$$Z_{Ltr} = \beta^2 Z_L$$

It may be very helpful to note that:

$$\beta Z_L + Z_{sh} = \beta(Z_L + Z_{ser} + Z_{sh})$$

CHAPTER 8

UHF RFID Protocols

8.1 What a Protocol Droid Should Know

Every communications process is based on agreements about certain conventions or agreements about how messages are to be sent, and what they mean. A communications *protocol* must address questions like:

- **Medium**: what is the medium by which messages are to be exchanged? People use the media of speech, writing, and pantomime to communicate directly with each other. Machine-to-machine communication can be based on electrical signals carried by a cable, light in a silica fiber, ultrasound, or radio waves.
- **Message format**: speech can be formatted in English, Swedish, Japanese, Hindi, or any of the wonderful panoply of languages that have arisen over the millennia since humans invented the ability to harangue their spouses. Each language has a vocabulary of phonemes and words, and a grammar describing how these elements are to be combined.
- **Medium access**: in some cases a particular medium — a wire — is dedicated to a particular communications process, and there is no possibility of contention, but many media are shared. The audible medium is shared when a group of people congregate to talk; some means must be arranged to allocate the medium so that individuals can be understood. In informal situations, a person usually waits until no one else is speaking and then attempts to talk; if they collide with another person with similar intent, both go silent and wait for a random time to try again. This scheme, dressed up as *carrier-sense multiple access with collision detection*, is the basis of the shared-medium part of Ethernet networking, used in almost every local area network in the world. When people gather in more formal settings, one individual may take control of the right to speak, and periodically poll the group for those who wish to take the floor, granting them rights as she sees fit.
- **Context and interpretation**: even if a message is received without error and the words and sentences deciphered, the meaning of the contents must be established by reference to a context in which the exchange takes place.

Each of these basic protocol elements must be defined for any communications system, and in particular for any RFID system. The choices that are made in defining an RFID protocol are shaped by the need to minimize the demands on the limited power and computational

© 2013 Elsevier Inc. All rights reserved.

ability of the tags. These requirements are particularly onerous for passive tags, where very little power is available (see chapter 5). However, even battery-powered tags must be frugal with their limited stored energy, and specialized protocols, or specialized subsets of existing protocols, are used to minimize the number of exchanges and the energy cost of each one.

As we have previously noted, once one makes the choice of using radio communications, the means that can be employed are heavily constrained by regulatory authorities. The designer is limited to specific frequency bands, maximum power levels, bandwidth, and channel residence within the bands. Key medium properties from an RF point of view are bandwidth, interference levels, propagation characteristics, antennas, and circuit requirements.

A medium-range UHF RFID protocol meant to operate in the US doesn't have too many choices. In most cases operation will be in either the 902−928 MHz or 2.4−2.483 GHz unlicensed bands. The lower band defaults to 500 kHz channels, and the upper to 1 MHz channels. Operation under European (ETSI) guidelines is a bit more complex, but basically requires the reader to operate in 200 kHz channels between 865 and 868 MHz. Recall from chapter 3 that the data capacity of a transmission channel is determined by the bandwidth and signal-to-noise ratio. A channel 500 kHz wide is sufficient to provide for one to two hundred kbps with no special measures, so either band is suitable for many RFID passive tag applications, which tend to be low-rate, even using the inefficient modulations that passive tags are stuck with. Both US bands are exposed to interference from many existing unlicensed devices, including cordless telephones and wireless local-area networks, though the proliferation of WiFi and Bluetooth hardware in recent years makes the 2.4 GHz band a tougher interference environment. Many other sources of interference also exist, including intentional radiators in nearby bands (like cellular basestations) and unintentional radiators (like poorly grounded spark plugs). The properties of some of the sources likely to be encountered in the 902−928 MHz US ISM band are summarized in Figure 8.1.

Other readers are a key source of interference in densely-populated environments, as we noted in chapter 6. In addition to the various physical configuration solutions, protocols can make the reader interference problem loom smaller or larger, depending on the overlap of the spectrum of tag replies and reader transmission. If the two are kept separated in frequency, it may be possible to filter out other reader transmissions in the receiver and leave only the tags. This approach is employed in EPCglobal Class 1 Generation 2 Dense-Interrogator operation (section 5 of this chapter).

Since passive RFID is a short-range technology, the propagation issues of interest mostly have to do with indoor propagation and obstacle tolerance. Radio waves can get through obstacles in three ways:

- **Direct penetration**: many dielectric materials, like dry paper or cardboard, dry wood, non-conductive plastics, most textiles, and glass, are substantially non-absorbing and

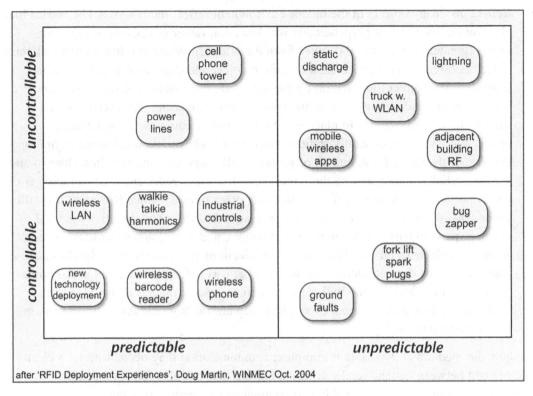

Figure 8.1
Sources of interference in the 900-MHz band, and their general properties.

have modest refractive indices (2−4) for 900-MHz radio waves. Such materials are sometimes known as **RF-lucent**. Radiation incident on lucent materials suffers modest reflections due to refractive-index mismatch − a refractive index of 3 causes a loss of about 3 dB per interface. Absorption is negligible. So many common materials that are solid obstacles for visible light are of moderate to negligible consequence for 900 MHz radiation. In contrast, metals reflect essentially all the radiation that falls upon them. Water, with a dielectric constant of around 80, also reflects almost all of an incident wave, and absorbs most of the rest. They are **RF-opaque**.

- **Diffraction**: visible light has a wavelength of around 0.5 micron − about 1/2000 of a millimeter. Human-sized objects are thousands or millions of wavelengths across, so to use light seems to travel in straight lines. However, 900 MHz radio waves have a wavelength around 32 cm − the length of a piece of letter-sized paper. Typical indoor objects are only a few wavelengths across. RF-opaque objects still cast shadows, but they are diffuse and not monotonic. A typical object a few wavelengths across casts a shadow about 10−15 dB deep, with a relatively shallow region − **Poisson's Bright Spot** − roughly in the geometric center of the shadow.

- **Reflection**: many objects in the indoor environment reflect radio waves. Dielectrics like glass reflect modestly at perpendicular incidence but rather effectively at glancing angles (greater than about 70 degrees from the normal). Water and metal are excellent reflectors. Signals can bounce off a reflective object and illuminate a region that is shadowed from direct illumination by the reader antenna. Reflected waves add to or subtract from the direct wave from the reader antenna, causing the received signal strength to vary from place to place in a fashion that is complex and not readily predictable, even in a generally static environment. This variation is known as *fading*. Because of the limited link budget of passive RFID tags, the tags are often close to and at least partially illuminated by the direct beam from the reader antenna; so fading is of less importance in passive RFID than in many other radio systems. However, it is still significant. When a battery is available, or a large tag with good directivity is used, ranges expand to tens or even hundreds of meters, and fading and shadowing become dominant influences on link functionality. Reflections particularly from the floor, cause read zones to be discontinuous, with tags read at (say) 10 meters and not at 9. People are wonderful reflectors, and will cause tags to be read or missed at the edge of the read zone as they move around, even when they are far from the direct beam from the reader antenna to the tag.

So the radio medium at 900 MHz is complex; communication may occur without a clear line of sight between tag and reader antenna, but the link can also fail sporadically. The protocol cannot assume a reliable and continuous connection, or simple monotonic changes in signal strength or link quality.

The use of 900-MHz radiation and unlicensed operation basically limit us to 6−10 dBi of reader antenna gain. Combined with the propagation difficulties alluded to above, the relatively broad beams created by such low-gain antennas imply that localizing an inquiry to any specific physical region will not be possible. Protocols must assume that any tags in a given general area will be illuminated; if only some of these tags are the object of a given inquiry, the protocol must provide means for addressing the inquiry only to the relevant subset.

The choices that a given protocol makes for each of the protocol elements collectively form an attempt to meet a number of contradictory requirements. Before plunging into an examination of how specific protocols work, let's take a look at the tools a protocol designer has at hand and the tradeoffs they face in using those tools.

Symbols and Modulation: Recall from chapter 3 that to convey any information a radio signal has to be modulated, and the translation between binary bits and a particular sequence of modulation constitutes the symbol vocabulary available on the channel. Passive tags have no means of extracting frequency or phase, so reader modulation is limited to changes in signal amplitude with time. Active tags have a local oscillator and phase modulation can be used. Since a passive tag also extracts its power from the reader signal, whatever choice of modulations and encoding is used must ensure a high average reader transmit power over a time comparable to the time a tag can store energy. The desire to keep power high most of the time leads to symbol sets in which RF power is on except for small low-power gaps, and the desire to inventory tags rapidly means that the symbol duration is short, but the combination of these two circumstances implies that the spectrum of the transmitted signal is wide: typical passive reader symbols are very inefficient users of bandwidth.

The reverse link from tag to reader, examined in some detail in chapter 5, can employ either amplitude or phase modulation at the tag, but with no assurance that the corresponding change in the signal at the reader will be similar in nature. Increases in backscattered signal power may lead to a decreased reader signal; changes in phase at the tag may lead to changes in amplitude at the reader. Only the fact that a transition has occurred is likely to be detected reliably, so all passive tag encoding schemes must be variants of frequency-shift keying.

The spectrum of a frequency-shift-keyed signal generally peaks around the frequencies employed. For example, if the tag can change its state either at 100 kHz or 200 kHz, the backscattered spectrum will have much of its energy in peaks displaced 100 and 200 kHz from the carrier; normally, both the sum and difference frequencies will be present, though their amplitude may vary depending on the (uncontrollable) phase relationships of the overall signal. As we noted in chapter 4, one of the biggest challenges a reader faces in receiving a passive tag signal is the noise from its own signals, which is concentrated close to the carrier frequency. Therefore, there is some advantage to making the frequency of the tag transitions high, so that the information is far from the carrier and not swamped by transmit leakage noise. The sharpness of the spectrum is dependent on the number of cycles at a given frequency, so the cleanest tag spectrum is produced when a single bit contains many tag cycles; but this means that it takes longer to transmit a bit for any given tag cycle time. Tag speed, receiver noise, and data rate must be traded off in the choice of the tag cycle time. We will see a clear example of this phenomenon in the Miller-Modulated Subcarrier scheme employed in EPCglobal Class 1 Generation 2: higher Miller indices provide better interference rejection at the direct cost of lower tag data rates and thus slower inventories.

Packet Format and Command Sets: Most digital communications are based on *packets* of data: discrete chunks of a few tens to a few thousands of data bits, accompanied by standard headers and optional tails that provide information about the contents and purpose of the packet to the system. A simplified example is shown in Figure 8.2. Packet headers, often known as *preambles*, serve several important functions. The preamble usually contains some fixed sequence of symbols that help the receiver to recognize the beginning of a packet and

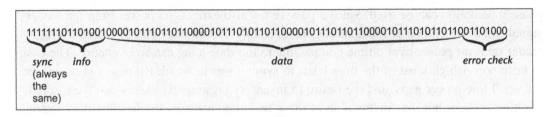

Figure 8.2
Schematic depiction of a data packet.

become synchronized with the clock of the transmitter; the latter is helpful to ensure that the instants at which the signal is sampled correspond to the appropriate moment within each transmitted symbol. Preambles also often contain some indication about the type of packet and the number of bits, so the receiver can know when to stop. The trailing bits appended to the end of the packet are often used to check for errors. Some or all of the packet may be encoded using one of a large number of schemes that allow for very efficient detection and correction of errors. Data or packets may be interleaved — transmitted out of time sequence — to guard against bursts of errors resulting from interference or fading. The data can be encrypted to provide security against interception by an unauthorized listener.

The peculiar limitations of passive tags place constraints on packetization. Passive tags have limited logic and memory resources, so if packets are used they should either be amenable to interpretation as each bit is received, or be short so the memory required to store the received bitstream is small. Convolutional and block codes demand rather more resources than the tags have, so any error checking must be very simple. Encryption if used must also be very simple to implement.

Two very simple means of error checking, widely used in passive RFID, are *parity checks* and *cyclic redundancy checks* (CRC). A parity check is extremely simple: the bits in a chunk of data are added up, and one additional bit is sent whose value is determined by whether the sum was odd or even. A parity check requires only a single flip-flop to calculate and so is practical even for a power-limited RFID tag IC. However, parity checks are rather inefficient and not very powerful. In order to avoid adding a lot of extra data, a parity bit should be appended only after several bits have been sent. The parity check will detect all single-bit errors in the dataset, but all multiple-bit errors in which an even number of bits are flipped will not be detected.

Cyclic redundancy checks are almost as easy to implement as parity checks and much more effective. A CRC is essentially computed by dividing the number represented by the data by a smaller known number and taking the remainder. Since bit errors change the value of the data (possibly by a large amount, depending on the bit that gets flipped), they will also in general change the value of the remainder. The probability of randomly choosing a number with the same remainder becomes quite small when the divisor is large; for example, a 16-bit CRC provides a probability of roughly 1 in 65,000 of a correct result being produced by noise. Varying the number of bits in the error check allows the designer to trade off robustness of error detection for speed and simplicity: 5-bit, 8-bit, and 16-bit CRC's are common. CRC calculations can be carried out using shift registers (flip-flops connected in series) with some of the output values fed back, and thus can be implemented using only a handful of gates and very little power.

In conventional communications systems, packet formats are usually invariant with respect to the data carried by the packet, to make sure that protocols are relatively independent of their application. However, in passive RFID, time and computational costs are so important that this practice is generally not followed. Instead, special packets and special symbols are used to convey commonly-encountered commands, to minimize the amount of time the reader spends talking and maximize the number of tags that can be read with a given amount of time and energy.

Medium Access Control: Who gets to talk when? In most conventional communications systems, a separate discovery process allows nodes to become aware of what other nodes they might wish to converse with, and then messages can be routed in an efficient fashion based on the existing table of possible contacts. Such a scheme is way too slow, power-hungry, and complex for a population of mobile passive tags. In most applications, the reader has no way to know how many tags are listening until it reads them. Instead, methods must be found to allocate the right to reply to tags in real time as they are counted. The selection of a single tag with which to communicate from a population of nominally identical tags is called *singulation*.

Two basic solutions to the singulation problem are the *binary tree* and *Aloha* protocols. A binary tree protocol exploits the fact that (hopefully) every tag is associated with a unique identifying number. If we picture all possible tag IDs as being leaves on a twofold-branching (inverted) tree, we can see that each ID is at the end of a unique path through the tree, and that at each step in the path, the number of possible IDs that could be at the end is reduced two-fold (Figure 8.3).

We can construct a polling procedure that will access every possible tag ID, by exhaustively examining each possible node in the tree. That is, we start by asking for all tags whose first digit is 0, thus automatically addressing only half of all possible tags. If no tags respond, we need proceed no further on this half of the tree. Along each branch where a tag response is found, we bifurcate further, eventually achieving a condition where only one tag is responding, so that its ID can be obtained without interference from other tags. Furthermore, the time taken should be roughly logarithmic in the number of nodes if the leaves are either sparsely populated or highly concentrated, since each time we find a node with no tags on it, we are able to eliminate all the leaves attached to that node. When we have a large tree with

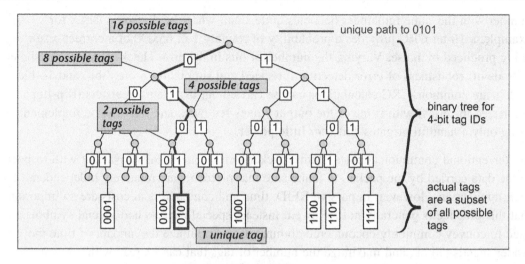

Figure 8.3

Simple 4-bit binary tree showing the unique path from the root to the leaf corresponding to ID 0101.

only a few tags, we will find empty nodes very high up in the tree and quickly eliminate most of it, concentrating only in those regions where tags are present. Only in the case where the leaves are densely populated and uniformly distributed over the number space will we need to survey a large part of the tree, but this will never happen for typical tag ID spaces: a 64-bit tag ID allows about 1.8×10^{19} unique tag IDs, insanely larger than the number of tags that could be placed within the read zone of a single antenna.

It is also possible to intentionally define a uniformly-populated but much smaller tree by associating a random number with a tag; this random number could be generated by the tag, or we could use (for example) the CRC calculated on the tag ID, which is a much shorter number with reasonably random properties. The advantage of such a procedure is that it avoids the common circumstance where a population of tags shares all but a few bits of their ID, and thus responds in concert to most binary-tree queries, causing multiple collisions and making it difficult for the reader to navigate the tree. A random singulation ID guarantees that the tree is sparsely and uniformly populated, so that along most paths more than a couple of nodes deep only one or a few tags will be present. The cost is that the random numbers are not guaranteed to be unique, and the process will occasionally fail when two tags take the same random number.

Binary tree procedures are simple to implement and reasonably fast. One important disadvantage is that binary tree traversals work best when tags are uniformly present and responding to reader throughout. For example, it is most efficient to traverse the tree by sending bits in sequence, without the laborious and slow elaboration of where the bit is in the tree. That is, it's pretty easy to tell the tags: "START A TRAVERSAL: ok,

$1 - 0 - 1 - 0 - 1 - 1 - 1 - 1$", at which point we are 8 nodes deep with 8 bits transmitted. However, a tag entering the process in the middle has not the slightest idea where we are and must therefore wait until the end before joining. Instead, we can send the whole path each time: "OK, TAGS, WE ARE AT THE NODE 10101111, and all tags whose next bit is a 0 can respond." "OK, TAGS, NOW WE'RE AT 101011111..." This process allows a tag to respond to a query addressing a path it is on, even though it did not hear the first few nodes. However, in either case, a late-arriving tag may be on a path that has been marked by the reader as being empty, and thus may be ignored. These problems can be addressed by traversing the whole tree every time a tag is counted; the EPCglobal Class 0 protocol incorporates a very efficient means of performing complete tree traversals for every tag ID, but the efficiency is achieved at the cost of a prior synchronization process that still prevents tags from joining an inventory in mid-stream. The Class 0 protocol, because it assumes bit-by-bit interactions between tag and reader, also limits tag speeds to no faster than reader speeds. This is a problem because tags usually have more information to transmit (their ID) than the reader, and reader spectra, being transmitted at much higher power levels, are of much greater concern to regulators than tag scattering.

Some of the limitations of binary tree MACs can be addressed by the clever use of randomness. Aloha protocols are based on early networking research conducted at the University of Hawaii (hence the name), and are based on the idea of random access to a channel. In basic Aloha, a station transmits a message whenever it has a message to send, and waits for an acknowledgement. If no acknowledgment is received, it is presumed that the message was lost due to error or collision with another contending station. The station waits a random delay time, and then retransmits the message. Aloha protocols are decentralized and thus scalable to large populations of stations, and are very efficient when the offered traffic is a small percentage of the total traffic that can be carried on a channel.

A very common variant of this procedure, *slotted Aloha*, restricts the start of a transmission to specific time slots; slotted Aloha is somewhat more efficient when there is a lot of traffic to send and it comes in well-defined sizes, because once a station 'captures' a slot, there is no possibility of a collision between two stations until the beginning of the next slot. However, for slotted Aloha to work efficiently, the probability that a station attempts to transmit in a slot must be adjusted to keep the offered traffic load moderate so that every slot is not lost to collisions. In the immensely popular Ethernet (IEEE 802.3) networking protocol, when a station does not receive an acknowledgement for a transmission, it waits for a random period – the *backoff* time – that increases exponentially each time the transmission is attempted and fails. In this fashion, stations adjust their behavior to the load on the network without any need for central coordination. In the case of an RFID reader, expecting tags to remember how many times they have attempted to transmit their ID and calculate an appropriate backoff is rather ambitious. Instead, tags are usually equipped with a simple counter of some sort that is incremented or decremented by commands from the reader, so that any complex traffic management is performed by the reader rather than the computationally-challenged tags. To be efficient, the reader must do a good job of adjusting the likelihood of tag responses, for which purpose the reader needs to be able to distinguish between the three cases of no tag response, a single tag response, and a collision (multiple tags responding in a single slot).

8.2 Days of Yore

In the happy times at the dawn of UHF RFID, life was simple and problems were few.

Well, actually, the author lived through those times and can certify that there was no dearth of difficulties to surmount in every field of human endeavor. What was in short supply was the fast, ultra-low-power CMOS circuitry we have now come to take for granted. Designers of passive RFID systems had to make do with minimal functionality, and created simple protocols. Let's look at a few.

We've already encountered some of the earliest work in the field, due to Koelle and coworkers at Sandia Laboratories (Figure 2.6, repeated below for convenience as Figure 8.4), but we're now in a much better position to appreciate what these early folks accomplished.

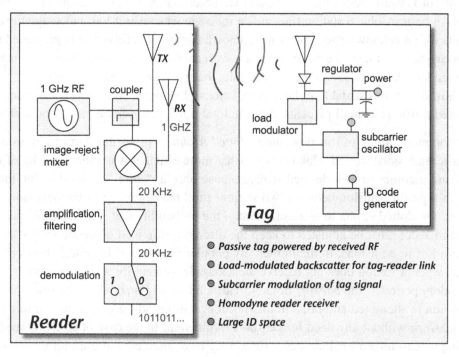

Figure 8.4
Early scheme for UHF passive tag, after Koelle et. al., Proc. IEEE, 1975.

In this scheme the reader transmitter did nothing but send a continuous-wave (CW) signal to power the tag. The tag used a diode to generate DC power from the RF signal. As soon as the DC power supply was sufficient, the tag started backscattering an ID code. This is a *tag talks first* scheme, since the tag sends a message without waiting for the reader to instruct it. (This is a good thing, since the reader doesn't talk at all.) The tag used a *subcarrier* oscillator to flip the state of the tag antenna 20,000 times per second — that is, the subcarrier frequency was 20 kHz. The ID code in this case modulated the amount of subcarrier modulation of the tag load rather than the frequency: this is an *amplitude-shift keyed, subcarrier-modulated* uplink (Figure 8.5). The tag is read-only (though these early tags did have a provision for modulating the subcarrier frequency based on the local temperature, thus supporting sensor integration). While not relevant to protocol issues, it is fun to note that Koelle's group used an image-reject mixer in the receiver to make the result insensitive to the absolute phase of the backscattered signal.

Recall that the small tag signal must be combined with other reflections, so that the amplitude and phase modulation of the resulting signal may not be simply correlated with the changes in radar cross-section at the tag. Thus amplitude modulation, even of a subcarrier, may not produce the desired signal at the reader. The problems of an amplitude-modulated return link were soon understood; subsequent developments quickly moved to frequency-shift-keyed tag modulation. During the 1980's, commercial standards were developed for identification of shipping containers and railcars: AAR S918 and ISO 10374. These standards are still tag-talks-first in the sense that the tag begins transmitting its ID once it has powered up. The tag symbols are based on frequency-shift keying: symbols use both 20 kHz and 40 kHz modulation of the tag antenna load impedance. The actual symbols employ a sort of *Manchester* coding (a term we will encounter again in connection with ISO 18000-6 below): the beginning of a binary '1' symbol is modulated at 40 kHz, but the frequency transitions in mid-symbol to 20 kHz. A binary '0' starts at 20 and transitions to 40 kHz. Like the Sandia tag, an AAR tag simply sends out all its data again and again once it is powered up.

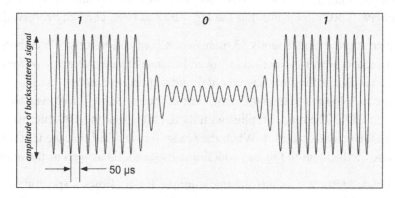

Figure 8.5
Amplitude-shift-keyed, subcarrier modulated symbol. Baseband signal (amplitude of the RF backscattered signal, with the 1 GHz component removed) shown here.

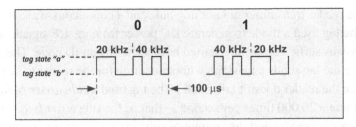

Figure 8.6
AAR S918/ISO 10374 tag symbols.

The tag contains 128 bits of data, rather more than in the Sandia work. Bits 126 and 127 violate the rules for Manchester symbols above, and are used to mark the beginning of a dataset (frame) — a trick we'll also encounter again. The ID encodes certain features of the object being marked; for example, in the case of a railcar, the ID number contains the car length, the number of axles, and the side on which the tag is mounted.

By the early 90's, automobile toll tags were coming into wide use. Toll tags are usually semi-passive, so avoiding unwanted activation is important. In addition, unlike a railroad (where only one car is in front of the reader at a time), a tolling station must deal with a number of tagged cars present simultaneously, and so have some ability to deal with medium allocation. For example, California Title 21 is a ***reader talks first*** protocol, in which the reader not only transmits a CW signal, but can also send data, using Manchester-coded amplitude-shift keying at 300 kbps. The reverse link symbols are simple frequency-shift keying: a binary '0' uses antenna modulation at 600 kHz and a binary '1' is modulated at 1200 kHz; the data rate is also 300 kbps. (The high data rates are required to support toll reading on fast-moving automobiles in toll road applications.)

The data consists of a header, some bits describing the tolling agency with which the tag is associated, and a unique 32-bit transponder ID. The data also includes a 16-bit ***cyclic redundancy check*** (CRC) to ensure that the tag's ID has been correctly received.

To start a conversation, a reader sends 33 microseconds of continuous '1' symbols, followed by a 100 microsecond power-off period (allowed because these are battery-powered tags). The reader then sends a polling message, containing a short ***preamble*** — a set of bits that is always the same to help the tag recognize the start of the message — followed by an agency code and a 16-bit CRC. The reader replies with its ID, after sending 100 microseconds of '0' symbols to help the reader sync to it. When the reader has verified that the CRC checks, it sends an acknowledgement to the tag, containing the tag's ID as well as the reader ID.

Since the acknowledgement contains the tag's unique ID, it allows a specific tag to know it has been read. Once this ACK is received, the tag must refuse to reply to another polling message with the same agency code for 10 full seconds. This provides a simple sort of collision resolution: once a tag is read, it goes silent so that other tags can be heard.

These early protocols developed some key aspects of a passive tag protocol:

- Long range using radiative coupling at UHF frequencies
- Amplitude-shift-keyed reader symbols
- Tag power harvesting from incident RF
- Frequency-shift-keyed tag symbols with antenna load modulation
- Packet-based communications with coding violations and standard preambles to mark the start of a packet or frame
- Cyclic redundancy check bits for error detection (but not correction)
- Persistent quiet states for simple collision mitigation

However, in addition to relatively high tag cost, they lacked some important capabilities for more sophisticated applications:

- Medium access control adequate for dealing with a large number of tags in the read zone
- Multiple tag states and reader commands to allow segments of tag memory to be read
- Ability to modify the tag memory contents
- Variable data rates to adapt to differing operating conditions

With the advent of RFID as a tracking technique for a wider variety of less-expensive assets, more sophisticated communications protocols were quickly proposed and demonstrated to support more varied requirements and take full advantage of rapidly-advancing integrated circuit technology. Let us now examine these more recent approaches to specifying the interactions of tag and reader.

8.3 EPCglobal Generation 1

We briefly reviewed the history of EPCglobal in chapter 2. In the formative days of the AutoID Center, two new protocols were developed, denoted Class 0 and Class 1. These new protocols attempted to simultaneously provide improved forward and reverse link coding, medium access control, and tag state management, while being amenable to implementation in very simple, low-cost ICs. Neither of these protocols were completed or ratified by EPCglobal, but both saw widespread more-or-less compliant commercial implementation in the period 2002–2006. Though Class 0 and Class 1 tags have been displaced by Class 1 Generation 2 tags, they remain of considerable theoretical interest as examples of approaches to solving the problems faced by a UHF passive tag protocol. Therefore we shall briefly examine both.

8.3.1 EPCglobal Class 0

The Class 0 category was defined as tags that are factory-written read-only, though in practice field-rewriteable tags are commercially available. Tags with both 64-bit and 96-bit

Electronic Product Codes were envisioned and saw commercial implementation, although at the time of this writing 64-bit tags are (happily) mostly preserved in retrospective collections and 96-bit tags are the norm. Matrics (since sold to Symbol Technologies and thence to Motorola) and Avery Dennison sold substantial commercial quantities of Class 0 tags; Matrics and Impinj produced chips implementing the protocol.

Reader symbols for Class 0 are shown in Figure 8.7. The symbols are pulse-length-encoded and amplitude-shift-keyed. The binary '0' and binary '1' symbols are chosen to provide continuous high average power to the tag; a random sequence of 1's and 0's would result in an average transmitted power of about 65% of the peak (CW) power of the reader. The special symbol **null** is encountered rarely (it is used to launch tree traversals and induce certain state changes in the tag), and thus its impact on average transmitted power is minimal even though the average power during a null is quite low (about 1/4 of the CW power). The parameters shown are typical of US operation, and provide a raw reader data rate of about 80 kbps.

Tag symbols are depicted in Figure 8.8. Here the symbols are defined by the frequency of transitions of the tag between different scattering states; the exact nature of the states is not critical as long as the radar cross-section differs sufficiently in amplitude or phase to produce a backscattered signal. The symbols are conceptually similar to the 20/40 kHz frequency-

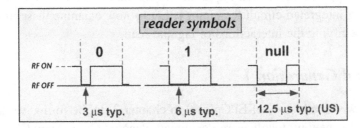

Figure 8.7
Class 0 reader symbols, depicted as baseband levels. Actual transmitted symbols consist of a high RF (900-MHz) amplitude ("RF ON") with low-amplitude or zero-power excursions ("RF OFF").

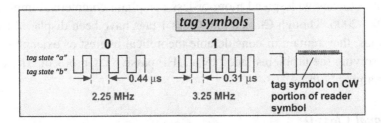

Figure 8.8
Class 0 tag symbols; tag states "a" and "b" may be any states between which the radar cross-section differs in amplitude or phase.

shift-keyed tag symbols employed in early UHF protocols, but the actual frequencies employed are higher and are not harmonically related. The tag backscattering is performed during the CW portion of the reader symbol. This is made possible by the choice of reader symbols and the protocol structure. Reader symbols are defined by the duration of the low-power gap; as soon as power is restored, it is in principle possible for the tag to decode the symbol received. Furthermore, the protocol is structured as a bit-by-bit query-response rather than a packet transmission followed by an extended tag response. Therefore the tag IC can be designed to decode the reader bit and decide on its response immediately after completion of the gap, allowing modulation to occur during the end of the reader symbols.

This interesting scheme has strengths and weaknesses. The system is in effect full-duplex: tag and reader transmit data effectively simultaneously, so the net data rate is twice the reader data rate. In order for such a scheme to work in the presence of the large baseband transients in the reader receiver that result from modulated transmit leakage, it is necessary to use a relatively high frequency for the tag transitions. In this case, the tag signal is displaced by 2−3 MHz from the carrier, so it is relatively easy to filter out the slow transmit leakage signals and recover the tag signal. (It is, however, worth noting that many dedicated Class 0 readers use a bistatic configuration, as described in chapter 4, to minimize transmit leakage. No amount of filtering will help if any part of the receiver stage reaches saturation due to transmit leakage transients.) The use of a tag signal displaced from the reader carrier by such a substantial offset makes it relatively easy to design a sensitive receiver, and Class 0 tags typically display excellent and apparently forward-link-limited read range, despite the relatively early IC technologies used. The author's experience suggests that ranges of 5 meters could readily be obtained with a monostatic reader, and 8−10 meters was achievable with a bistatic configuration, using commercial tags.

Using such a large tag spectral displacement also carries significant disadvantages. Under European regulatory recommendations, in which tag radiation is explicitly included, Class 0 tags could be expected to radiate unacceptable power levels outside of the allowed bands for RFID operation. Even in jurisdictions in which regulatory issues do not arise, but where a narrow band is available for RFID operation, the tag backscattered signals will be primarily or exclusively out of band. The tags are therefore exposed to interference from legal, possibly licensed, radiators whose operation may be beyond the influence or control of the RFID user. In US operation, out-of-band and reader-reader interference are also substantial concerns since the tag signals will extend outside the ISM band when operating at near-edge channels, and the binary symbols cross channels. For example, for a reader operating at 915 MHz, the tag signals are at 917.25 and 918.25 MHz; a reader transmitting on either the 916.75−917.25 or 917.75−918.25 MHz channel will block the tag signal unless the reader transmissions are very well confined to the center of the channel, which is often not the case. Readers at 912 and 913 MHz will also downconvert to the tag frequencies at baseband. Thus interference susceptibility can be as much as four times as high as a more conventional near-carrier tag signal.

Class 0 employs an interesting variant of the binary tree for medium access control. The scheme is depicted in a simplified form in Figure 8.9 through 8.13. The reader starts tree traversal with a special sequence. When tags hear this sequence, all tags backscatter the first bit of their ID. If the reader hears only a 0 or only a 1, it echoes that bit, by implication traveling down that branch of the tree. If it hears 0 and 1 symbols, it can choose to echo either bit at random. Each tag that hears the bit it just sent backscatters the next bit of its ID; if a tag hears the opposite bit from that which it sent, the tag goes mute until the next traversal. When the reader has echoed all the bits in a tag ID (four bits in this very simplified version), there should be only one tag left responding, and all the bits of its ID have been read. Thus, the tree is traversed one time for each tag that is present, and no empty branches are traversed. The tags need only interpret one bit at a time and require minimal memory; the reader's job is also relatively simple as it need not keep any records of which nodes contained what tag replies, but merely be guided by the tag response to each bit.

This method of proceeding down the tree is relatively easy to implement and efficient in terms of symbols transmitted. However, several problems arise from the procedure. Because the receiver must detect tag replies within a few microseconds of the end of a reader symbol, it is indispensable to use tag symbols well-separated in frequency from the reader symbols (so that the reader transients can be readily filtered), which as we noted causes out-of-band and interference challenges. The procedure as described results in the reader transmitting every bit of each tag's ID. Since a reader transmits at a high power level, it is easy to intercept; if a clear line of sight is present a listener with a directional antenna can capture the reader signal from several kilometers away. When the reader echoes all the tag bits, the tag IDs can therefore also be readily intercepted, which may reveal information a person or company would prefer to keep private.

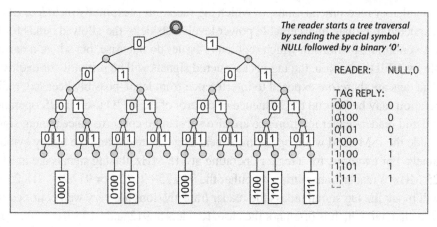

Figure 8.9
Simplified Class 0 MAC scheme, step 1.

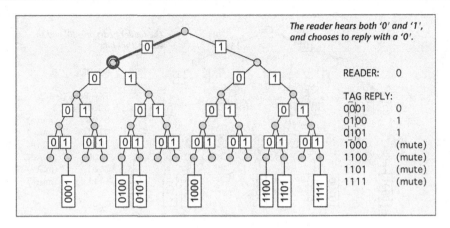

Figure 8.10
Simplified Class 0 MAC scheme, step 2

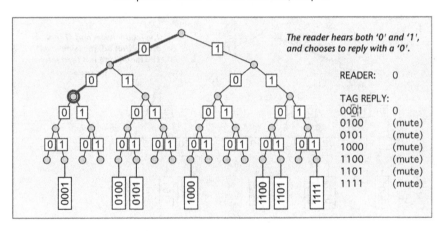

Figure 8.11
Simplified Class 0 MAC scheme, step 3

A more subtle problem arises as a consequence of the interactions between tags that we mentioned in section 4.2 of chapter 7. When many tags with ID numbers that start with the same set of bits are placed in a dense array, they will all backscatter together (until we have proceeded far enough into the tree for the ID numbers to differ). Since the power received by a single tag is influenced by the presence of other tags, not only will some tags be shadowed, but the depth of the shadow will vary as the other tags modulate the state of their antennas. An example of this phenomenon is shown in Figure 8.14 In experiments conducted with Dan Kurtz of WJ Communications, the author observed tags mistaking the change in power due to such backscatter modulation for a **NULL** symbol from the reader, causing them to believe mistakenly that a new traversal had begun and re-enter after having become mute. If such a tag 'captures' the traversal by being echoed by the reader, it will inevitably fail to be read,

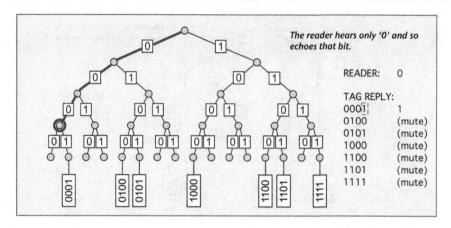

Figure 8.12
Simplified Class 0 MAC scheme, step 4

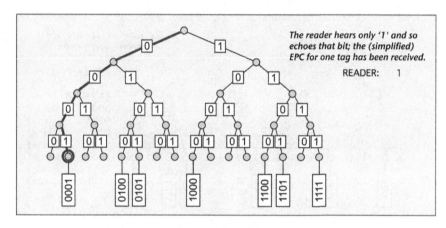

Figure 8.13
Simplified Class 0 MAC scheme, step 5. Images for steps 1–5 used by courtesy of RFID Revolution, LLC.

since the tag and reader disagree about which ID bit is being transmitted. The reader will terminate the traversal when the ID and CRC bits have been received, not realizing that the bits received are actually from two partial IDs, and will find an invalid CRC and reject the read. When this happens frequently, reading efficiency is greatly impaired.

Finally, a procedure based on the tag's unique ID can't be used if the tag has not yet been programmed and carries only a default ID value (e.g. all-0's).

In order to mitigate some of these problems, the standard provides alternative bases for singulation. The reader can choose to singulate based on any of three ID numbers. ID0 is a random number generated at the time a traversal begins; ID1 is a 64-bit random number stored in the tag at the time of manufacture; ID2 is the tag's unique ID (typically an ***electronic product***

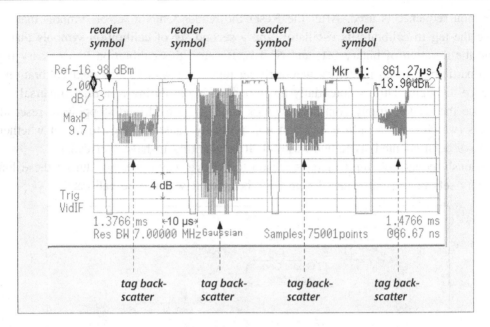

Figure 8.14

Received power vs. time for an antenna placed near an array of Class 0 tags with similar IDs, being used to singulate. Note relatively slow 25-microsecond symbol times are used here. Image used by courtesy of WJ Communications.

code, EPC). If either ID0 or ID1 are employed for singulation, most of the security and collective modulation problems are solved or greatly ameliorated. (The choice of which number to use is made with the rather obscurely-named **SetNegotiationPage** command.) The disadvantages are that because the new ID's are random, there is some chance that two tags will have the same random number. In this case, one or both will fail to be counted. In addition, the tag ID is not read until a tag is singulated. (In this case the reader does not echo the tag ID bits but sends random data to cloak the tag information.) This means an additional step must take place after each tree traversal to complete the inventory, reducing the net count of tags per second.

In addition to the special symbol **NULL**, which is used to initiate tree traversals and also to induce certain state changes in the tags, there are explicit reader commands. Commands are of fixed length (8 bits) with a single parity bit; in addition to **SetNegotiationPage** there are commands to reset the flags that indicate that a tag has been counted, force a tag to be dormant or mute, a **Read** command to read the tag ID, and a **Kill** command. **Kill**, when accompanied by the correct 24-bit password, is supposed to render the tag IC inoperable. The tag's unique ID2, normally an EPC, is protected by a 16-bit CRC.

The tag state diagram is rather complex (Figure 8.15). A tag is Dormant when it is powered up by a **Reset** (which is just 400–800 microseconds of CW from the reader. The high-frequency tag backscatter symbols require an accurate time base, so a special calibration sequence is used. After the **Reset** the reader sends a set of symbols that enable the tag to calibrate its oscillator, and a second set of calibration symbols that define the duration of binary 0,1, and **NULL**. Because this calibration is necessary for a tag to participate in an inventory, late-arriving tags must wait for the next calibration sequence to begin. This is potentially an important issue: during a normal traversal sequence the reader simply continues to look for tags to read, and stopping to reset all the tags (which takes a millisecond and will probably cause the tags to forget whether they have been counted) is rather wasteful. It is tempting to have the reader continuously count tags until there is reason to believe it is done, but during these long inventory sets no tag can enter the process, because it will not be calibrated.

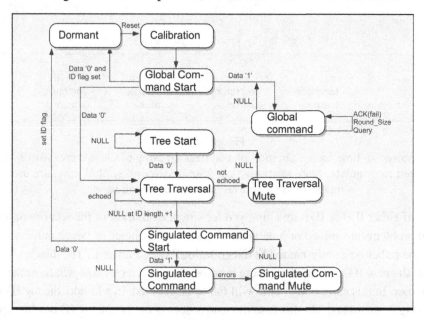

Figure 8.15
Simplified Class 0 state diagram.

A fairly typical path of a tag through the state diagram is depicted in Figure 8.16. After calibration, all calibrated tags will normally go into Global Command Start. Although several commands are possible, the one command typically issued is **SetNegotiationPage**, which tells the tags whether ID0,ID1, or ID2 will be used for singulation. The reader then issues a **NULL** to move all the tags back to Global Command Start. From there, the reader sends a binary '0'. Tags whose ID flat is set − that is, tags that remember they have been counted, possible in a previous inventory operation − move back to Dormant, but the other tags enter the Tree Traversal state, and proceed to send their ID bits. As long as a tag's ID bits are echoed, it remains in Tree Traversal; if the reader sends a different bit, the tag moves to Tree Traversal Mute and waits for the next **NULL** to rejoin. Once all the ID bits

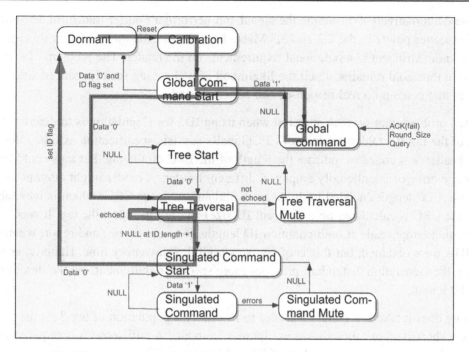

Figure 8.16
Typical path of a tag through the Class 0 state diagram during an inventory operation.

have been sent, the reader issues another **NULL**, and the (presumably unique) tag that has been singulated can be addressed individually. The same NULL pulls tags that were not singulated (and therefore transitioned to Tree Traversal Mute) back to Tree Start, from whence the same binary '0' that causes the singulated tag to flip its ID flag starts the other tags into a new tree traversal.

One of the interesting problems that arises in implementing the standard is to know when you're done. How does the reader know when a tag is trying to reply? How does one distinguish between noise and signal? One possible approach is to continue until a certain number of total tag reads produce CRC errors, or alternatively until a certain number of consecutive reads are in error. When the error criterion is reached, the reader turns off transmit power and executes a frequency hop; once the new channel is stabilized, the process beings again. In this approach, complete binary-tree traversals may occur based only on noise, and one would expect that on occasion (roughly one of every 65,000 reads) a CRC would check against the data simply at random. That is, we would see a tag that is not present: a ***ghost tag***. Ghost tags can be a problem in applications where readers can be expected to perform tens of thousands of reads per day and a real tag may only be read a few times, so it may not be realistic to require that a tag be read five or ten times consecutively in order to be considered real.

One could alternatively oversample the signal and perform a Fourier transform, to verify that backscatter power in the 2.2 and 3.3 MHz channels exceeds the noise, but this approach imposes more stringent computational requirements on the reader. The problem of deciding on a noise threshold remains, albeit the likelihood of a ghost tag read is reduced since traverses that contain no real responses can be terminated.

A related problem that was encountered when using ID2 for singulation is to determine the length of the tag ID. Older versions of EPCglobal's tag data specification assumed that the EPC's header was coded to indicate the length of the EPC on the tag, but tags could be encoded in error or intentionally employ a different header. A reader might attempt to reader the EPC length encoded in the header, find an erroneous CRC (which is inevitable if in fact the CRC is calculated on a different ID size) and fail to report the tag. It was possible to attempt reads at both common ID lengths (64- and 96-bit) and report whatever valid ID's were obtained, but this is of course wasteful of inventory time. Happily, as we will see, the Generation 2 standard provides more specific requirements on the description of the ID length.

How long does it take for a Class 0 reader to inventory a population of tags? At the relatively short symbol times shown in Figure 8.7, about 1.4 milliseconds is required for the tag to send and reader to echo 96 bits of ID and 16 bits of CRC. This corresponds to a raw read rate of around 700 tags per second. However, read rates in the field are often slower. Many commercial readers used a longer symbol time (25 microseconds) to get better signal-to-noise performance and narrower reader spectra; in this case, the peak achievable read rate is reduced to around 350 tags/second. The fact that late-entering tags will be missed implies that it is necessary to restart the protocol (that is, to issue a RESET and calibration sequence) fairly frequently. A more substantial source of overhead is the need for many readers to communicate with a host over a serial or networked connection after one or a few read attempts are completed; this communication can take tens of milliseconds. The result is that actual peak read rates are generally in the range of 200 tag/second for a moderate population of tags, and the number of times a given tag can be read in one second may be as low as 10 for real commercial readers. This low attempt rate has significant impact on the ability of a reader to read fast-moving tags in a conveyorized environment. In Class 0 operation it is a good practice to provide a sensor to ensure that a tag inventory attempt is always launched when a tagged object is properly positioned with respect to the reader antennas.

Because the Class 0 protocol document described Class 0 tags as being factory-written, no provisions for writing new data to the tags were specified. However, in practice it is often the case that one needs to be able to write an EPC to a tag. Therefore, field-writeable tags were produced, and since the commands and memory organization were not specified in the standards document, different vendors used different, mutually incompatible approaches. Happily this oversight has been resolved in the second-generation Class 1 standard.

8.3.2 EPCglobal Class 1 Generation 1

Class 1 tags are passive tags able to backscatter a unique ID to a reader. The Class 1 category assumed that tags could be re-written at least once, and in fact all commercial implementations the author is aware of permit repeated programming of the tag ID. Tags with both 64-bit and 96-bit Electronic Product Codes saw commercial implementation, although again at the time of this writing 64-bit tags mostly remain in old tag collections. Alien Technologies and Rafsec, among others, sold substantial commercial quantities of Class 1 tags.

Tags are to support **LOCK** and **KILL** commands. The **KILL** command is protected by an 8-bit password, and a timeout period after a **KILL** attempt to try to thwart dictionary attacks on this very short key. (However, the timeout isn't much protection, and Class 1 tags can hardly be considered secure against **KILL** attacks.) Unlike Class 0, the Class 1 protocol employs a fairly conventional packetized half-duplex protocol in which the reader sends a complete command packet, and then tags transmit a complete reply.

The Class 0 and Class 1 standards documents are resolutely incompatible, specifying different tag symbols, medium-access control, command sets and state diagrams. The one exception is the reader symbology (Figure 8.17). The Class 1 reader symbols are qualitatively similar to the Class 0 symbols, both being pulse-duration-encoded amplitude-shift-keyed data. The alternate symbol set for Class 1 readers is nearly identical to that used for Class 0 binary data, although the special **NULL** symbol is not used in Class 1. The default or "base" set of reader symbols is qualitatively similar but uses shorter pulse durations and thus provides slightly higher average power to the reader, at the cost of a slightly wider transmitted spectrum for the same data rate. The raw reader data rate at the default timing is about 70 kbps.

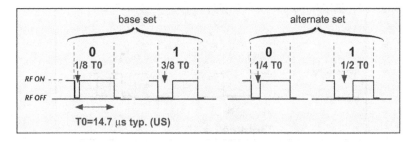

Figure 8.17

Class 1 reader symbols, depicted as baseband levels. Actual transmitted symbols consist of a high RF (900-MHz) amplitude ("RF ON") with low-amplitude or zero-power excursions ("RF OFF").

The sequence of communication for most commands is shown in Figure 8.18. A transaction gap longer than any other intentional interruption in RF signals the beginning of a command. The gap is followed by 64 microseconds of continuous RF transmission, after which the reader transmits the symbols corresponding to a command and corresponding parameters. (More information on the contents of these fields will be found below.) A terminating binary '1' signals the end of the command frame (EOF); after a sync interval during which the tags may reflect on the advisability of speaking up, another binary '1' initiates the time period during which tags may respond to the command. Unlike Class 0, each reader command contains its own startup and synchronization, so tags have a new chance to enter an inventory at every transaction gap. On the other hand, the protocol is half-duplex, with readers talking for a millisecond or so followed by tag replies, so for the same reader data rates it is somewhat slower than a Class 0 exchange.

The approach to tag modulation, sometimes known as F2F, is quite different from Class 0, in accord with the use of a packetized approach. The symbols are depicted in Figure 8.19.

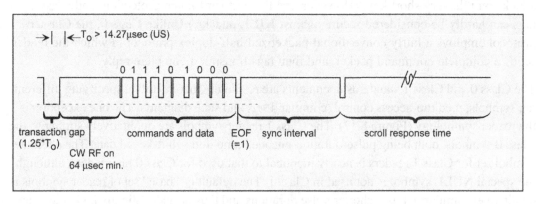

Figure 8.18
Class 1 reader transmission sequence.

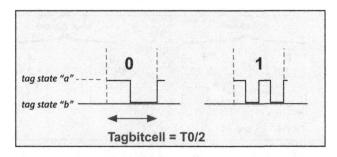

Figure 8.19
Class 1 tag symbols; tag states "a" and "b" may be any states between which the radar cross-section differs in amplitude or phase.

A state transition occurs at the edge of every symbol. When a binary '0' is transmitted, a single additional state transition is present in the center of the symbol; when a binary '1' is to be sent, three additional transitions (corresponding to a doubling of the frequency of the fundamental) are present. Although the approach is basically frequency-shift-keyed as in all backscatter coding, the frequencies employed are much lower than those in Class 0. The tag data rate is about 140 kbps when the default T0 value is used. Higher tag data rates make sense, both because tags usually have a lot of data (the EPC) to send, and because regulation of tag transmissions is generally less stringent than the corresponding rules for readers due to the much lower power levels used. The peak values of the tag spectrum will be around 140 and 280 kHz displaced from the carrier frequency for a 140 kbps return data rate; the latter frequency is at the edge of a 500-kHz US ISM-band channel.

The Class 1 protocol's approach to medium access control is much less specific than that of Class 0. Class 1 provides some specialized commands to facilitate a binary tree approach to collision resolution, but does not specify any particular way to navigate the tree, and alternative commands for accessing the EPC of a tag are available. Thus there are several rather different ways to use a Class 1 reader, appropriate depending on the number of tags one expects to encounter simultaneously in the read zone.

The simplest approach is to ignore the possibility of collisions. A reader can simply and repeatedly issue the command **ScrollAllID**, causing any tag that hears the command to backscatter its ID and CRC. The approach is sometimes known as *Global Scroll* mode. This scheme unsurprisingly works just fine when only one tag is in the read zone, and it is relatively fast. When more than about three tags are present and their response is of comparable magnitude (i.e. when they are at similar distances from the reader antenna), collisions will nearly always occur and this approach is very ineffective. However, a slight elaboration does remarkably well with small tag populations. The reader issues a ScrollAllID command until it obtains a readable reply with valid CRC; it then issues a **Quiet** command with the ID just received as the argument. The tag that successfully replied is placed in the Quiet state and no longer responds (at least until it loses power). This frees the medium up for the next-most-powerful or next-luckiest tag. Each time a tag is read, it is placed in the Quiet state, so that the remaining population of unread tags shrinks and collisions become less likely. This approach works reasonably well for tag populations up to around 6–8 tags, particularly if circumstances are such that the tags are likely to receive differing power levels, so that the reader can proceed from strongest to weakest response. It is very simple to implement, but the reader needs to transmit the whole EPC of each tag that must be sent to Quiet, and do so using the reduced data rate available to the reader, so this approach is around 3 times slower than Global Scroll.

Finally, the reader may use the **PingID** command, which is designed to aid binary tree navigation. A **PingID** command provides a filter consisting of a starting location and a bit pattern; this filter is known as the **mask**. Tags whose EPC matches the mask bits at the relevant location respond to the **PingID**; non-matching tags do nothing. After issuing the command, the reader uses isolated binary '1' bits to define eight reply slots, known here as bins; the scheme is depicted in Figure 8.20.

A tag chooses a slot to reply in based on the three bits of its ID following the mask bits (Figure 8.21), so simply by observing which slot a reply occurs in, the reader may deduce the next three bits of the ID of the tag or tags replying.

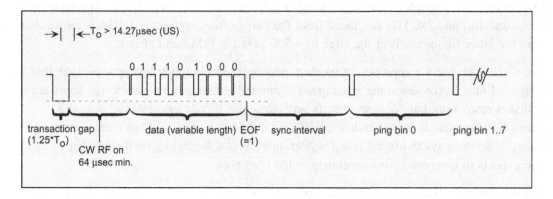

Figure 8.20
The PingID command is followed by bin markers defining eight reply slots. The detailed data structure of the Ping command is explained below.

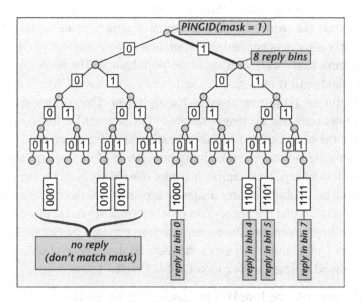

Figure 8.21
Simplified example of binary tree navigation using the PingID command.

The tag reply actually contains the next 8 bits of its ID, so in principle the reader could read those bits, if no collision occurred, and jump ahead through the tree. However, since no error checking of the last 5 bits is provided, this is not entirely reliable, and some algorithms simply note the presence of a reply and step three bits deeper into the tree.

Since there are essentially no restrictions on the contents of the mask in PingID, reader designers have considerable leeway on how they choose to navigate the binary tree. For example, if EPCs of interest are believed to all come from a given manager number and have the same model number (SKU), the mask can always start at this point, thus very much reducing the scale of the binary tree to be searched at the cost of eliminating any tags in other parts of the tree from consideration. A reader may also choose to follow a tag's bin replies all the way through the EPC, or short-circuit the process at some point and issue a **ScrollID** command, which causes a replying tag to backscatter its complete ID. If more than one tag was actually replying in a given bin (and therefore sharing the same mask bits), a collision may result and the CRC verification will fail. Just as in Class 0, the reader faces the possibility that a number of tags with nearly identical EPCs may be present in the read zone; to ensure a uniformly populated tree segment, the reader can use the CRC bits, which should be randomly distributed, to perform singulation. An abbreviated sequence using the optional command **PingScroll** can be used to cause a tag to immediately reply in a bin with the remainder of its ID, and then continue on to the next bin without reissuing a command and mask, while the (presumably counted) tag transitions to the Quiet state. A reader can choose to issue a scroll command after each ping stage, on the assumption that the time wasted in a collision is modest if collisions are rare enough. The reader can declare the process done when a final ping command on the lowest previously-occupied level of the tree garners no replies.

Whatever approach to navigation is employed, the procedure is likely to be unfriendly to tags arriving mid-inventory: if they happen to lie in a portion of the tree that has already been searched, they will not match a filter value (except by chance) and will not reply until the next tree traversal is initiated.

The state diagram for Class 1 operation is a bit simpler than Class 0 (Figure 8.22). Tags power up and enter the Awake state unless they have a persistent timer telling them that they were recently counted, in which case they become Asleep. In cases where tags might have been left in the Asleep state in a prior inventory but it is desired to count all the tags present in the read zone, the reader should issue a **Talk** command before performing any other inventory actions, to ensure that all tags are awake. Tags do not automatically transition to Asleep after replying with their ID, except when they are replying to a **PingScroll** command and hear the appropriate Bin signal after their reply. The protocol

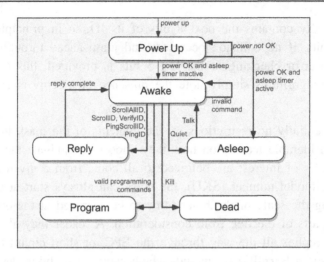

Figure 8.22
Simplified state diagram for Class 1 tag.

calls out explicit programming commands **EraseID** and **ProgramID**, and specifies a memory organization for the tag EPC and CRC. Tags can be programmed directly from the Awake state without being singulated, which is useful for programming tags that have not received an ID and are thus awkward to singulate. The command **VerifyID**, which causes any unlocked tag that hears it to return its CRC, ID, and password, is also useful in dealing with tags that have not yet been programmed, or tags that may have been programmed in error. Some readers will also report the results of a **VerifyID** command without validating the CRC; this permits low-level debugging of tag responses that would otherwise be suppressed when CRC validation fails.

The general arrangement of the reader command bits is shown in Figure 8.23. All commands start with a preamble of 20 binary '0' symbols to help the tag recognize and synchronize with the reader packet. All commands are 8 bits in length and protected by a single parity bit. Most commands use a mask, and must therefore define the starting location and length of the mask as well as the data; each block is protected by a parity bit.

The measured baseband symbols encountered in a typical reader-tag exchange in global-scroll-like operation are shown in Figure 8.24. Each command is preceded by the preamble (spinup) of 20 binary '0' symbols. The reader begins an inventory with a **Talk** command to ensure that tags are Awake, and then issues **ScrollAllID**. The single tag responds with a preamble of 7 binary '1' symbols and a binary '0', followed by the CRC and EPC. The whole process, running at 70 kbps for the reader and 140 kbps for the tag,

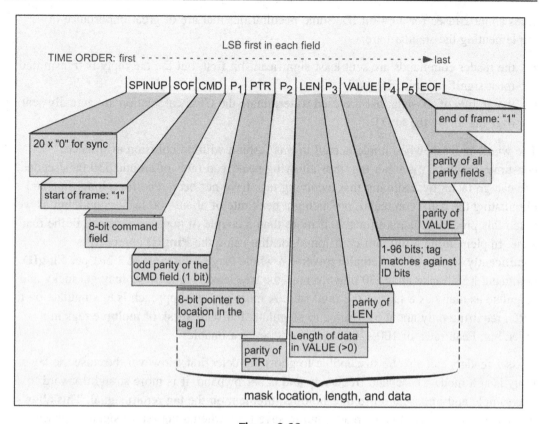

Figure 8.23
Class 1 reader command structure.

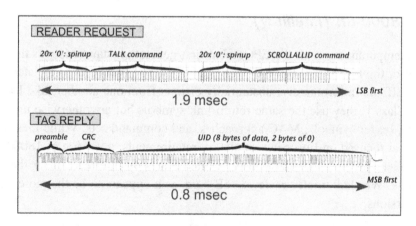

Figure 8.24
Example of the exchange between a commercial Class 1 tag and commercial reader.

takes about 2 msec for a 64-bit ID. Some peculiarities that are of great importance in implementing the standard are:

- the reader commands are sent least-significant-bit first, but the tag reply is transmitted most-significant-bit first; and
- the 16 bits of '0' data that are used to terminate the CRC calculation are actually sent over the air by the tag (!).

The whole process, when implemented in this fashion with no collision avoidance, consumes around 3 msec per tag, thus allowing peak read rates of around 330 tags/second. We can go faster by assuming that incoming tags have not been read for a long time and eliminating the **Talk** command, producing a peak rate of about 500 tags/second, but as we noted this procedure is questionable if more than a couple of tags are expected in the read zone. Implementing a full anti-collision algorithm using the **PingID** approach is significantly slower. If we actually traverse a whole (say 64-bit) tree at 3 bits per **PingID** command it will take about 30 msec to read the tree leaves, though we may get lucky and singulate as many as 8 tags in the final step. A more prudent approach is to singulate on the CRC, requiring only about 5–7 msec to singulate, but with a risk of multiple tags in a given bin. Peak rates of 100–200 tags/second are reasonable.

Class 1 readers can also be susceptible to ghost tag detection. However, because the tag reply is at a modest baseband frequency and is narrowband, it is more straightforward to oversample and impose a noise threshold requirement on the tag return signal. This allows the reader both to avoid ghost reads, and to save time when a tag return signal is absent, or degrades partway into a read (which will happen when the forward link power is marginal; the IC runs out of stored power partway into backscattering its EPC).

8.4 ISO 18000-6B (Intellitag)

Roughly contemporaneous with the EPCglobal first-generation standards was the development of the ISO 18000-6A and −6B standards. The two parts of the standard, −6A and −6B, are unfortunately substantially distinct from one another, like EPCglobal Class 0 and Class 1: they use the same return-link symbols but are otherwise incompatible, with differing reader symbols, MAC approaches, and command sets. While these standards were somewhat focused on European regulatory requirements, a version of 18000-6B has also been implemented commercially in the United States by Intermec, under the trade name Intellitag. We shall briefly review −6B here, as it appears to be the more widely used of the two versions.

18000-6B is a packetized, reader-talks-first standard. The reader symbols are Manchester-encoded ASK, with data rate of 10 or 40 kbps (Figure 8.25). Manchester coding uses a state transition in the middle of a symbol time; the direction of the state transition (low-to-high or high-to-low) indicates the identity of the bit. In Manchester coding, the RF power is in the "low" state half of the time when the reader is modulating. The power delivered to the tag can be traded against the ability of the tag to see the reader symbols by varying the depth of modulation; the protocol allows either 15% modulation (lots of power, poor symbol definition) or 99% modulation (prominent symbols but 50% reduction in forward-link power).

The tag symbols are FM0, which is also used in 18000-6C (EPCglobal Class 1 Generation 2); we introduced FM0 in section 4 of chapter 3, and will have more to say about shortly, as it is also used for the tag coding in EPCglobal's second-generation standard. The tag datarate is 40 kbps, so each symbol lasts 25 microseconds. Tag transmissions are preceded by a preamble composed of 8 binary '0' symbols, followed by the sequence 1_v-0_v-0-1-1_v-0, where the subscripted v indicates a violation of the normal FM0 rules (Figure 8.26).

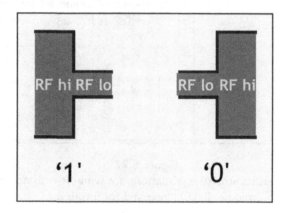

Figure 8.25
ISO 18000-6B reader symbols.

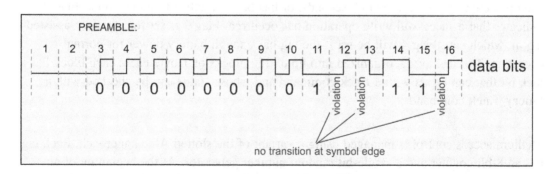

Figure 8.26
ISO 18000-6B tag packet preamble. The tag is silent for at least two symbol times after a reader command. Violations are symbols with no state transition at a symbol edge.

Violation symbols lack a state transition at the end of the symbol, as is normally required by FM0.

A reader packet (Figure 8.27) starts with 400 microseconds of CW power to ensure that tags are ready. The reader then sends one of two (currently) possible preambles, both containing symbol rule violations to help the tag uniquely identify the preamble. Reader packets are protected by a 16-bit cyclic redundancy check (CRC), as are tag replies.

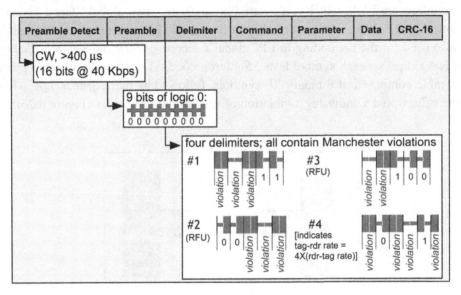

Figure 8.27

ISO 18000-6B reader packet structure. Violations are symbols with no state transition in the middle. RFU = "reserved for future use".

Tags are required to maintain eight 1-bit status flags, though only four are currently defined. Flag 1 is the **Data Exchange Status Bit** DE_SB, and is set to 1 unless the tag has been powered off for more than 2 seconds, or has been initialized by the reader. Flag 2 indicates that a successful write operation has occurred. Flag 3 is set for a battery-assisted tag, in which case Flag 4 will be set if the tag has enough battery power for normal operation. Tag memory is organized into up to 256 blocks of 1 byte each; each block has a lock bit that can be set with a **Lock** command and whose status can be checked with a **Query_Lock** command.

Medium access control is managed using a variant of the slotted Aloha approach. Each tag has an 8-bit counter and a single-bit random number generator. At the beginning of an inventory, all tags set the value of COUNT to 0 and transmit their unique identifying number. If the reader is able to read a number, or sees no reply at all in the slot, it sends a **Success** message. When tags hear the **Success** command they decrement their count by 1.

If the reader detects a collision between tags, it sends a **Fail** message. If the reader sees a good-looking response with a CRC that doesn't check, it may send an optional **Resend** command that causes the tag to backscatter its ID again. When tags hear a **Fail** message, if their COUNT is non-zero they increment it by 1, If COUNT = 0 and the reader sends **Fail**, the tag generates a random bit and increments COUNT if the bit is 1; otherwise it tries sending again. The net effect is that when there are only a few tags present, COUNT will be 0 or close to 0 and the tags will send their UIDs. When many tags are present, frequent collisions will drive COUNT to larger values until good reads or empty slots start to occur, at which point tags will be able to again count down to a reply.

The tag states include *Power-off*, *Ready*, *ID*, and *Data_exchange* (Figure 8.28). Tags are in the *Power-off* state until they power up (who would have thought?). Group selection commands are used to place the desired subset of tags to be inventoried in *ID*. Tags in the *ID* state are participating in an inventory, in essence trying to be identified by the reader. Tags that have scattered their ID can transition to *Data_exchange* upon receiving an appropriate command with their ID in it. A tag whose ID number is known *a priori* can be moved directly into the *Data_exchange* state.

The mandatory commands include eight variations of **Group_Select**. Each Select-type command includes a starting address, byte mask, and eight bytes of mask data. Starting at the starting address, each byte in memory is assigned one of the eight bits in the byte mask. Each byte is multiplied by the mask bit (so that if the mask bit for a given byte is 0,

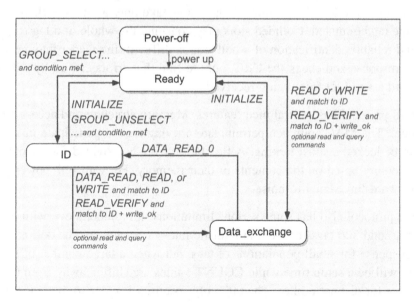

Figure 8.28
Simplified tag state diagram for ISO 18000-6B-compliant tag.

that byte is taken to be all-0). The masked bits are then combined into a 256-bit binary number which is compared to the number supplied in the mask data; various commands allow tags to be selected for an inventory if their masked memory is the same as the data, different from, larger than, or smaller than the data. The tag flags can also be used for selection operations; they are treated as a single byte and masked bitwise by the byte mask.

Tags must implement the **Success** and **Fail** commands described above in connection with managing tag inventories. In addition, they must implement **Initialize**, which returns a tag to the *Ready* state and resets Flag 1. The **Read** command is accompanied by a unique ID; a tag with that ID transitions to *Data_exchange* and backscatters 8 bytes of memory starting at the address cited in the command. **Write** allows the reader to write to a single byte of memory, unless that byte is locked; the tag replies with an error code if a write to a locked byte is attempted. A special command, **Query_Lock**, is provided to establish the lock states of bytes in memory.

How long does it take to read a tag? If we ignore any selection operations consider only the actual counting operation, presuming that enough time has passed so that the tags' COUNT values allow mostly collision-free processing, we can estimate inventory speed. At 40 kbps it takes about 1 millisecond for the reader to transmit a **Succeed** or **Fail** command. The tag replies with 8 bytes of data (a 64-bit ID) and 16 bits of CRC in about 2.6 milliseconds. The whole process thus ideally takes about 3.6 milliseconds, allowing roughly 275 tags/ second to be counted. Actual performance will not be this fast: in general we must expect some fraction of empty slots and collided slots in a large tag population, due to the stochastic nature of the MAC algorithm. Empty slots take only a bit more than the 1 ms needed for the tag command. Collided slots may consume the whole of a tag response time, since fast and reliable identification of a collision is difficult: the reader may need to listen to the whole response and check the CRC, only to find that errors caused by a collision have prevented a valid ID from being received.

The 18000-6B protocol has several nice features. Memory structure and access are specified in the protocol. The MAC approach permits late-arriving tags to participate in an inventory. Memory can be locked against writing. A flexible scheme is provided for selecting subsets of tags to inventory based on the contents of their memory. Forward-link power can be traded for forward-link signal-to-noise.

However, the protocol also has some serious limitations. It is rather slow, with an ideal peak rate of around 300 tags/second and realistic rates rather lower. The design of the MAC allows fast response for small populations of tags, but when a large number of tags is present there will be a setup time while COUNT values are shifted away from 0 to allow enough spacing for distinct replies. The tag reply spectrum is close to the reader carrier signal (roughly 40 kHz away), making it challenging to filter out phase and amplitude noise from reader transmit leakage. Multiple collocated readers will interfere with one another unless they happen to enforce stringent transmit spectral masks (which is not required by the protocol document). Singulated commands require the reader to send the unique ID of a tag, and have no encryption, so an eavesdropper can readily intercept tag IDs and other detailed information. Lock commands are not protected by a password; instead it is assumed that the memory will be locked prior to placing the tag in service. Finally, there is no provision for killing a tag.

8.5 ISO 18000-6C (EPCglobal Class 1 Generation 2)

ISO 18000-6B, Class 0, and Class 1 tags and readers saw substantial commercial deployment and helped enable the early stages of RFID implementation for many supply-chain vendors and retailers. These first-generation tags and readers validated the idea that simple, inexpensive ICs and low-cost tag antennas could provide acceptable UHF performance. However, the initial protocols suffered from significant limitations some of which we have alluded to in discussion:

- Both EPCglobal protocols employ variants of binary-tree-based collision resolution and are thus unfriendly to late-arriving tags.
- None provides any link-level security during programming operations.
- All have difficulties maintaining unique sessions with tags that lack compliant EPCs or UIDs (like tags that have not yet been programmed).
- The relationship between tag and reader data rates is inflexible.
- There is no control of the reader transmit spectrum and no flexibility to adapt the reader and tag spectra to minimize interference.
- The EPCglobal protocols are susceptible to phantom (ghost) tag reads.

By the end of 2003, these limitations were sufficiently apparent that resources in EPCglobal were re-allocated from working groups supporting the first-generation protocols to a second-generation standard that could address them.

The Generation 2 standard (hereafter abbreviated Gen 2) is different in most respects from the first-generation standards. Several of the most important enhancements are:

- Flexible tag data rates
- Spectral control of reader and tag transmissions to minimize interference
- Separate protocol control bits with explicit declaration of the EPC length
- Use of Aloha-based adaptive collision resolution with a readily-variable number space (the *Q-protocol*)
- Random-number-based logical sessions allowing singulation in the presence of identical or absent EPC's
- Multiple persistent flags supporting quasi-simultaneous inventories from different readers

- Variable-length commands for inventory speed improvement
- Explicit specification of memory maps, lock and permalock provisions, and programming procedures
- Link cover coding for secure tag programming
- A compliance and interoperability testing procedure defined by EPCglobal

Gen 2 is robust and flexible, generally producing improved read performance both in single-user and dense-reader operation. Gen 2 has substantially replaced earlier protocols in many passive RFID applications. Let us try to understand how the protocol achieves this superior performance. We first insert a brief digression providing some context for the Gen 2 standard within the envisioned EPC system.

8.5.1 EPCglobal Network Protocols

The physical layer standards were conceived as part of a broader infrastructure for moving commercial data, summarized in Figure 8.29. The underlying idea is that each physical object in a supply chain is given a unique identifying number, defined by a Tag Data Specification and accessible through tag physical layer interfaces. Information about that object is stored in databases at subscriber organizations and exchanged using query and discovery services.

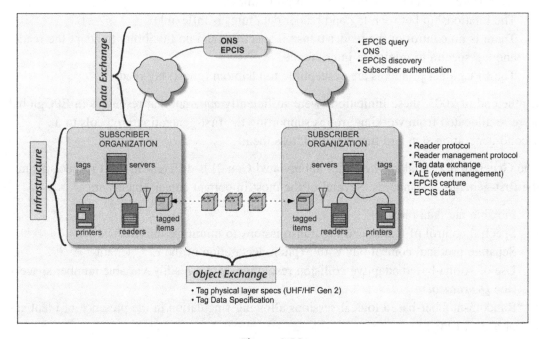

Figure 8.29
Example overall data architecture for auto-ID.

It's not clear just how much of this architecture will be implemented and how much tag data will remain within proprietary or private data networks. However, the overall functions shown here are relevant to any scheme where the tag stores only an identifying value: in order to be useful, the tag data must point to information stored elsewhere.

8.5.2 The Electronic Product Code

The founding concept of the AutoID Labs, a unique identifying number for every object (see chapter 2), is embodied in the Electronic Product Code or EPC. This code identifies the object to which the tag is attached (*not* necessarily the tag itself). A tracking tag's most fundamental task is to provide this code to a reader. Before discussing the protocol, we will take a quick look at the number that it is meant to move.

The EPC can take a number of differing forms and sizes, determined by the value of an 8-bit header. Two 96-bit examples are shown in Figure 8.30. The general ID format contains a 28-bit manager, intended to be an organization or corporation assigned a unique number by EPCglobal (a privilege for which the organization must pay a fee!). The object class is a 24-bit identifier, assigned by the organization, describes the model or type of object. The serial number, 36 bits here, allows for about 68,000,000,000 uniquely identified objects. An organization that runs out of numbers for a single model of a product can consider itself very successful indeed. The formal definition of the EPC distinguishes between the pure identifier and the actual number stored in the tag, which can contain physical-layer-specific information (like error checking) that is stripped off when the pure identifier is forwarded to a database.

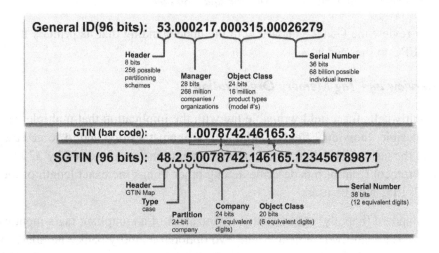

Figure 8.30

96-bit general ID and SGTIN variants of the Electronic Product Code. *Wal-Mart RFID Supplier Briefing 11/4/04.*

The SGTIN is intended to be a method of mapping the existing GTIN (bar code) identifiers used in commerce into an EPC. The GTIN header indicator digit is mapped together with the Item Reference into the Object Class field of the SGTIN. The company is mapped into the company field. A new Type field classifies the sort of object that is identified, and the partition describes how bits are allocated between company and class. The serial number concept does not exist in GTIN's, where every object of the same type has the same bar code; it is assumed that the company organization will assign a unique number to each physical object it identifies. Note that since decimal digits are being mapped into bits, some of the bits are wasted. To identify 10 digits we need 4 bits, only 3 1/3 being "used"; the same applies for larger decimal numbers.

The process may be slightly less comprehensible but more entertaining when expressed in verse:

> **The GTIN Song**
> *To convert a boring GTIN to a spiffy EPC,*
> *Thereby showing bar codes have transformability,*
> *First we write a Header with hex zero and hex 3,*
> *Then a Filter Value, not a pure identity.*
> *The Partition tells you how to read the remainder of the bits*
> *Encompassing the largest serial number that can fit.*
> *But first an Item Reference and a Company ID,*
> *And cyclic checking at the end to get redundancy.*
> *If you've followed the instructions thoroughly and well,*
> *Your EPC can indicate a cracker, bike, or bell.*
>
> *If you find this poem less useful than an application note,*
> *You're an engineer, and you can't rhyme for squat. So there.*

We can now review the Gen 2 standard with the understanding that its primary task is to provide the EPC to the reader.

8.5.3 Overview and Tag Memory Organization

Gen 2 explicitly calls for a field-writeable tag with the implication that multiple write cycles are possible. Individual fields can be locked against writing and in some cases reading, and the tag can be killed. Lock and Kill operations are protected by 32-bit passwords. Protocol Control bits describe among other things the exact length of the EPC, so any length EPC may be used.

The Gen 2 standard (happily) specifies the organization of a compliant tag's memory (Figure 8.31). There are two obligatory and two optional memory banks, numbered in binary (that is, bank '10' is decimal 'bank two'). Bank 00 contains (at least) the 32-bit KILL and ACCESS passwords. Bank 01 contains, in addition to the tag's EPC, a 16-bit **Protocol Control** (PC) word, describing the length of the EPC, as well as some optional

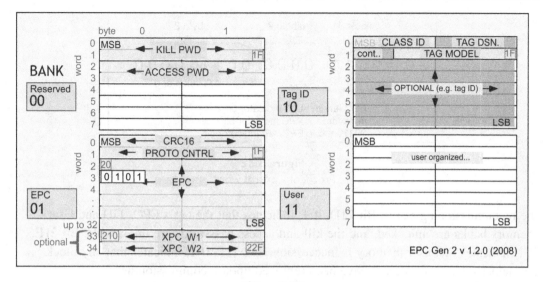

Figure 8.31
Gen 2 tag memory organization and terminology.

information about the tag, and the CRC16 used for error checking of the EPC value. Note that in a Gen 2 tag the CRC is calculated by the tag, and need not be written to memory by the reader. The optional Extended Protocol words XPC W1 and W2 support recommissioning of a tag (that is, reusing it after it has been "killed"), and simple sensor applications. The optional Tag ID bank 10 provides for identifying information related to the tag, distinct from any object to which it might be attached. Such a capability is useful for tracking tag IC manufacturing and tracking tag inventory. Finally, optional User bank 11, whose organization is not constrained except for the numbering of bytes and words, is available for any application-specific data.

The structure of the Protocol Control word is shown in Figure 8.32. The first five bits describe the EPC length in words. For example, the value shown, 00110 in binary or 6 in decimal, describes a $6 \times 16 = 96$-bit EPC. The next bit is the User Memory Indicator, 1 if user memory is present. The XPC indicator bit, if set to 1, indicates the presence of an Extended Protocol Word. Finally, the EPC/AFI indicator bit tells the reader if an ISO-compatible Application Format Indicator is to be found in the second byte of the PC word.

The XPC words are appended in memory after the EPC but are transmitted before it. Since the longest EPC length code possible is 11111 (decimal 31), in order to properly describe the length of a transmission including up to two extended protocol words and the EPC, the EPC length can be no more than 29 words, or 11101 in binary. Since 29 words provide more than 10^{139} unique identifying numbers, more than sufficient to track every baryon in the universe (if you could tell them apart), that doesn't seem like a major impediment to progress. The first XPC word uses the three least-significant bits to indicate

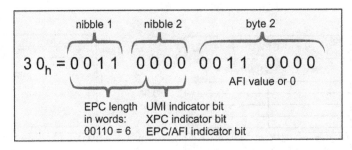

Figure 8.32
Protocol Control word.

the recommissioning status. Bit 21D, if 1, indicates that the tag's EPC, TID and User memory banks are unlocked, and the kill and access passwords are unreadable. Bit 21E, if 1, indicates that user memory is inaccessible. Bit 21F, if 1, indicates that any block permalocks that were present have been removed upon recommissioning. The most-significant bit, if 1, indicates that XPC_W2 is present and non-zero.

Locations in memory are specified by *extensible bit vectors* (EBV),which allow memory banks to be of arbitrary size; the banks are not limited to the 8 words depicted in Figure 8.31. In the EBV scheme, the first byte of an address is divided into an extension bit and seven data bits. If the extension bit is 0, the data bits contain the complete address. If the extension bit is 1, at least one more byte is to be appended to the address data. Since that byte also has an extension bit, the address can be arbitrarily long. This flexibility is achieved at the cost of the loss of 1/8 of the capacity of the address bits, since one bit of every byte is devoted to extension control even when no extension is required. Two examples of EBV addresses are depicted in Figure 8.33. In the top example, the extension bit is 0 and the value is simply that of the remaining bits, decimal 5. In the bottom example, the first byte's extension bit is 1, so the remaining bits of the byte are multiplied by $2^7 = 128$, and the next byte is examined. Since its extension bit is 0, the data in the remaining bits finishes the EBV, and the value is $128*5 + 3 = 643$.

The status of a particular tag in inventory operations is managed by five one-bit flags: four session flags S0 through S3, and a Select flag SL. The state of these flags can be set by the versatile **Select** command. They are used to constrain the subset of tags to which inventory operations apply, and to maintain the status of a tag with respect to up to four quasi-simultaneous inventory operations. We shall have more to say about this system as we proceed.

8.5.4 Reader and Tag Symbols and Coding

Gen 2 is a packetized, reader-talks-first protocol. The reader symbols are the amplitude-modulated, pulse-interval-encoded (PIE) symbols we introduced in chapter 3 (Figure 8.34). A binary '0' consists of a power-on interval followed by a power-off interval of equal duration. The

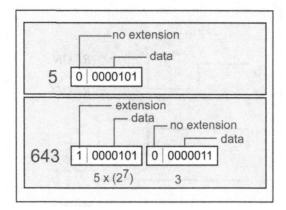

Figure 8.33
Examples of an EBV address.

total length of a binary '0' defines the time interval *Tari*; the pulsewidth *PW* is half of Tari. A binary '1' uses the same pulsewidth at the end of a longer power-on interval; the duration of a '1' can be as short as 1.5 Tari or as long as 2 Tari. Standard values of Tari are 6.25, 12.5, and 25 microseconds, corresponding to symbol rates of 160, 80, and 40 kbps.

The Gen 2 standard defines three different operating categories for readers, with corresponding limitations on the transmitted spectral width. The first, *single-interrogator* operation, imposes no requirements on the reader transmission beyond those asserted by the relevant regulatory authority. *Multiple-interrogator* operation, designed for those cases where the number of simultaneous collocated readers is modest compared to the number of available channels, places some constraints on the transmitted spectrum sufficient to minimize interference in adjacent or second-adjacent channels. *Dense-interrogator* requirements are designed to allow successful tag reading even when every available channel is occupied by a reader transmitter.

The limitations on bandwidth are expressed in terms of *spectral masks*, which show the maximum power a reader can transmit in each frequency range relative to the carrier frequency f_c. Spectral masks for multiple- and dense-interrogator operation are shown in Figure 8.35.

In both masks the limitation is placed on the integrated (total) power in a specified frequency range, relative to the total power in the center (intended) channel. The limits are expressed in dB relative to the power in the center channel, *dBch*. Frequency offsets for the multiple interrogator mask are expressed in terms of channels relative to the intended channel; in US operation, these channels would generally be 500 kHz wide. Because the channel width is fixed for a given regulatory region, multiple-interrogator requirements can in principle be met by simply reducing reader data rates, presuming reasonable symbol filtering. In contrast, dense interrogator frequencies are specified in terms of the inverse of

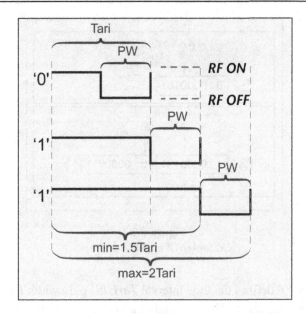

Figure 8.34
Gen 2 reader symbols.

Tari, the duration of a binary 0, so the width of the mask scales with data rate and conformance cannot be achieved by changes in rate.

The spectral mask is depicted in more familiar terms as a function of frequency offset in Hz in Figure 8.36, for the specific case of Tari = 25 μs. In this view it is easy to see that the main power in the signal is constrained to lie within a 100-kHz-wide region centered on the carrier frequency.

Dense reader operation is particularly important under European or Asian regulations, where relatively little spectrum is available. For example, under ETSI 302 208 (where Tari = 25 microseconds is most likely to be employed), the 865–868 MHz band is divided into 200-kHz-wide channels. If readers are operating in every channel, the emission masks will abut each other, and the power of a compliant interfering reader within the neighboring channel will only be down by about 30 dB. This is insufficient for interference-free operation, as a tag signal is typically 50–60 dB below the reader signal. However, if readers are limited to alternate channels, the tag signal can be centered 200 kHz from either reader signal, with a 100-kHz wide region in which the reader power is reduced by 60 dB (Figure 8.37). This is sufficient to allow reasonable tag reception even in the presence of modulated reader signals. We shall see shortly how Gen 2 arranges for the tag spectra to be appropriately offset from the carrier.

The bandwidth of the transmitted spectrum for amplitude-modulated PIE symbols is set by the width of the feature PW, which is fixed. We looked at simplified ideal spectra for PIE symbol streams in chapter 3; an ideal spectrum has a sharp feature at the frequency offset

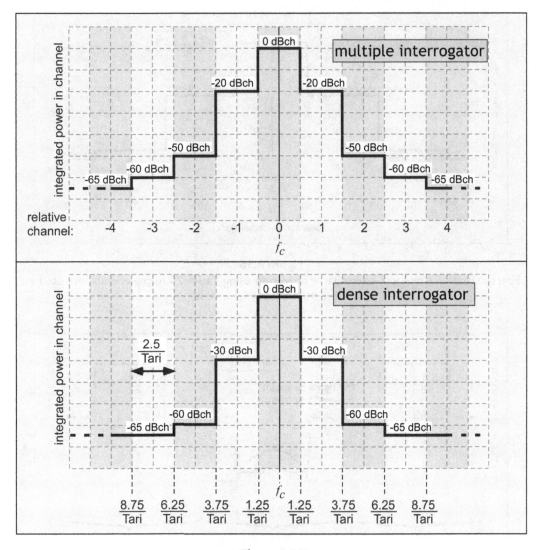

Figure 8.35

Spectral masks for multiple- and dense-interrogator operation. See text for explanation of terminology.

corresponding to a binary '0' symbol ($f = 1$/Tari), and substantial power out to about twice that frequency. The extent of the spectrum outside these fundamental limits is highly sensitive to the extent to which the edges of the pulse are smoothed. A very smooth transition from high to low RF power provides a narrow spectrum, but the link budget is somewhat reduced, because the transmitted power only attains a zero value briefly at the bottom of the smoothed pulse; if the sampling time is slightly offset from this point, the depth of modulation is reduced.

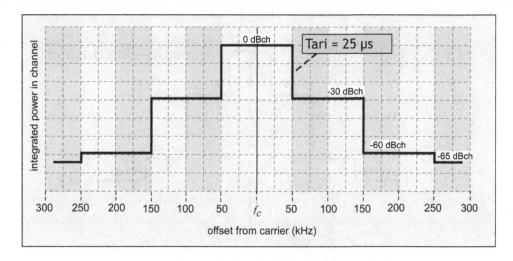

Figure 8.36
Spectral mask for multiple-interrogator operation using Tari = 25 microseconds, depicted vs. offset from the carrier in kHz.

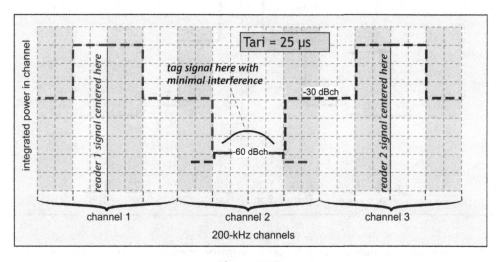

Figure 8.37
Spectral masks for dense reader operation in alternating 200 kHz channels.

Examples of measured RF power vs. time for smoothed (filtered) and unsmoothed (unfiltered) Gen 2 symbols are depicted in Figure 8.38. The unfiltered single-interrogator symbols (top) show very sharp edges; power drops by 40 dB in a period of about 3 microseconds. The tag can sample anywhere within the pulse and will detect a very low power level. In contrast, the filtered multiple-interrogator symbols (bottom), which employs a *finite-impulse-response* (FIR) filter to ensure very smooth changes in power

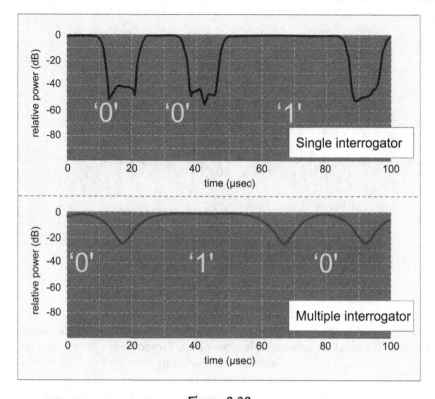

Figure 8.38
Measured transmit power vs. time for reader using unfiltered symbols (single interrogator mode) and filtered symbols (multiple interrogator mode); Tari = 25 microseconds.

levels, have very gradual transitions between high and low RF power. The lowest power level reached, at about 20 dB below the peak power, is still quite sufficiently suppressed to provide a good signal to a tag, but the smoothed transition means that if the sampling time is displaced by (say) half the pulsewidth, the apparent depth of the pulse is reduced from 20 to 10 dB; the corresponding voltage signal is 1/3 of the peak voltage rather than 1/100 as is the case for the unfiltered symbol, and is obviously more susceptible to misinterpretation due to noise and offsets. Filtering reduces spectral width but also affects link budgets, particularly for passive tags with imprecise clocks and limited signal processing.

The measured spectra for unfiltered (unsmoothed) PIE symbols – "single interrogator" operation – and optimally filtered symbols for multiple interrogator operation are depicted in Figure 8.39. It is quite obvious that the price of the extra link budget of unfiltered symbols is a very wide spectrum, with substantial power as much as 1 MHz from the carrier even for very modest 40 kbps data rates. Single-interrogator readers may interfere across multiple channels, and are inappropriate for use in critical

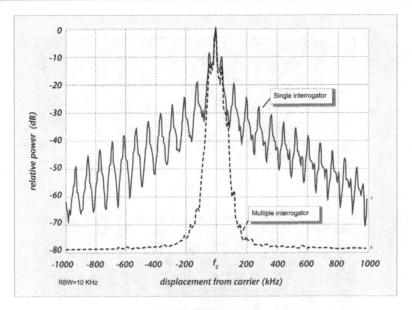

Figure 8.39

Measured transmitted signal spectrum for a commercial reader in single interrogator (unsmoothed) operation and multiple interrogator (smoothed) operation, both using Tari = 25 microseconds.

dense-reader areas except at very low duty cycle. The multiple-interrogator operating mode has a much narrower spectrum with essentially no radiated power farther than about 100 kHz from the carrier. However, by reference to Figure 8.36, we can see that this is not good enough for dense interrogator compliance: the power in the signal must be 30 dB below the carrier level for displacements of only 50 kHz from the carrier, a very stringent requirement that will at best be marginally achieved by even the multiple interrogator signal.

Smoothing a symbol by itself is can thus provide limited benefits in spectral width. The Gen 2 standard provides two alternative methods for reducing the width of the transmitted spectrum. Both approaches allow the reader to use bandwidth more efficiently than amplitude modulation alone, while behaving from the tag's point of view as a conventional amplitude-modulated signal. We examined these modulation approaches in some detail in section 4.1 of chapter 4; the reader may wish to briefly revisit that section to recall these matters to mind. The first, ***phase-reversal amplitude-shift keying*** (PR-ASK), is a variant of binary phase-shift keying, and is very similar to duobinary data transmission. The second, ***single-sideband*** (with carrier injection), removes one of the sidebands that would normally be present in an amplitude-modulated signal. Both these techniques achieve roughly a factor of two improvement in bandwidth for a given data rate.

SSB modulation in this context requires that the carrier frequency be offset from the nominal channel center during reader modulation and then returned to the center frequency during CW transmission. While this sort of operation is possible with modern digitally-synthesized transmitters, it is relatively complex. PR-ASK is simpler to implement, produces a symmetrical spectrum, and does not require any special offsets. However, there are a few subtleties to be aware of in using it. Since the pulse feature (PW of Figure 8.34) is the result of the transition of the transmitted signal between positive and negative phases, the time for the transition must be carefully managed in order to produce the correct pulse width. As we shall see shortly, the beginning of every reader transmission also contains a special delimiter symbol that is always the same duration, irrespective of the data rate; this may require special provisions in PR-ASK. Finally, any offset correction that is used in the receiver must account for the fact that even if the signal returns to full CW power before a tag response is to be detected, it may be at the opposite phase from that at which the reader transmission began.

A closeup view of measured spectra for single, multiple, and dense interrogator operation using the same Tari of 25 microseconds is depicted in Figure 8.40. It is clear at this scale that the single and multiple interrogator spectra are substantially identical in the core region within roughly 40 kHz (= data rate) of the carrier. Filtering only affects the spectra outside this region: a 40 kHz sine wave is necessary to form the binary-'0' symbol of duration Tari,

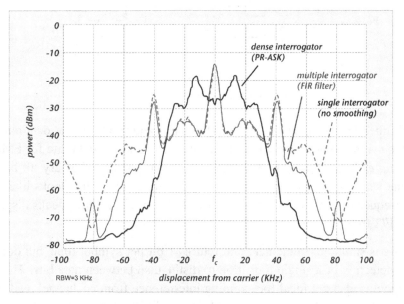

Figure 8.40
Measured transmit spectra for a commercial reader in single interrogator, multiple interrogator, and dense-interrogator modes, all with Tari = 25 microseconds.

even when optimally smoothed. It is difficult to fit this spectrum into a mask extending only 50 kHz from the carrier. (The reader may find it interesting to compare these measured spectra with the calculated spectra in e.g. figures 3.11 and 3.12. The differences are due to the finite resolution of a real spectrum analyzer, and averaging over a longer stream of symbols for the measured data.)

The dense interrogator spectrum (implemented in PR-ASK) is obviously quite distinct from the other two. The prominent peaks corresponding to the binary-'0' feature are displaced by slightly more than 10 kHz instead of 40 kHz. At 50 kHz from the carrier the spectral power is reduced by about 50 dB from the peak power, so this spectrum is amenable to robust compliance with the spectral mask. The power at the carrier frequency is considerably reduced relative to the modulation peaks, as is always the case using phase modulations; since the carrier contains no information this is a desirable result.

Gen 2 tag symbols and uplink signaling are remarkably flexible, and inevitably quite complex. The default operating mode, known as *FM0*, is reasonably straightforward (Figure 8.41). The tag changes its backscatter state at the edge of every symbol. A binary '0' has an additional state transition in the middle of the symbol; a binary '1' doesn't. All transmissions end with a 'dummy' binary '1' symbol, and always end on the same ('low') state; if the final '1' bit ends in this state there is no transition after the final dummy bit (and it could be said not to exist!).

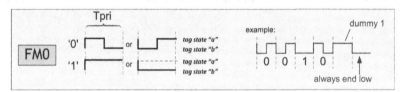

Figure 8.41
Default G2 tag signaling (FM0).

The symbol time, denoted **Tpri**, is the same for 1-bits and 0-bits, so the data rate is the inverse of the symbol time. The symbol frequency, equal to the data rate for FM0, is known as the **backscatter link frequency** (BLF). It is clear that a string of binary '0' bits looks like a clipped sine wave with a frequency of BLF, and a string of '1' bits looks like a clipped sine with a frequency of BLF/2, so we'd expect FM0 spectra to have peaks displaced about BLF and BLF/2 from the carrier.

FM0 signaling is simple to implement and returns 1 bit per symbol time, but depends on the accurate detection of a single transition to distinguish between data bits. The spectrum of the tag signal is also not ideal for avoiding interference from cochannel readers. For example, referring to Figure 8.37, in order to avoid interference from dense-interrogator readers, we could set the BLF to about 200 kHz, putting the spectrum from binary '0' symbols in the region where the reader power is minimized. However, the spectrum of a stream of binary '1' symbols would then be located about 100 kHz from the carrier, in the region where modulated reader power is only down by 30 dB, and quite susceptible to cochannel interference. (Remember what we're worried about is a different reader operating on the same channel; the reader that is trying to read the tag is transmitting CW power when the tag is backscattering, and thus emits a very narrow spectrum, as long as phase and amplitude noise are small.)

In order to provide more flexibility for noise vs. data rate tradeoffs and spectral management, the Gen 2 standard defines a second, closely-related approach to tag signaling: *Miller-modulated subcarrier* (MMS) encoding. In MMS, the data bits are first encoded in (just for fun) the opposite fashion of FM0: that is, a binary '1' is given a state transition in the middle of a symbol time, and a binary '0' is not. Let's call this the baseband encoding. Not only are the symbol definitions different from FM0, but instead of there being a state transition at every symbol edge as in FM0, in the baseband coding of MMS there is *no* state transition at the symbol edges between consecutive 1's, or between a 1 and a 0. However, there *is* a state transition at a symbol edge between two consecutive 0's.

The baseband coding is then multiplied (or added modulo 2, as you prefer) by a square wave containing M cycles in every baseband symbol, where M can be 2,4, or 8. The result is a square-wave-like signal with periodic phase inversions; if this signal were the RF transmission, we would say that it is binary-phase-shift-keyed. A simple example with M = 2 is shown in Figure 8.42.

It is important to note that MMS does not change Tpri or the backscatter link frequency. MMS encoding with M = 2 means that two cycles of the subcarrier are needed for every bit: the data rate is half of what it is using FM0 at the same link frequency, and of course in general the data rate is BLF/M. An example showing how the same data is coded at differing Miller indices is depicted in Figure 8.43. Note that, like FM0, MMS reader transmissions also have a binary '1' symbol appended at the end of the data.

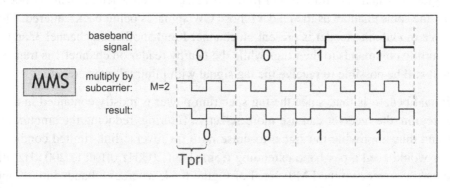

Figure 8.42
Miller-modulated subcarrier example, with M = 2.

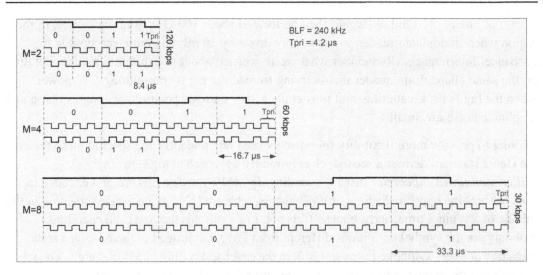

Figure 8.43
Example of encoding the same datastream using MMS with Miller indices (M = 2,4, and 8).

With increasing values of M, Tpri, the state cycle time for the tag, is fixed, but the number of cycles composing a single-bit symbol is equal to M, so the rate of transmission of data is reduced. At the same time, the spectrum becomes more narrowly centered around the major peaks, which are displaced by the link frequency from the carrier. An example of this effect is shown in Figure 8.44 for a string of about 160 random symbols for FM0, or 40 MMS symbols.

Thus, by using MMS with (for example) M = 4 and BLF = 200 kHz, we can put almost all the tag backscattered power in a narrow region 200 kHz above and below the carrier; for example, referring to Figure 8.37, all the tag power would fall in the center of channel 2 if the tag is excited by a reader transmitting in channel 1. If all the readers are dense-interrogator-compliant, and placed in alternate 200 kHz channels (as in Figure 8.37), the tag power will lie in a region in which very little reader power is present, even when there are collocated readers on the same channel as the reader whose CW signal is being backscattered. That is, even when a modulating reader is present on channel 1 and another on channel 3, and both are transmitting commands to their tags while the nearby reader on channel 1 is transmitting CW, it will still be possible to receive the tag signal with minimal interference.

An additional benefit is that, since the tag spectrum power is mostly contained in a narrower region, the receiver can use more selective filtering, reducing the amount of noise entering and thus improving the signal-to-noise ratio for reverse-link-limited conditions. A receiver would need a passband extending from about 20 kHz offset to 200 kHz offset from the carrier to receive the FM0 signal of Figure 8.44, whereas a bandwidth of about 120 to 140 kHz is needed to capture essentially all of the FM0 signal — that is, the SNR improvement scales with the Miller index, 8 in this case.

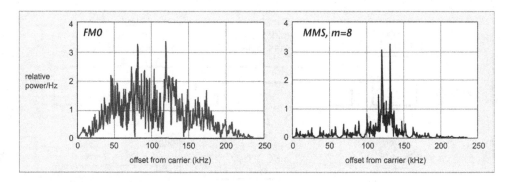

Figure 8.44
Comparison of tag response spectra for random finite string of symbols; BLF = 125 kHz for both.

In addition to the choice of whether to use FM0 or the three variants of MMS, the backscatter link frequency BLF can vary from 40 to 640 kHz. The data rate is the ratio BLF/M, so the tag data rate can range from 5 kbps to 640 kbps. The same data rate can be obtained with differing spectra and link characteristics by specifying differing Miller indices and link frequencies. This flexibility provides the user with the ability to adapt to differing conditions, but it also makes both the design of the tag ICs and the software to manage communications complex relative to earlier standards.

8.5.5 Packet Structure

Gen 2 is a packetized, reader-talks-first protocol, like Class 1 Generation 1. The reader chooses the forward-link and reverse-link parameters, and communicates these choices to the tags using the opening symbols of each packet, which are known as a ***preamble*** when the reader is initiating an inventory operation, or a ***frame sync*** when any other command is sent. Preambles set tag (uplink) parameters for the remainder of an inventory session. Frame syncs contain only the reader timing information, to help tags stay synchronized during successive operations. The baseband representation of these sequences is shown in Figure 8.45. Each sequence begins with a ***delimiter***. As we alluded to previously, the delimiter is always 12.5 microseconds long, regardless of the data rate employed by the reader or tag. This makes the delimiter relatively easy for a tag to detect, but can be awkward when PR-ASK is used: since pulses are the result of the path of the transmitted signal in phase space passing through 0 during transitions between +1 and −1 constellation points, the rate of the transition between the two states must be changed to send the delimiter except when Tari = 25 microseconds.

After the delimiter, the reader sends a binary '0' defining the value of Tari, followed by the special reader-to-tag calibration symbol ***RTcal***. As shown in the inset to the figure, the total duration of RTcal is equal to the sum of the duration of a binary '0' and a binary '1'

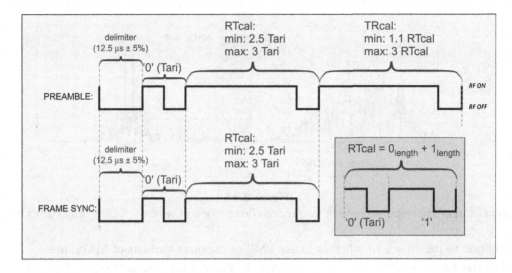

Figure 8.45
Preamble and frame sync sequences. Inset shows that RTcal = sum of duration of binary '1' and binary '0' symbols

symbol. RTcal is allowed to be from 2.5 to 3 times as long as Tari, implying that a binary '1' symbol can be from 1.5 to 2 times as long as Tari. (A longer symbol results in a lower reader data rate but is easier for the tag to distinguish from a '0'.)

The preamble, used only at the start of an inventory operation, contains the additional tag-to-reader calibration symbol *TRcal*. TRcal itself is not useful until the reader also sends the parameter *divide ratio* (DR), which accompanies the command containing the TRcal symbol. The tags that hear the command can then determine their link frequency:

$$BLF = \frac{DR}{TRcal} \tag{8.1}$$

The divide ratio can take the values 8 or $64/3 \approx 21$. Some example values for these parameters are given in Table 8.1. The tag data rate is BLF/M. So, for example, the following parameter sets all produce a tag data rate of 320 kbps:

$$TRcal = 67 \ \mu s, DR = 64/3, \text{ and } M = 1$$
$$TRcal = 33 \ \mu s, DR = 64/3, \text{ and } M = 2$$
$$TRcal = 25 \ \mu s, DR = 8, \text{ and } M = 1$$

Because the spectrum is mainly determined by BLF and to a lesser extent M, this scheme allows the user to adapt the location and width of the backscatter spectrum somewhat independently of the data rate. However, recall from chapter 4 that the reader receiver filters the received signal to remove as much noise as possible; ideally the receive filter is well-matched to the bandwidth of the signal to be detected. So it doesn't make any sense to

Table 8.1: Example Values for Tag-Reader Calibration Symbol
Duration TRcal, Link Frequency BLF, and Tag Cycle Time Tpri.

Divide Ratio	TRcal (μs)	BLF (kHz)	Tpri (μs)
64/3	33	640	1.6
	67	320	3.1
	83	256	3.9
	225	95	10.5
8	17	465	2.1
	25	320	3.1
	50	160	6.3
	200	40	25

allow the user of a reader to change the tag data rate unless the reader's receiver is also designed to make the appropriate changes in the receiver filter's passband. While it is possible to make adjustable baseband filters using modern active filter circuitry, it is cheaper to use conventional passive components (capacitors, resistors, and possibly inductors) to form the filter. Inexpensive readers may be designed to operate only in a narrow range of tag data rates; changing the requested rate in the software may produce poor results due to receiver filter mismatch with the tag signal. A user shouldn't be able to change these settings unless the reader designer has ensured that corresponding filter changes will take place!

Tag replies, in turn, are preceded by a preamble and optional *pilot tone* (Figure 8.46). The preamble is a sequence of binary symbols 1-0-1-0 with a *violation*: a symbol that does not obey the normal rules for FM0 symbol formation, and is used only in the preamble. The violation helps to uniquely identify a preamble and distinguish it from any other tag data stream. However, the violation symbol, which persists at one value for longer than any normal data symbol, has a slightly different spectrum with more energy at low frequencies. Baseband filtering in the reader receiver must be specified with the violation symbol in mind; if it is too narrow, the transient response of the filter can create a spurious zero-crossing in the middle of the violation character, making it look like a conventional 0−1 sequence and causing the reader to fail to recognize the preamble.

The pilot tone is particularly simple, being just a series of identical symbols, and helps the reader to identify the start of a tag reply. The pilot tone is not guaranteed to be unique: an EPC or other tag data could also contain 12 binary '0' symbols in a row. However, as we shall see below, tags and readers operate under fairly strict timing requirements, so the period in which the reader searches for a pilot tone is not likely to be occupied by other symbols (at least when only one reader is operating in a given vicinity!). Whether the pilot tone is used or not is determined by another reader command parameter, *TRext*. A pilot tone can also help the reader detect collisions: if a pilot tone is recognized, but no valid preamble is detected after it, the reader can guess that a collision occurred. We'll take a closer look at collisions a bit later in the discussion.

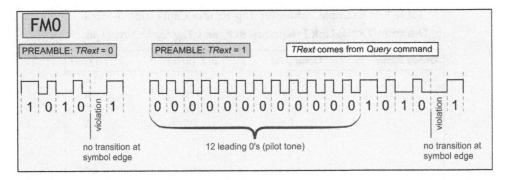

Figure 8.46
Tag-to-reader packet preamble when using FM0 encoding.

When a tag is using MMS, the preamble and pilot tone are slightly different. The default preamble contains 4 M cycles of the tag, equivalent to four consecutive binary '0' symbols. Then the sequence 0-1-0-1-1-1 is sent using whatever value of M is current. NO violation symbol is used. If TRext is set to 1 by the reader, the tag starts with 16 M cycles (16 consecutive '0' symbols, each M cycles long) instead of 4 M.

8.5.6 Medium Access Control

Instead of the binary tree variants employed by the first-generation standards, the Gen 2 MAC approach is based on a slotted Aloha variant, originally known as the *Q protocol*. The basic scheme is:

- The reader specifies the number of slots in the inventory *round*
- Each tag randomly chooses a location to reply within the round
- The reader issues short commands to mark the beginning of each slot within the round
- If a tag has chosen that slot, it replies with a random number
- If the reader can decipher the number and acknowledge it, the tag sends its EPC
- The random number exchanged between tag and reader also allows the reader to maintain a unique logical *session* with that tag, independently of the tag's EPC, or even whether it has a unique EPC. With the aid of this number, the *handle*, the reader can read from and write to the tag's memory, and perform other operations unique to that tag.

Let's look in more detail at how this works. Recall each tag has four 1-bit flags corresponding to four *sessions*. To initiate an inventory, a reader issues a **Query** command pertaining to one of the four possible sessions (Figure 8.47). The Query command contains a number of parameters, some of which control which tags will participate in the subsequent inventory. We will cover these in more detail later, so for the present we can assume that a certain subset of tags hearing the **Query** command will consider themselves participants. The command also contains a numerical parameter, Q, which is of such

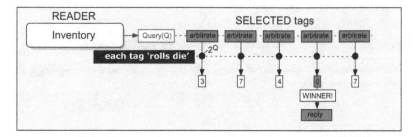

Figure 8.47

An inventory operation starts with a Query command; the example shown assumes Q = 3. Tags start in the arbitrate state; see Figure 8.52 for details of tag states.

interest that the protocol as originally proposed was named after it. Q specifies the space in which the tags are to randomly distribute themselves, and thus also the number of slots in the upcoming inventory round. The first step a participating tag takes is to create a random number whose value is between 0 and $(2^Q - 1)$. This random number specifies which slot the tag will respond. The number of slots available needs to be of the same order as the number of tags in the read zone: if there are (say) sixteen tags, 512 slots are a waste but one slot isn't nearly enough. A complete set of 2^Q slots constitutes an inventory **round**.

The use of an exponentially-expanding number space allows the single compact parameter Q to span a large range of possible tag population sizes. Q can take on values from 0 to 15, so only 4 bits are needed to specify its value, but this allows the protocol (at least in principle) to work smoothly with anything from 1 tag in the field to $2^{15} = 32,768$ tags. Unlike the ISO 18000-6B COUNT protocol, the Q protocol can be quickly adapted for a wide variety of tag populations, and can benefit from *a priori* knowledge of the size of that population on the part of the user.

If a tag's random number is equal to 0, the tag responds in this (first) slot; otherwise, it records the value of this number in its **slot counter** and waits. The tag that rolls a 0 generates a new 16-bit random number RN16, and sends a short packet containing only the preamble (and pilot tone, if applicable) plus the RN16 value (Figure 8.48). This short packet has no error checking (that is, no parity bits or CRC), because the reader is going to echo the value back to the tag; if the value is wrong, the acknowledgement will simply fail and the inventory will continue. If the reader hears this reply packet in the right time period after the **Query** command, the reader sends an acknowledgement command, **ACK**, containing the same RN16. If the tag hears this packet, and the RN16 is the value it just sent, it replies with a much longer packet containing its protocol control bits, EPC, and CRC16. An example of measured tag signals for such a sequence is depicted in Figure 8.49. At this point, the reader can choose to access this specific tag, to read from or write to various parts of the tag memory. The RN16 value exchanged between tag and reader allows them to create a unique logical session, even if many other tags are in hearing of the reader, and even if the tag has no other unique identifying information.

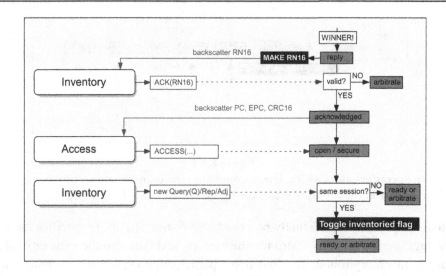

Figure 8.48
A tag whose slot counter is 0 replies to the reader with a random number. See Figure 8.52 for tag state details.

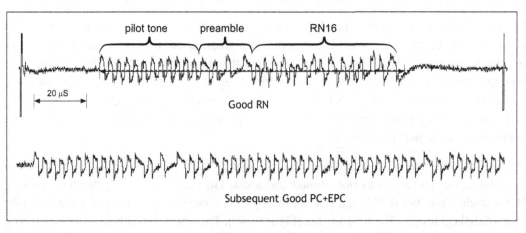

Figure 8.49
Example of the baseband signal resulting from a tag reply, followed by an EPC.

If on the other hand, all the reader needed was the EPC of the tag, it can issue a new **Query** command, or (much more likely) a **QueryRep**. The **QueryRep** command is used to signal the end of a slot, and thus is used very frequently. To maximize throughput, the **QueryRep** packet is very short: after the pilot tone/preamble, the command is just two bits of binary '0', followed by two bits describing which session the command applies to. When a tag that has just sent its EPC (and is thus in the acknowledged, open, or secure states) hears a **QueryRep** in the session it has been operating in, it flips the flag corresponding to that session and returns to waiting for a new inventory. Other tags decrement their slot counters by 1 and start the process again. If no tag's slot counter is 0, there is no reply in this slot, and the reader must issue another **QueryRep**. If a tag's counter was at 1 and decrements to 0, it replies in that slot, and the process starts over again.

If more than 1 tag replies, there will be a collision — a garbled or indecipherable transmission. The reader may (or may not) be able to recognize the collision and distinguish it from an empty slot. An example of a collided response is shown in Figure 8.50. Because the pilot tone and preamble are identical between the two tags, and both are timed with respect to the end of the previous reader command, the pilot tone and preamble remain decipherable (if perhaps a bit distorted) even though more than one tag is talking. However, the random number values (RN16's) in the responses are different, so the result is a mess.

Being able to unambiguously detect a collision, and distinguish it from an empty slot and a valid tag reply, is very useful to the reader, because after it finishes going through the inventory round of 2^Q slots it needs to figure out what to do next. When it finishes a round, it has five options:

1. Increase the value of Q by 1 and do another round.
2. Leave the value of Q alone and do another round.
3. Decrease the value of Q by 1 and do another round.
4. Issue a new Query command with an arbitrary value of Q.
5. Stop.

The first three options are accomplished by issuing a **QueryAdjust** command, which maintains all the other parameters of the current inventory operation unchanged.

In general, the reader has no *a priori* means of knowing that all accessible tags have been counted. If most of the slots in the previous inventory round contained no reply, the value of Q probably ought to be reduced. If most of the slots in the previous inventory contained collisions, the value of Q ought to be increased. If around 30–50% of the slots contained decipherable replies, with the remainder being a mix of empty and collided slots, the value

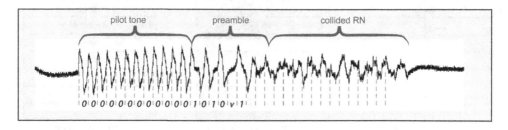

Figure 8.50
Example of the baseband signal resulting from a collision between two or more tags. Dotted lines show symbol timing extracted from pilot tone.

of Q was about optimal for the tag population; it might be prudent to repeat the same inventory, or optimistically assume that many tags were counted and reduce Q by 1 before trying again. Finally, if Q was previously a small value (e.g. 1 or 0) and no responses were seen, it is likely no tags are present and option 5 can be chosen. To make a good decision on the matter, the reader needs to know how many slots contained 0, 1, or 2 or more replies. It is also very helpful for throughput to be able to distinguish a good tag reply from a garbled tag reply, because when the reader attempts an **ACK** with what turns out to be a bad RN16, it must then also listen for a valid reply with PC, EPC, and CRC, and having failed to find one issue a **NAK** command to let the tags know that no tag was read. This process takes a lot longer than simply issuing a **QueryRep** at the end of an empty slot.

If all the tags to be addressed started in a specific state – for example, with the relevant session flag set to state A – we should ideally complete this set of inventory rounds with all the session flags flipped to state B. Each tag in the read zone should have responded at least once and been heard and recorded; after being successfully read the flag should have been flipped. Therefore, one ought to be able to validate that all tags have been counted using a procedure like (Figure 8.51):

- Count all the tags whose flags are initially in state A;
- Repeat the process targeting tags in state B
- If the same tags are found, all tags have been counted; otherwise repeat.

However, in practice the author has not found the world to be quite so simple. In addition to the obvious problem of some tags finding themselves in local fades and thus not receiving enough power to turn on, it appears that sometimes a tag will "leak" through the inventory: that is, a tag will flip the session flag even though the reader never counts it. There are at

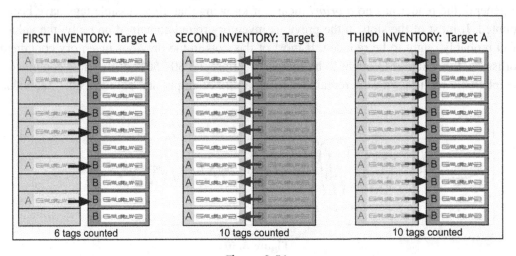

Figure 8.51
Example of a sequence of inventories with alternating target flag values.

least a couple of ways this can happen. After a tag has sent its EPC, if it sees a **QueryRep** command it assumes it was successfully read and flips its session flag. A **QueryRep** command has no error checking and is short. A tag receiving the reader only marginally might mistake the first two bits of another command, such as **NAK**, for the '00' indicating a **QueryRep**, and erroneously flip its flag. Another mechanism has to do with the structure of the tag reply. The Protocol Control word that is sent before the EPC tells the reader how long the EPC is; we will discuss the details below in connection with the Gen 2 tag memory map. A reader using a relatively slow processor (which might be the case for a handheld or other battery-powered reader) may not be able to decipher the PC word in time to prepare itself properly for the EPC, and so may simply read the PC word and then issue another request to the tag for the EPC, having in the intervening time decoded the received data and prepared to capture the proper number of bits. However, since the command following the first transmission of the EPC was not **NAK**, the tag may properly assume it was counted and flip its flag; if there is a bit error in the second transmission of the EPC, this assumption will be incorrect, but the tag doesn't know. In the author's experience, to count a large number of tags, it is more efficient and more effective to perform a series of inventories using the same target value but exploiting the option to use sessions with flag values that persist through successive inventories. We'll take a look at session persistence shortly.

The Gen 2 MAC approach is fast, flexible, and relatively easy to implement. By changing the starting value of Q we can operate in a Global-Scroll-like mode (Qstart = 0 or 1), appropriate to a fast-moving conveyor with only one box near the reader antenna at any given time, or use a large Q value to deal with tens or even hundreds of simultaneous tags in the read zone. Any tag that hears a **Query** command can participate in the subsequent round, so late entry is not a problem. Tags are not singulated using their EPC, so the reader never sends the EPC value over the air, and tags with a zero-value EPC or a non-EPC-compliant ID can still be singulated. The RN16 mechanism provides a way to maintain a logical session between the reader and a specific tag regardless of the state of the tag's EPC. Since only one or a few tags reply in any given slot when Q is anywhere near a good value, and reader and tag transmission are separated in time, tag-to-tag interactions are unlikely to disturb operation of the protocol. Since the reader looks for the pilot tone and preamble before declaring a valid tag reply, and the reader must find an RN16, ACK, and then see a valid reply before even attempting to read the EPC and check it against the CRC, the probability of finding a *ghost tag* — a false tag read in noise — is vanishingly small.[1]

---

[1] One can still construct scenarios in which a tag is incorrectly read without detection. For example, a real tag can reply to an ACK and be interrupted by a burst of interference that causes more than 16 bits to be garbled; in this case, there is a tiny chance that the CRC will randomly match the resulting noisy EPC. Such a scenario requires interference bursts infrequent enough to allow the RN16 handshake to proceed but frequent enough to disturb on the order of $2^{16} \approx 65{,}000$ tag responses in a period of interest.

Thus the G2 MAC solves many of the outstanding problems encountered in the first-generation protocols.

8.5.7 States and Commands

Gen 2 tag states are shown in Figure 8.52. The simplest path through the state diagram proceeds from *Ready* to *Arbitrate* upon receipt of a **Query** command. When the tag's slot counter reaches zero it transitions to *Reply*; if a valid **ACK** is received the tag moves to *Acknowledged*, and then returns to *Ready* at the next slot boundary.

The most commonly-used commands are summarized below:

Query   This command launches a new inventory round. The 4-bit command code is 1000. There are 22 bits total (after the pilot tone and preamble). Parameters:

*   DR: this is the divide ratio we discussed above in connection with the uplink symbols. See Table 8.1 and the associated discussion. DR is employed along with the duration of the calibration symbol TRcal to determine the backscatter link frequency, and by implication the data rate. Since there are only two allowed values (8 and 64/3), this parameter only requires 1 bit.

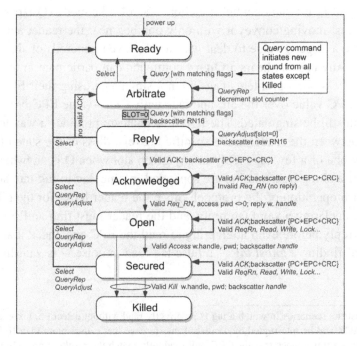

Figure 8.52
Gen 2 state diagram.

- M; the Miller subcarrier modulation index. See Figure 8.42 and discussion thereabouts. TRcal, DR, and M together determine the backscatter data rate and the tag backscatter spectrum. There are four possible values so 2 bits of the command are devoted to M.
- TRext: if this is set to 1, tag transmissions are preceded by a pilot tone of 12 binary '0' symbols.
- Sel: The Query can apply to only tags whose Select flag is set, only tags whose Select flag is cleared, or any tag. Two bits are required since there are three possibilities.
- Session: Each **Query** command applies to one of four sessions. (That is, when a tag gets counted, this parameter sets which flag gets flipped.) Two bits are needed for the four sessions.
- Target: Each session flag can be in either state A or B. For the session specified by the Session parameter, only tags whose flag is in the Target state respond to the **Query** command.
- Q: Q specifies the size of the upcoming inventory round. Since Q can range from 0 to 15, four bits are needed to specify it.
- CRC5: A 5-bit cyclic redundancy check is appended to the **Query**, since it is a long and relatively complex command. A tag does not respond to a **Query** whose CRC fails verification.

QueryRep   This is the most commonly-used of all commands and so is very short: only 4 bits. Two bits are the command code '00', and the other two determine to which session this QueryRep applies. There are no other parameters.

QueryAdjust   This command launches a new inventory round, with all parameters unchanged save the value of Q. This 9-bit command has the 4-bit command code '1001' and two parameters:

- Session: The session to which this adjustment applies. A tag only responds to a **QueryAdjust** if the session is the same as that of the most recent **Query**.
- UpDn: The value of Q can be increased by 1, unchanged, or decreased by 1. Three bits are devoted to this parameter even though there are only three possible values; this provides simple error checking, since the values are selected so that a single bit error in a valid choice always produces an invalid parameter value, which is ignored by the tag. Thus it is fairly unlikely that the tag will adjust Q incorrectly.

ACK   The reader sends this to acknowledge that it has received an apparently valid RN16 and echo the value. It has the 2-bit code '10', followed by the 16-bit random number the reader thinks it received from a responding tag. There is no error checking, since the tag is already comparing the value in the ACK to the number it sent.

NAK   Not-an-ACK, with the relatively long moniker '11000000'. A **NAK** tells a replying tag that its EPC was not successfully received. No flags get flipped. The tag that responded does not choose a new slot value; it will have to wait until the next inventory round before again replying.

ReqRN   Once a tag has been singulated, further exchanges with it are based on a second random number, called the tag's *handle*. The reader obtains this value by Requesting a Random Number: **ReqRN**. This command can also be used to obtain random numbers for encoding reader transmissions. Encoding is achieved by bit-by-bit exclusive-or (XOR) of the data to be sent with the RN16 value; this is known as *cover coding*. The 8-bit command code '11000001' is followed by the tag's 16-bit RN or handle (if one has already been sent), and a 16-bit CRC for error checking. The tag replies with a new RN16 value, also sent with a 16-bit CRC.

Read   This command can read from any memory area that is not locked against reading. The command code is '11000010'. The length of the command is not fixed, because the memory address to be read is described by an extensible bit vector (EBV), as described in Figure 8.33, but it is at least 56 bits long, and thus would hardly be appropriate as part of a high-speed inventory process. Parameters are:

- Bank: any of the four memory banks; two bits.
- WordPtr: the starting address of the memory segment to be read, as an EBV.
- WordCount: the number of 16-bit words to be read. Since this is an 8-bit parameter, up to 255 words can be read.
- RN: this is the handle of the tag to be read.
- CRC16: error check on the command.

Access   The **Access** command is used to request a tag handle, or an RN16 for cover coding. When the tag has received a valid **Access** command it transitions to the *Open* state (if the tag's Access password is 0) or to the *Secured* state if the password is non-zero and correctly provided. *Open* and *Secured* both allow data to be read and written, but only in the *Secured* state can a tag's memory be locked. Parameters are:

- Password: The 32-bit password is sent cover-coded, and in two 16-bit halves in consecutive **Access** commands.
- RN: the handle or RN16 of the tag being addressed.
- CRC16: error check on the command.

Write   The default command for writing to tag memory works one 16-bit word at a time. The syntax is very similar to that of **Read**, except that the WordCount is replaced by the 16 bits of cover-coded data. For good security the reader should request a new RN16 code key prior to each **Write** operation.

We have left the **Select** command for last although it is often the first command issued, because of its complex usage and syntax. The idea behind the command is to provide a flexible means by which to select a subset of all possible tags for inventory.

Successive **Select** commands can be combined, allowing complex Boolean operations based on the contents of the tag memory to define the desired subset. The syntax of the command is summarized in Figure 8.53. The Target parameter specifies which of the five flag values is to be modified based on this command. There are eight possible Action codes, and each code has a different effect depending on whether the target is a session flag or the SL flag, and whether a given tag's memory data matches Mask bits or not. The Mask is a bit sequence of up to 256 bits to be matched to a segment of the tag memory. The relevant segment is specified by the MemBank, Pointer (indicating the start of the relevant bit sequence), and the Length. Note Bank 00 cannot be used for **Select** commands. Finally, the Truncate parameter can be used to cause the tag to issue an abbreviated EPC during the subsequent inventory, if it is set to '1' on the last **Select** command before the inventory begins.

Because the flag states are preserved between commands, sequential **Selects** can be used to implement Boolean operations. For example, we can choose tags with a specific vendor (Manager) code and either of two model numbers (SKU's) by using two **Select** commands, each containing the same Manager but different SKU's in the mask, and each targeting the SL flag with action "assert" for matching and "deassert" for non-matching tags (that is, action code '000'). The Boolean AND can be implemented for non-contiguous memory segments using successive "assert" or "set" actions for matching tags with no action for non-matching tags. An exclusive-OR (XOR) results from using "toggle" action codes.

The simplest **Select** operations are the most reliable. Complex sequences of commands are allowed by the standard, but the user should remember that the more commands are used, the more likely it is that a tag will fail to receive at least one member of the sequence.

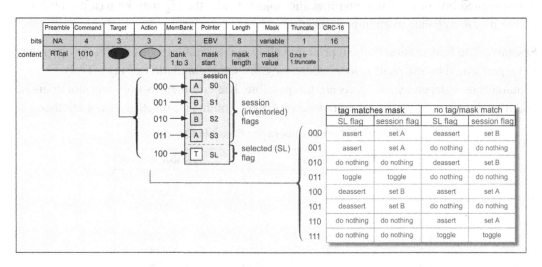

Figure 8.53
Syntax of the Select command. Assert sets the SL flag, and Deassert resets it. Toggle inverts the state of the flag.

8.5.8 Normal Operation and Key User Parameters

Starting Q: In normal operation of the Gen 2 protocol, tags power up into the *Ready* state. Although the standard is written implying that at least one **Select** command will precede any inventory operation there is no requirement that this be done, and inventories can be launched immediately upon power up. The reader then issues a **Query** command. The first Query is a step into the unknown for the reader, which has no *a priori* means of knowing what's waiting to be read. However, this may not be the case for the user. It is often true that the rough number of readable tags is known prior to the beginning of an inventory. A conveyor may be set up so as to allow only one or two tags to be present in the read zone at any given time. A forklift load passing through a portal may have a known number of tagged cases and a single pallet tag. If the user knows about how many tags are present, they can short-circuit a lot of work by the reader by specifying the starting value of Q, which is usually provided as a parameter in the reader software. The ideal value of Q is around 1.5 times as large as the number of tags, but since only powers of 2 are available, the starting value can't be chosen very precisely, and needn't be that accurate (see Table 8.2. A starting value of 5 or 6 will work fine for anything from around 25 to 100 tags, both because the protocol is robust and because the reader will adjust the value of Q based on the results of the first inventory or two. However, a starting value of 0 or 1 will not work very well with a large reader population; the reader may have a hard time seeing anything intelligible in the mess of collisions that results and decide it is done without reading any tags. Similarly a starting value of 6 or 7 in a high-speed conveyor environment is likely to result in a series of empty inventory rounds; if the read zone is limited by (say) a misoriented box or a challenging metallic object inside, the tag may be past the reader before it gets a chance to attempt a response.

Sessions: The four session flags in principle allow quasi-simultaneous inventory operations to be performed by different readers addressing the same population of tags. Truly simultaneous independent sessions are not possible, because there is only one slot counter, and it is always reset whenever a tag hears a **Query** command, whether or not another

Table 8.2: Slot Numbers for Possible Q Values.

Q	Number of Slots	Q	Number of Slots
0	1	9	512
1	2	10	1,024
2	4	11	2,048
3	8	12	4,096
4	16	13	8,092
5	32	14	16,384
6	64	15	32,768
7	128		
8	256		

inventory was in progress in a different session. Truly interpenetrating inventories would constantly reset tag slot counters, causing tags to leak out of an inventory round and making it difficult for the readers to adapt their Q values. To make use of this capability it is also necessary for the competing users to create some sort of convention about how the sessions are allocated amongst various readers, since nothing is accomplished if two readers address the same session flag.

Of perhaps equal importance in managing inventories is *session persistence*. In general, the state of a tag IC is preserved while power is supplied, but lost (except for its non-volatile memory) when power is lost. (Recall that power will generally be lost when the reader executes a frequency hop, and may also be lost due to local fading as the tag, reader, and other objects in the vicinity move around.) However, the Gen 2 standard requires that the session flags have persistence, and that the different sessions have differing properties in this regard. All inventory flags default to A when a tag powers up after an extended siesta. The state of the Session 0 flag persists indefinitely when tag power is on, but the state of the flag is lost immediately when tag power is lost. Session 1 has a fixed and limited persistence even when tag power is on; if the flag has not been refreshed by an inventory operation in more than the persistence time (up to 5 seconds) it will reset the flag to A, unless an inventory is actually in progress when the persistence timer runs out. Sessions 2 and 3 persist indefinitely when tag power is available, and for at least 2 seconds after power is lost. (The author has observed actual persistence times of up to 60 seconds on commercial Gen 2 tags in sessions 2 and 3.) The Select flag state also persists for at least 2 seconds when power is lost.

Table 8.3: Tag Flag Persistence Behavior

Flag	Persistence: Tag Power On	Persistence: Tag Power Lost
S0	Indefinite	none
S1	$500 \text{ ms} < t < 5 \text{ s}$	$500 \text{ ms} < t < 5 \text{ s}$
S2	Indefinite	$t > 2 \text{ s}$
S3	Indefinite	$t > 2 \text{ s}$
SL	Indefinite	$t > 2 \text{ s}$

Persistent sessions can make a considerable contribution to inventory performance when a large number of tags is present in the read zone, and fading is not negligible (so that not all tags are readable at a given moment). Let us imagine that we wish to count (say) 100 tags. We provide a reasonable starting Q value of 7 and tell the reader to inventory the tags. The reader may execute a series of inventory rounds with (for example) Q = 7,7,6,5,4,3,2,1,0, finding no reply on the final round and terminating that series; the set might consume 100 msec. Some subset − say fifty − of the tags present are read. The reader then executes a frequency hop and starts the process again. Since the flag state has been lost, the fifty tags that were counted previously are quite likely to participate again and be counted again − a useless waste of reader time and resources if all we need is the fact of the count and not its frequency. In the course of a second or two we might count many of the tags five or ten times in order to achieve one count of the small number of tags with poorer response. If in contrast we perform the same operation in Session 2, each tag that is counted remembers this fact. As long as the Query target is unchanged (say, A), subsequent inventories will address only tags that have not been inventories. As we proceed, the population of participating tags will become smaller, and their chance to reply larger. No time is wasted re-counting tags, and the whole countable population will be rapidly surveyed.

The author has found that both the time required to count a large tag population, and the number of tags that are missed or marginally counted, are substantially reduced when persistent sessions are employed. Because of the small but finite possibility that a tag can "leak" through an inventory, flipping its flag despite not having actually been counted, it is also a good policy to carry out alternate series of persistent inventories with different targets. That is, one might count for 500 ms or 1 second with a session S2 and target A, and then count for an additional 500 ms with the same session but target B. Any tags leaking through the first set are discovered in the second set, and the chances of an exhaustive count are improved.

Because tag states are persistent, possibly for seconds or tens of seconds, inventory results can be history-dependent. This fact is important to note in applications where consecutive reads of the same tag may have differing goals or consequences. For example, if a case tag is read on a conveyor to record the fact that the case is present, and then to be read a few seconds later after the case has been sorted onto another belt, if a persistent session is used and both stations employ the same session and target, the case may be missed due to flag persistence. Choosing an alternate target risks the same consequence if the time between reads is comparable to the persistence time. When history matters it is prudent to select non-persistent sessions for inventory.

Reading and Writing: Because tag memory organization is specified by the Gen 2 protocol, reading from and writing to tag memory can be expected to be independent of the tag IC vendor. Reading from memory is fairly straightforward, except to note that it is possible to lock bank 00 (the password bank) against reading as well as writing. Once this has been done there is no way to recover the password values. We will discuss locking in a bit more detail below.

Writing a new EPC to a tag requires a protocol control (PC) word, which was introduced in Figure 8.32, repeated below for convenience as Figure 8.54. The first 5 bits of the PC word describe the length of the EPC in 16-bit words. A 96-bit EPC is 6 words long. For a default EPCglobal tag, this is the only non-zero portion of the PC word. Thus, for a 96-bit EPC, the protocol word in hexadecimal notation is 30_h, as shown in the figure. If bit 8 is set to 1, the PC word includes an ISO-compliant Application Family Identifier (AFI). The AFI value is assigned depending on the general class of use for the object the tag identifies, and is

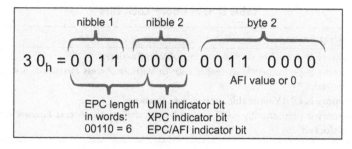

Figure 8.54
Protocol Control (PC) word structure.

generally used as a vendor-independent filtering mechanism (i.e. as part of a **Select** mask) to include only tags pertaining to a certain type of use. The extended protocol control indicator bit will be set to 1 to indicate more elaborate capabilities, some of which are described in section 6 below.

Using the default **Write** command, the bytes of the PC and EPC are written into memory words 1 through 7 of bank 01 (for a 96-bit EPC). Note that this requires issuance of one command for each word written. Memory word 0, containing the CRC, is calculated by the tag each time it powers up, and thus there no need to write a value there. The optional **BlockWrite** command writes multiple words in a single command, without using forward-link cover coding. At the time of this writing not all commercial tags support **BlockWrite**. The related optional command **BlockErase** allows more than one word of memory to be cleared in a single command.

Locking Memory: Gen 2 has a sophisticated tag memory security system. All Gen 2 tags have a 32-bit Access Password stored in the Reserved memory bank, 00. When a tag is singulated, it enters one of two states. If the Access Password is all zeros, the tag goes to the *Secured* state. If the password is non-zero, the tag passes to the *Open* state. An *Open* tag can be *Secured* by providing this **Access Password** to the tag. A *Secured* tag allows more control over its memory than an *Open* tag. The exact privileges that are granted depend on the lock states (to be described shortly) of the different memory banks. Only *Secured* tags allow changing of these lock states.

All tag memory banks can be locked, which is useful to prevent unauthorized access to tag memory. For example, customers can be kept from changing the EPC of a retail item in order to modify its price. The *Access* and *Kill* passwords, both of which reside in the Reserved memory bank, can be locked independently. There are therefore 5 total regions that can be locked: *EPC*, *TID* and *User* memory banks, and the **Access** and **Kill Passwords**. Each of these regions have 2 bits to control their lock state: *Pwd* and *Permalock*. Each region can therefore be in one of four lock states, as shown in Table 8.4. For the *Access*

Table 8.4: Memory Lock Bits

Pwd	Permalock	Description
0	0	Memory is writeable with or without *Access Password*.
0	1	Memory is permanently writeable with or without *Access Password*. It may NEVER be locked.
1	0	Memory is ONLY writeable with *Access Password*.
1	1	Memory is permanently NOT writeable with or without *Access Password*. It may NEVER be unlocked.

and *Kill* passwords, locking restricts writing *and* reading. For the other regions (*EPC*, *TID* and *User* memory banks) locking only restricts writing. Once *Permalock* has been set for a region, the lock status for that region can never be altered.

The **Lock** command provides a mask that allows the user to set each of the lock/permalock bits for all the banks and the two passwords. Once a permalock bit is set, it can of course not be reset.

It is important to note that the lock status of the tag memory is explicitly specified as unreadable. Lock status can only be inferred as a result of attempts to read from or write to memory segments.

Kill Command: Tags can be killed if they have a non-zero *Kill* password; a compliant *Killed* tag is required to no longer respond to any command, unlike some earlier tags in which the kill command merely erased memory. The *Kill* and *Access* passwords are both 32 bits long, so there is little danger of a dictionary attack succeeding on individual tags. Like **Access**, the **Kill** command sends the password in two 16-bit halves, each protected by cover-coding with a 16-bit random number. Tags can support recommissioning, where memory is unlocked (for overwriting) instead of being killed.

Providing unique *Kill* and *Access* passwords to individual tags represents a substantial password management problem. In many cases it can expected that large classes of tags will share the same passwords, and the labor required to mount a dictionary attack might be more readily justified. For example, a parallel assault on 100 tags at 10 tries/second would crack a password in about a month. Preliminary results from researchers at the Weizmann Institute also suggest that power analysis attacks to extract passwords may be possible on individual tags; see Further Reading for more information.

8.5.9 Protocol Performance and Link Timing

How fast can a tag be read? In order to answer this question, we need to take into account the peak data rates and the amount of data to be sent over both directions of the link. We also need to take into account certain delays built into the protocol. These delays are shown Figure 8.55. The first specified timing limitation is imposed upon the reader; it must wait at least twice RTcal after a command with no reply (like a **Select**) before issuing another command. Since RTcal is roughly twice the duration of an average reader symbol, this delay corresponds to only about four reader bits and has a modest impact on long commands like **Select**; in addition, it doesn't arise for commands like **Query** where a reply is expected, so we can ignore it in our peak performance analysis. After a **Query** or **QueryRep**, the tag must wait at least T_1, which is the larger of RTcal or 10 cycles of the FM0 carrier. Which is it? The shortest value of RTcal that is allowed for the shortest value of Tari (6.25 microseconds) is 2.5 Tari = 15.6 microseconds. The fastest BLF is 640 kHz, corresponding to a cycle time Tpri of 1.56 microseconds, so that — surprise! — the two constraints are identical. The tag should reply within 15–16 microseconds after the **Query** command is complete. If there is a reply, the reader must acknowledge it within no more than 20 Tpri = 32 microseconds. If everything goes smoothly, the tag will again wait 15 microseconds after the **ACK** to reply.

Assuming the minimum Tari and maximum BLF (with divide ratio 64/3), the preamble for each reader command takes 67 microseconds; a frame sync takes a bit less (34 microseconds) since there is no TRcal symbol. During full-speed operation the reader will be sending only **QueryRep**, **ACK**, and when things don't work out, **NAK**. Treating a typical bit as 1.25 Tari and multiplying by the command length in bits, we can estimate the

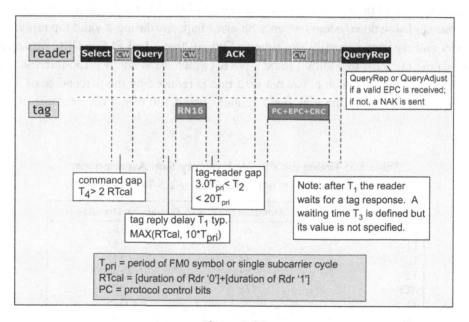

Figure 8.55
Gen 2 link timing summary.

duration of the reader commands. Assuming no pilot tone and FM0, we can estimate the time taken for the tag to reply. Putting everything together (Table 8.5), we get the total time for one successful slot to be about 581 microseconds, with most of the time taken by the **ACK** command and the long tag reply. This time corresponds to a maximum possible rate of 1720 slots per second.

Can the protocol really run this fast? Even in the ideal case, the use of a random MAC protocol means that it is not possible for every slot to be filled with exactly 1 tag. Let us, for example, assume that every other slot is empty. When a slot is empty, the reader issues a **QueryRep**, waits for T1 plus the duration of the preamble, and then some hardware-dependent time (we'll guess 9–10 microseconds for the sake of a round number) to decide that no tag has responded and issue another **QueryRep**; that means an empty slot consumes 100 microseconds. In this case, the actual tag read rate will be about 1470 tags/second.

In the real case there are some substantial obstacles to sustained operation at this speed, some having to do with the protocol itself and some with hardware issues. Except in the case when only one tag per inventory attempt is guaranteed, the reader doesn't know exactly how many tags are present, and can't choose a perfect value of Q; we must expect some slots to contain collisions. In most collisions the reader will still be able to extract a preamble and (erroneous) RN16. The reader will then **ACK** with an invalid random number, to which neither tag will reply. This will consume an extra 200 microseconds, representing a quite substantial slowdown if collisions are at all frequent (as must sometimes be case while we are adjusting the value of Q). That's the good news. A much more substantial slowdown occurs when a bit error happens during a valid tag reply. The reader and tag spend one full slot time to find out that a bad EPC + CRC was received, after which the tag must in addition issue a NAK (which takes another 130 microseconds or so) before it can start a new slot. The net read rate is reduced by the percentage of reads containing bit errors. Since the long reply contains about 130 bits, a bit error rate of less

Table 8.5: Timing for a Single Inventory Slot. Assumptions:
Tari = 6.25 ms, BLF = 640 kHz, RTcal = 2.5 Tari, TRext = 0

Step	Duration (μs)	Cumulative Duration (μs)
QueryRep	66	66
T1	16	81
Tag RN16	36	117
T2	31	148
ACK	175	323
T1	16	339
Tag PC + EPC + CRC	211	550
T2	31	581

than 10^{-3} is required to keep the tag reply error rate below 20%. To reduce the number of bit errors, one ought to employ a lower backscatter link frequency and higher value of M (to narrow the spectrum of the tag reply, and thus the receiver baseband filter width), but of course this reduces the tag data rate and slows reads.

The timing requirements imposed on the link are also challenging, particularly for handheld/ battery-powered readers with limited computational capability. At full link speed the reader has 31 microseconds to determine whether a valid preamble was received, demodulate the RN16, validate that it wishes to send an **ACK**, and formulate and transmit the command. (These times can be stretched a bit by clever software design, demodulating the bits as they are received and constructing the reader command at the same time.) For a low-cost 500 MHz processor and 5x oversampling, assuming four computations per sample, that gives us around 30 processor cycles per computation, not including memory access and conditional execution times. If a low-power 100 MHz processor is all we have, we're down to 6 cycles per sample. So while it's quite possible for a reader to operate at the full rate, it is demanding of computational resources, and limits the sophistication of data processing algorithms that can be applied.

The use of dense interrogator modes puts further constraints on the data rates, not related to the limitations of the hardware but to regulatory conditions. In order to place most of the tag spectrum at the boundaries of a channel (US operation) or in an alternate channel (ETSI operation) the backscatter link frequency is constrained to equal the desired offset from the carrier: around 240 kHz for the US, or 200 kHz for Europe. Furthermore, in order to make sure that most of the spectrum power is away from the carrier, a Miller index of 4 or 8 is desirable, further reducing the actual tag data rate. Finally, the reader data rate must also be limited to keep the transmitted spectrum narrow.

Fortunately, the flexibility allowed by the protocol lets readers adjust tag and reader data rates for optimal performance given the resources available and the read range desired. Readers the author has worked with have demonstrated peak rates of 200–500 tags/second for populations of 50–100 tags in the read zone, under FCC conditions. This is much faster than the values for first-generation tags in comparable circumstances.

8.5.10 Concluding Remarks

UHF tag protocols have come a long way since the early work at Sandia Laboratories. The Class 1 Generation 2 protocol demands a rather sophisticated tag IC, but in exchange it provides a number of benefits:

* Unique tag sessions based on 16-bit random number handles, independent of the tag's unique identifier;
* Simple encryption for memory write operations;

- Long passwords for memory lock and tag kill;
- Support for late-arriving tags;
- Ghost (phantom) tags are nearly eliminated;
- Dense populations of collocated readers can operate simultaneously with minimal interference;
- Flexible, persistent sessions allow efficient sequential inventories with minimal re-reads of tags already counted;
- Protocol-specified memory maps and terminology ensure interoperable memory reads and writes;
- Flexible ability to lock memory banks temporarily or permanently.

The flexibility that makes the protocol powerful also makes it relatively complex to implement; a Gen 2 chip requires about five times as many transistors as a Gen 1 chip. This means that while absolute prices for the ICs may continue to fall, they will always be more expensive than a simpler protocol would require.

If the Gen 2 standard has a substantial area of weakness, it is in security and privacy. The EPC itself can be read by any compliant reader, regardless of lock states. There is no provision for reader authentication or hash security for the EPC; if the EPC itself contains any information of value to an attacker it is easily obtained. Cover coding is reasonably secure if the tag transmissions can't be detected and the reader uses a new random number to encode every packet, but this is hardly absolute security. The author has intercepted tag transmissions from about 5–7 meters using a 13-dBi Yagi-Uda antenna and a spectrum analyzer; with more appropriate equipment it is likely a tag near a reader could be heard from some tens of meters away, or through a wall or two. Current tag IC's are also vulnerable to power-sensing attacks, making the passwords susceptible to extraction, particularly if (as is likely) a large number of tags share a password. While these limitations are of modest import in standard supply-chain tracking, for which Gen 2 was primarily designed, they represent challenges for security-based applications (e.g. in anti-counterfeiting agents for pharmaceuticals and other goods), and in applications where privacy is important (such as library loans or consumer item purchases). It seems reasonable to anticipate that future passive tag protocols will increasingly incorporate security and privacy protections, as continued progress in IC technology makes more logical capability available at the same DC power.

8.6 ISO 18000-6 Extensions

The EPCglobal Generation 2 standard was adopted with minimal changes by the International Organization for Standardization (ISO) as 18000-6C. For basic passive UHF tags, the two standards are substantially identical. Some interesting extensions were added to 18000-6C in 2010 and will be reviewed briefly below.

8.6.1 Total (18000-6D)

The first, formally 18000-6D, is the definition of a random-backoff-based inventory tag, known by the acronym for its operation as TOTAL (Tag Only Talks After Listening). This protocol is based on earlier proprietary products from iPico, a South African company. Part D does not define any new activity for the reader; at its simplest, it can be implemented using a reader that only sends CW signals. TOTAL tags monitor the received RF power level in their band of interest (typically set mainly by the passband of the antenna rather than a true narrowband filter). If the power level exceeds a threshold, the tag wakes up and transmits a message after a random delay. The message may include a tag ID (TID), data, and a simple sensor page. If the tag then hears a modulated signal with a command that it recognizes, it can execute the command; otherwise, it goes quiet. The states are summarized in Figure 8.56.

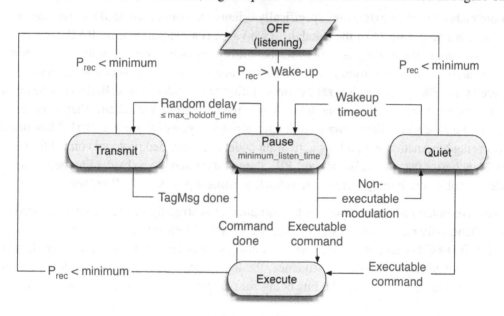

Figure 8.56
TOTAL state diagram.

The maximum holdoff time is at least 30 ms (allowing 33 delay slots per second). The minimum holdoff is 5 ms the first time the tag awakens, but shortens to 125 s thereafter. The wake-up timeout after hearing a non-executable command is 25 ms.

Data transmission uses subcarrier modulation with a subcarrier frequency of 512 kHz and data rate of 256 kbps. There are two options for modulation: Miller subcarrier with $m = 2$ may be used, but a pulse-position-modulation scheme, in which a pulse is placed in either the first or third of four available time slots, may also be used. Data is organized in 64-bit pages, with a page preamble and error check. If Protocol Control bit 57 is high, a sensor page is included.

Random backoff is a simple and robust strategy for collision resolution, used in shared-medium Ethernet, 802.11 wireless communications, and many other contexts. It works very well when the offered traffic is small enough to ensure that about two-thirds of the possible timeslots are not occupied and poorly when most timeslots are filled.

The 18000-6D protocol supports both passive and battery-assisted tags. Awakening only when RF power is detected also suits battery-assisted designs that need to conserve battery energy. However, the simple threshold detection mechanism will cause false wakeups whenever radio transmitters are common in or near the target band.

8.6.2 Battery-Assisted Tags

A more elaborate set of extensions specifically oriented to battery-assisted passive tags is also included as section 11 of the standard. Two variants are described. BAP-PIE uses the same pulse-interval encoding as 18000-6C, but adds provisions for sleep states to conserve battery energy, and a **Flex_query** command extension to support the needs of battery-powered tags. Flag persistence is set by timers rather than reader power. BAP-Manchester is a more extensive set of changes, including a new reader modulation scheme that is more suited to the case where the reader is no longer supplying power to the tag. BAP-Manchester supports tag hibernation with a long activation code, and extended protocol control for sensor data. Manchester tags must also support PIE, though they may do so by cycling between modes. A tag whose battery has died falls back to being a passive 18000-6C tag.

As we have noted previously, passive-tag modulation is strongly influenced by the need to simultaneously talk to and power the tag. Pulse-interval encoding is used in Gen 2/18000-6C because the reader power is always on at least half the time, even during a long string of binary 0's. As a consequence, the local "DC" level of a PIE-modulated signal varies with time, depending on the bits being sent (Figure 8.57). A simple, low-power

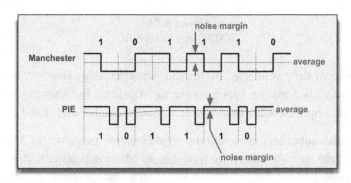

Figure 8.57
Comparison of Manchester and PIE modulation, showing locally averaged DC level.

approach to demodulating the reader signal uses a comparator with an R-C filter on one input. The comparator compares the instantaneous signal level to the local average signal to locate the high and low parts of the transmitted signal. When such a scheme is used for PIE, the noise margin is reduced as the local average value rises toward the high level during long strings of binary 1's.

In Manchester modulation, a rising edge in the middle of the symbol denotes a 0, whereas a falling edge denotes a 1. An additional rising edge is inserted at the beginning of the symbol if needed to send consecutive 1's. Manchester modulation is one of many schemes that are DC-balanced: the local average over a few symbols is equal to the global average. Manchester is also more bandwidth-efficient than PIE. Manchester modulation is a better choice than PIE when the average power delivered to the tag is no longer a concern. The improved noise margin supports operation at lower received signal power from the reader, thus allowing for longer read range. Recall from chapter 3 (Figure 3.35) and section 8 of chapter 5 that a square-law diode detector can provide about 15−20 dB improvement in receive threshold relative to the power required by a passive tag.

ISO 18000-6 BAP tags support an INACT_T timer that starts on loss of RF signal. It is useful to include some hysteresis to avoid chattering: that is, once the incident RF signal falls below some power threshold RFlow and the timer starts, the RF signal must rise to a threshold RFhigh that is higher than RFlow before the timer stops. The timer may optionally be reset after each preamble or command. After a maximum inactive time, the tag falls into one of the three states. The first is Stateful Sleep, in which the flag persistence timers continue to run but the tag is otherwise inactive. The second is Stateful Low Power Listen, which applies to BAP-PIE tags. The third is the Hibernate state, which applies to Manchester tags. Finally, an optional global timeout enables tags to return to a low-power state in the presence of a high-power interferer that keeps the detected RF signal high when no reader is active.

The inventory flag behavior must be modified for battery tags, since power is never lost (or so one hopes). For a BAP-PIE tag, S0 persists until INACT_T is activated to emulate the passive-tag behavior. S1 lasts for 0.5−5 s after INACT_T is activated. Session flags S2 and S3, and the SELECT flag, persist for 2−20 s after INACT_T starts. A timer expiration during a Query does not cause the flags to revert.

A Manchester tag uses a rather different scheme. All the flags are treated identically and have a persistence set by the Activation command that wakes the tags, which can be from 0−4096 s. A complex set of Session Locking options is provided.

In addition to the existing Query command, a **Flex_Query** command is included. The **Flex_Query** command can specify the type of tag that should respond: passive, BAP, sensor, or a mix. The Simple Sensor function can be set on or off. The Miller index m can be as large as 64. The use of large Miller indices reduces the data rate, but improves the signal-to-noise ratio. Large values of m are used to improve the reverse link performance at long range, so the tag doesn't become reverse-link-limited (section 5 of chapter 3).

Making sure that a battery-assisted tag stays asleep except when a reader wants to read it, without using a bunch of power in the sleep state, is not trivial. The Manchester forward link includes a number of features that are intended to support low-power sleep with selective wakeup. The standard packet preamble for all commands includes a long 21-bit training segment, to ensure that the receive decoder has stabilized, and a 15-bit delimiter (Figure 8.58). The same training sequence, combined with the logical complement of the standard delimiter, is used to awaken the tag from Hibernation. A longer variant form, including interrogator information, is also available. The choice depends on a tradeoff between the additional power needed to decode the full sequence and the power saved by avoiding a wakeup when the "wrong" reader is calling. Note that the long read range possible with a BAP-Manchester tag means that it may hear many readers in one facility; keeping them straight will help avoid being constantly awakened by readers to which a given tag should not respond.

The use of a specific legal bit sequence as a wake-up code means that the wake-up code must **not** appear in ordinary data, lest hibernating tags be awakened by an exchange not intended for that purpose. To avoid this occurrence, all data is ***bit-stuffed***: whenever any data to be sent matches the first 14 bits of the activation sequence, the next bit is forced to be (stuffed with) a binary 0. That is, any sequence like:

01011 00100 0111*x*

(where *x* is any bit value) is stuffed to become

01011 00100 0111**0***x*

The tag must similarly recognize the 14-bit stuff sequence and unstuff (delete) the following 0 bit to recover the original data.

Figure 8.58
Preambles for preceding a command when the tag is operating (Standard) and waking the tag when it is hibernating (Activation).

Various modifications to existing commands, and new commands, are added, most of which are optional. The **Select** command adds an 8-bit short interrogator ID, again to help tags respond only to appropriate readers. **QueryAdjust**, **QueryRep**, and **NAK** also get an interrogator ID. In addition, **QueryAdjust** acquires the ability to set Q to any allowed value, rather than stepping it up or down by 1.

The new **Query_BAT** command includes a short interrogator ID, and a Type field, which includes sensor alarm, full or simple sensor capability, and battery-assisted passive tag types. A Simple Sensor response flag is also added. The backscatter link frequency BLF can extend all the way down to 25 kHz and up to 1920 kHz. The very low BLF value is useful to extend the reverse-link range for battery-assisted operation. The new commands **Next** and **Deactivate_BAT** are provided to moves tags into the Hibernate state.

The Extended Protocol Control scheme described in section 5 above is used here to support configuration and reporting information. The additional protocol control word W1 contains information on the sensor configuration (alarm, simple or full sensor), battery mode (passive or battery-assisted), and the three recommissioning bits described in section 5. Finally, an optional elaborate Tag Capability Reporting and Setting function is defined.

8.6.3 Sensors

Section 12 of the standard describes various sensor capabilities that may be included in a 18000-6C or -6D tag. Two types of sensors are defined: the Simple Sensor and the Full Function Sensor. Both are required to include a real-time clock, so that readings can be time-stamped. Timing is required to use Universal Coordinated Time (UTC), starting at midnight on January 1, 1970; an optional command is provided to synchronize tag timers. A Simple Sensor does not require user programming, and its data can be appended to the tag ID during inventory, if requested by the reader. The appended data, the Simple Sensor Data block, includes a bit of type information, a pass/fail or optionally an 8-bit sensor value, threshold information, and alarm status. A Full Function Sensor can capture varying data types, store the results, and deliver them to the reader upon request. A Full Function Sensor can also be programmed by the user.

The optional **HandleSensor** command provides a way for readers to communicate with sensor-capable tags. The HandleSensor command includes a 7-bit port number to allow the reader to transparently communicate with a specific sensor on the tag. A variable-length Payload field and a variable-length Response field allow flexible definition of commands to be passed into and out of the sensor, supporting proprietary capabilities and extensions.

Only certain specific sensor types are available for a Simple Sensor: temperature sensors with span 14 or 28°C, relative humidity, impact, and tilt sensors. The Simple Sensor block can be 32- or 48 bits long, though the 48-bit option is not yet defined in detail. The data block includes the type, the range — e.g., for the case of a temperature sensor, from −29 to +34°C in 14°C-wide blocks — and alarm information. The Sensor Data Block is either stored in User Memory (with the location of that storage being specified by a field in TID memory at word 0×26), or accessible via a port with the **HandleSensor** command. The **Flex_Query** or **Query_BAT** command can be configured to request that the data block be appended to the tag's ID as part of the reply to a reader ACK. The Sensor Data Block is **not** sent by default and thus isn't accessible with a conventional **Query** command.

Full Function Sensors use a *Sensor Directory System* (SDS), whose location is specified by TID memory word 0×22. The Sensor Directory System can also be used for Simple Sensors. For a Full Function Sensor, the SDS entry includes the sensor access type, sensor type and identifier, alarm, port number, and security and authentication information. Definition of the sensors includes conventions adopted from the IEEE 1451.7 standard.

8.7 Active Device Protocols

Protocols built for active devices, even low-power active devices, differ in significant ways from protocols specialized for RFID applications. In most cases, the links are symmetric or nearly so: both ends can send and receive data and commands. Bandwidth-efficient modulations, involving variations in carrier phase as well as amplitude, can be used, along with complex coding schemes that improve performance at low signal-to-noise ratios. In addition to the simple preamble and sparse data of a minimalist RFID reader or tag packet, active device protocols typically include source and destination addressing, type and encoding fields, optional link security, and physical-layer error checking. In addition, the protocols include elaborate means for joining a network (association) and leaving it (disassociation), supporting assignment of network addresses and discovery of key network parameters. All this complexity means overhead, and additional energy spent, both in transmitting and receiving the actual data, and in figuring out what to send and what has been received.

In this section we'll take a quick look at two popular protocols for active digital communication that have seen application in sensor and location networks: the IEEE standards 802.11, often referred to as Wi-Fi™, and 802.15.4, often referred to by the name of its upper-layer network protocols, Zigbee™. Both protocols are packetized, half-duplex general-purpose communications protocols that can in principle send and receive any sort of data. The now fabulously popular Wi-Fi was designed as a local-area-network extension, and that is how it was usually used. However, it has also been applied for sensor networks and active identifying devices; an example architecture for this sort of application was introduced in chapter 5 (Figure 5.32). The 802.15.4 standard was specifically designed for sensor networks and other low-power applications.

The 802.11 standard is built around a star topology: an Access Point (AP) organizes a local network that Stations (individual devices) can detect and associate with. The Access Point is presumed to be part of a larger local network, either through a wired or separate wireless connection. Peer-to-peer *ad hoc* networks are also allowed but have seen much less extensive deployment.

The 802.11 standard was conceived as a wireless extension of Ethernet networking, so it is built to carry Ethernet packets (Figure 8.59). An Ethernet packet contains a preamble and start-of-frame symbol, followed by some addressing information, the payload (usually from a higher-level networking protocol), optional padding, and an error check. Even if we ignore the preamble and start-of-frame, which might be removed if using a different link layer, the packet has a minimum of 18 bytes (144 bits) before any data is considered. The payload is likely to contain, for example, an IP or UDP packet, which has its own addressing, sequencing, and error correction information. Recall that, for example, the most frequent reader command in 18000-6C/Gen 2 requires only a preamble and four bits. Active protocols do bigger jobs and need more bits to do them.

The wireless standard adds a customized link control layer (802.2) and WLAN-specific medium-access control and physical layers to support the required wireless capabilities (Figure 8.60). Each layer requires additional support and thus additional bytes.

A typical 802.11 payload is a UDP packet (that is, a globally-addressed network packet with no network-layer acknowledgment). A UDP packet might be used to send some data from a sensor node to a server managing the sensor database. The use of UDP over IP provides a lot of power and flexibility − in principle, the server can be anywhere in the world, as long as it is reachable through the Internet − but at the cost of a lot of packet complexity. A representative result is illustrated in Figure 8.61. The physical-layer packet, at the bottom, consists of a preamble, header, and some data (the "MPDU"). The MPDU contains the MAC layer information: addressing, sequencing, and error check. The frame body of this packet is the SNAP packet (which is an extra layer present for historical reasons), embedded in which is the networking-layer UDP packet. The exact amount of overhead needed to send the data varies somewhat, but a reasonable estimate is that 90 bytes (720 bits) are required before any data gets delivered.

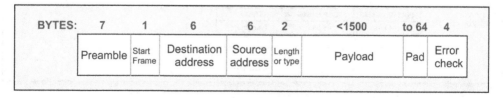

BYTES:	7	1	6	6	2	<1500	to 64	4
	Preamble	Start Frame	Destination address	Source address	Length or type	Payload	Pad	Error check

Figure 8.59
Basic Ethernet packet structure.

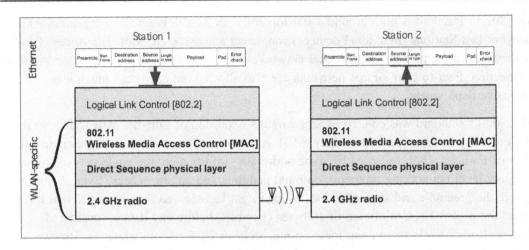

Figure 8.60
802.11 protocol layer stack.

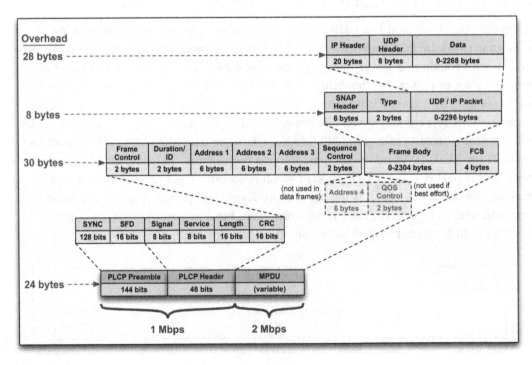

Figure 8.61
Typical UDP frame sent over 802.11 "classic" physical layer.

In the 802.11 "classic" direct-sequence physical layer, the final packet is sent using *quaternary phase-shift keying* (QPSK) at 2 Mbps, with the data being spread by an 11-chip Barker sequence, so that the signal uses a bandwidth of about 16 MHz. The classic PHY is not used much these days in conventional data networking but may still be convenient for low-power identification or sensor applications. At this rate, the packet described above, with a few bytes of data added, would take about 400 microseconds to send, followed by up to a few milliseconds waiting to get an acknowledgment.

An 802.11 device needs to associate with an Access Point before it can exchange messages. The association process involves detecting a beacon from the AP, making a request to join, and passing through any security hoops that are imposed by the AP. Once the device is a member of the network, it needs to go through a conventional DHCP process to get an IP address so that it can be seen at the network level. The whole process is complex and can take tens or hundreds of milliseconds, but fortunately it happens only infrequently. After joining the network, different constraints are imposed by different AP designs. Some require each station in the network to send a packet at some interval (e.g., every few minutes) to verify that they are still present; others may maintain a station on the active list until a transmission attempt fails. Depending on the application, a low-power sensor station could be set to send data every minute or two, thus automatically remaining active, or could go through the whole joining process again if data is sent very infrequently. The station may also periodically send inquiries confirming that its IP address has not been assigned to another station.

To conserve energy, an 802.11 Station can signal the AP that it will be inactive for some time period. The AP will buffer any packets addressed to that Station and announce their presence in the beacons that it sends periodically. The station can awaken when a beacon is due, check the beacon for a bit that will inform it a packet is waiting, and then send a message to retrieve the packet. The radio receiver can thus be turned off most of the time to conserve battery resources, while still supporting remote configuration and management when needed.

The energy cost of such an active device was discussed in chapter 5 (section 8). By using transmit power of 0−10 dBm (1−10 mW), an active 802.11 device can achieve open-area ranges of hundreds of meters, or indoor ranges of tens of meters (depending on obstructions), with much less energy use than a conventional Station transmitting 0.1−1 W, but still a lot more than a passive RFID tag. Active 802.11 devices benefit tremendously from direct access to the Internet and all the associated infrastructure but are bulky and expensive compared to a passive tag. Active 802.11 devices require substantial battery power, e.g., from a AA-sized cell to last for a year or more. They are not practical for tracking inexpensive consumer goods but may be very useful for locating valuable objects like people or expensive movable equipment.

The 802.15.4 standard was conceived as part of a ***personal-area-network*** (PAN) suite of standards, specifically for low-power, low-data rate applications. In addition to the star topology, in which a Coordinator device creates and runs a PAN, a peer-to-peer mesh networked architecture is allowed, in which devices in a PAN can communicate directly with each other without the intervention of the Coordinator.

Because this standard is targeted at low-power and low-data rate applications, standard Ethernet packets are not used, and the packet size is limited to a maximum of 127 bytes of data. Larger payloads must be split into multiple transmissions. A modified addressing approach is allowed in which each device in the PAN receives a short 16-bit address. In a star topology, addressing information may be compressed even more by assuming that a transmission is either sent from, or destined for, the Coordinator.

Several physical layers are defined for differing frequency bands and applications. The direct-sequence physical layer for the 902–928 MHz band uses QPSK modulation and a simple encoding scheme to provide a raw data rate of 40 kbps; 250 kbps is available in the 2.4 GHz band.

An example of the construction of a packet is shown in Figure 8.62. In order to make this closely comparable to Figure 8.61, an alternative networking layer that supports UDP packets over 802.15.4 links, 6LowPAN, is presumed to be used. The total overhead of about 26 bytes (208 bits) is about 1/3 of the overhead used in the 802.11 link, but because

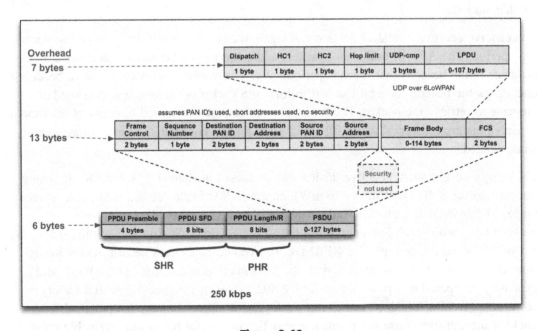

Figure 8.62
Typical UDP packet sent over 802.15.4 2.4-GHz physical layer, using 6LowPAN networking layer.

the link is slower, the packet will actually take longer to send. It's clear that the 802.15.4 structure provides better efficiency for small amounts of data. However, if a larger payload is needed, 802.15.4 must split it into multiple packets, each with its own overhead, vitiating the advantages of the more compact framing.

In both protocols, a considerable amount of computational infrastructure is also assumed. A device is presumed able to identify and join a LAN or PAN, get and retain the requisite addressing information, and respond appropriately to any signaling from the Access Point or Coordinator. Prior to transmitting, an 802.11 device listens to the relevant channel to ensure that it is not in use; if a packet is heard, the device must receive and decode a timing field, the *Network Allocation Vector* (NAV), that reserves the time needed to complete the current transmission. After sending a packet, the device's radio receiver must remain active to hear the acknowledgment from the AP; if no acknowledgment is received, the device must execute the backoff algorithm, find a clear channel, and try again.

Similarly, an 802.15.4 device in a PAN using a beacon must hear the beacon and establish what times are in use for contention-based channel access, which times are reserved for delay-sensitive communications, and which times are idle times, during which all devices can sleep. It must listen to the channel it wishes to transmit on to establish that no other device is using it before transmitting. After a transmission, the device will normally wait for an acknowledgment and execute a backoff and retransmission if no ACK is received.

To accomplish these relatively elaborate operations, some with real-time constraints, the networked device needs RAM, ROM, non-volatile memory for firmware, and a fully capable CPU with an appropriate register set. If the device is a sensor node, the CPU will also need to communicate with off-chip sensing devices using a low-power local serial bus like SPI or I^2C. Finally, a full radio transmitter with quartz crystal reference is needed to ensure regulation-compliant output, as well as a low-noise receiver with substantial RF gain, downconversion to baseband or an intermediate frequency, and more gain and filtering. That is how you get to the complex system shown in Figure 5.32.

With such a complex device, system, and protocol, no simple statements about battery life can be made. Battery life can be measured in years if the devices are off most of the time; battery life can fall to days if remote firmware updates or other power-hungry operations are frequently needed. Optimizations at the installation, device configuration, network configuration, and system management levels are all required to build a robust sensing or identification capability.

We note in passing that *ultrawideband* (UWB) devices have also seen some use in RFID applications, particularly for tracking location of people and valuable objects. UWB devices use a minimum of 500 MHz bandwidth in the 3−10 GHz range (there's also a sub-1-GHz variant). Signals may be a series of pulses or broadband orthogonal frequency-division multiplexing. In either case, a wideband signal supports accurate timing and thus location of the radiating device by time of flight. Time-of-flight location is more reliable than location by received signal strength, though both methods are subject to errors in the presence of obstructions and reflecting surfaces. The IEEE 802.15.4a standard supports UWB personal area networks with location and ranging.

8.8 Capsule Summary

All communications protocols are based on conventions agreed upon by the parties involved, specifying the medium of communication, the message format, access to the medium, and interpretation of the resulting data. Once the choice of electromagnetic transmission is made, practical and regulatory considerations limit the choices available for passive long-range RFID to a few frequency bands. These choices, and the limited computational capabilities of passive tags, constrain the properties of the wireless medium. Within those constraints, a protocol designer must choose what symbols to use, whether and how they are assembled into packets or frames, how to manage access for an unknown tag population, and what data to put on the tag.

Early protocols validated the use of frequency-shift-keyed tag symbols and amplitude-shift-keyed reader symbols, cyclic redundancy checks for error detection, and limiting tag data to identifying information. Collision resolution was limited to a persistent Quiet state for tags that have been counted.

The Auto-ID Laboratories' Class 0 and Class 1 protocols, even though never fully standardized, made significant contributions to the tool kit for tag protocols, adding sophisticated command sets and tag states, field-writeable tag memory, and binary tree medium access control for inventorying large populations of tags in a manageable time. ISO 18000-6B introduced the use of a simple slotted Aloha MAC algorithm, FM0 tag coding, and sophisticated tag subset selection.

Implementation of these early protocols also exposed important weaknesses and obstacles to wider use: inflexible data rates, incompatible memory implementation, intolerance of late-entering tags, ghost tags, lack of link-level security, reader-reader interference, and lack of a mechanism for talking to a tag independently of its unique ID. The Class 1 Generation 2 standard attempts to address all of these issues. Memory mapping and terminology is specified in the protocol. Gen 2 provides great (perhaps excessive) flexibility in tag and reader data rates, and special modulation and coding for interference mitigation. An enhanced Aloha-type MAC can adapt to varying tag population sizes while being friendly to late-arriving tags. Complex subset management is available using multiple sequential **Select** commands. Stringent multiple-step timing requirements make ghost tags unlikely. Tags use random numbers to manage unique logical sessions with a reader, irrespective of the contents of their EPC memory. Long passwords protect the memory-lock and tag-kill functions. Simple forward-link encryption makes interception more difficult. Enhancements support tag-talks-first inventory, battery-assisted tags, improved modulations, and sensor capabilities.

Existing wireless network protocols have been repurposed to support sensor applications, and new protocols have been designed for that purpose. In either case, the capabilities of the resulting networks greatly exceed those of passive or semi-passive tags but require extensive resources for device and system design, configuration, and management.

Future enhancements of passive protocols are likely to include enhanced authentication, privacy, and security functions.

Further Reading

General Communications Protocols

Understanding Data Communications (sixth edition), G. Held, New Riders 1999. *This is a nice introductory survey of the very broad field of communications, suitable for folks with minimal acquaintance. Some of the technologies discussed (e.g. ISDN and X.25) are now rather dated.*

Communications Networks, A. Leon-Garcia and I. Widjaja, McGraw-Hill 2000. *A rather more technical but very readable survey of communications networking. Focused on network architectures much more than the physical-layer concerns with which we have mostly concerned ourselves in the present volume.*

Digital Modulation and Coding, S. Wilson, Prentice-Hall 1996 (op.cit). *For the serious student; the fundamentals of signal modulation and detection, developed with considerably more rigor than we have employed.*

RFID Protocols: The Source Docs

If you are actually going to implement any of the protocols, no summary discussion — even the laborious Gen 2 examination above — will do the trick. You have to read the protocol documents. Reading protocols may also be indicated in cases of severe insomnia — the ISO 18000-6A/B document is in the same league as the US Federal Aviation Regulations in terms of the number of grams of caffeine needed to remain awake to the end.

Here are a few:

Association of American Railroads AAR S-918-00 adopted 1991, rev. 95,00. http://www.aar.org/. *This standard specifies the tags that are used on essentially all railcars in the United States. It is similar to ISO 10374, designed for use on shipping containers.*

Title 21, Automatic Vehicle Identification Equipment amended 1998. *Part of California state government transportation regulations; available from the State web site.*

Draft protocol specification for a 900 MHz Class 0 Radio Frequency Identification Tag, 2/23/03, http://www.epcglobalinc.org/standards/specs/. *As noted in the text, this draft was never completed or ratified, though substantially compliant tags and readers saw commercial deployment.*

Candidate Specification 860 MHz – 2500 Mhz – Class 1 RFID Air Interface, version 1.1 rev 1.02; *was previously available at http://www.epcglobalinc.org/standards/specs/ but is now missing in action. As noted in the text, this draft was never completed or ratified, though substantially compliant tags and readers saw commercial deployment.*

Class-1 Generation-2 UHF RFID Protocol for Communications at 860 MHz – 960 MHz, Version 1.2.0, October 23, 2008, http://www.epcglobalinc.org/standards/specs/. *This is the Gen 2 standard, substantially identical to the ISO 18000-6C standard.*

ISO/IEC 18000-6, Information Technology – Radio frequency identification for item management – Part 6: Parameters for air interface communications at 860 MHz to 960 MHz (Second Edition, 2010-12-01), http://www.iso.org/. *This document includes the older −A and −B variants (for whatever reason, some of the dullest standards documents I've ever waded through!), as well as the more recent −C and −D tag types, and specification of various battery-assisted and sensor-equipped options.*

IEEE Standard for Information Technology – Telecommunications and information exchange between systems – Local and metropolitan area networks – Specific requirements. Part 11: Wireless LAN Medium Access Control (MAC) and Physical Layer (PHY) Specifications. IEEE Std 802.11–2007, www.ieee.org. *A huge document about which numerous books (including one of mine) have been written.*

IEEE Standard for Information Technology – Telecommunications and information exchange between systems – Local and metropolitan area networks – Specific requirements. Part 15.4: Wireless Medium Access Control (MAC) and Physical Layer (PHY) Specifications for Low-Rate Wireless Personal Area Networks (WPANs). IEEE Std 802.15.4–2006. *Note that this is the physical layer standard that supports Zigbee networking. It is often referred to as Zigbee, but it isn't: 802.15.4 can also be used for other network protocols, including conventional TCP/IP communications.*

RFID Protocols: More Information

"An enhanced dynamic framed slotted ALOHA algorithm for RFID tag identification", Su-Ryun Lee, Sung-Don Joo, Chae-Woo Lee, The Second Annual International Conference on Mobile and Ubiquitous Systems: Networking and Services (MobiQuitous), 2005, p. 166–172
"Colorwave: a MAC for RFID Reader Networks", J. Waldrop, D. Engels and S. Sarma (MIT Auto-ID Center); Auto-ID Lab web site.
"The Reader Collision Problem", D. Engels and S. Sarma, IEEE SMC 2002
"Reading Protocol for Transponders of Electronic Identification System", C. Turner, and J. McMurray, US Patent 7,019,664
"Anti-Collision Methods for Global SAW RFID Tag Systems", C. Hartmann, P. Hartmann, P. Brown, J. Bellamy, L. Claiborne and W. Bonner, IEEE Ultrasonics Symposium 2004 p. 805
"A new medium access protocol for RFID networks with foresight," F. Baloch, D. Hoang, E. Sawan, and R. Pendse, Wireless Telecommunications Symposium (WTS) 2011

RFID Protocols: Security and Privacy

"A Lightweight Mutual Authentication Protocol for RFID Networks", Luo Zongwei, T. Chan, T, J. Li, IEEE International Conference on e-Business Engineering, 2005, p. 620
"Grouping Proof for RFID Tags", J. Saito and K. Sakurai, AINA 2005
"A feasible security mechanism for low cost RFID tags", Gwo-Ching Chang, International Conference on Mobile Business, 2005, p. 675
"A scalable and provably secure hash-based RFID protocol", G. Avoine, P. Oechslin, Third IEEE International Conference on Pervasive Computing and Communications, Workshops, 2005, p. 110
"A Study on Establishment of Secure RFID Network Using DNS Security Extension", YoungHwan Ham, NaeSoo Kim, CheolSig Pyo, JinWook Chung, 2005 Asia-Pacific Conference on Communications, p. 525
"RFID privacy: an overview of problems and proposed solutions", S. Garfinkel, A. Juels, and R. Pappu, IEEE Security & Privacy Magazine, **v 3** #3 p. 34 (2005)
"Social Acceptance of RFID as a Biometric Security Method", C. Perakslis and R. WolkProceedings of the International Symposium on Technology and Society, ISTAS 2005 p. 79

Active Protocols

"A comparative study of wireless protocols: Bluetooth, UWB, Zigbee, and Wi-Fi," J. Lee, Y. Su, and C. Shen, 33rd Annual Conference of the IEEE Industrial Electronics Society (IECON), November 5–8, 2007, Taipei, Taiwan

"Model-based design exploration of wireless sensor node lifetimes," D. Jung, T. Teixeira, A. Barton Sweeney, and A. Savvides, Proceedings of the Fourth European Conference on Wireless Sensor Networks, EWSN 2007, January 29–31, 2007

"Performance evaluation of IEEE 802.15.4 LR-WPAN for industrial applications," F. Chen, N. Wang, R. German, and F. Dressler, Fifth Annual Conference on Wireless on Demand Network Systems and Services, 2008, p. 89

Exercises

Packet structures and Medium Access Control:

1. The International Organization for Contention's (IOC) STAR (Slothful-Tag-And-Reader) Protocol requires that the tag transmit fifty '0' bits and a twelve-symbol preamble prior to sending its 96-bit identification code. A parity check bit is embedded after each 8-bit byte of the ID code. The ID code is followed by a 16-bit CRC. What percentage of the tag message is devoted to formatting and error checking instead of sending data?

   _____ %

   The IOC is organized into working groups. Working group RPD-1 (Rate Performance Disparagement) has tabled a proposal for providing tags with the option of replying with only the preamble, ID, and CRC, also eliminating the parity check digits. The working group is now stuck, since the British members believe that tabling refers to bringing a measure up for consideration, whereas the American members understand the term to mean that the proposal has been abandoned, the European members are on vacation for 36 weeks, and the Asian members are too busy making products to attend the meetings. Let's help them out. If the reader command causing a tag to reply with its ID requires 48 bits and is transmitted at half the rate of the tag reply, and a four-reader-bit gap is specified between reader command and tag reply and prior to another command, how much can throughput be improved by the tabled proposal?

   _____ %

   Is this improvement worth calling the European members back from the Mediterranean beaches?

   YES _____ NO _____

   It's winter, they're in the Alps _____

2. To the consternation of visually-impaired[2] Committee Chair Toulouse Track, all the 124 voting members have shown up at the meeting to vote on whether "tabling" should be interpreted according to the British, American, or Icelandic conventions[3]. Dr. Track decides to allocate the right to speak based on a slotted Aloha approach. He will ask each participant to choose a random number between 1 and 100 and write it down on the tablet of paper next to their glass of ice water. He will then flip a coin to decide whether to count up from 0 or down from 100, and call out numbers in the resulting sequence for each participant to speak. Each member will then have 1 minute to state their case favoring or opposing the resolution. If no one speaks for 10 seconds after a number is called, Dr. Track will go on to the next number. In the event that two or more members have chosen the same number, they will all speak simultaneously. Dr. Track will record that a collision occurred, draft a memo regretting the fact to be delivered to the IOC Intellectual Property Manager, Pat N. Pending, and go on to the next slot.

   What is the likelihood that there will be no collisions?

   100%_____   10% _____   5% _____   0% _____

   If there are ten 2-person collisions, three 3-person collisions, and one each with eleven people (the number '1') and nine people (the number '100'), how long does it take for Dr. Track to get through all the allocated slots, assuming a 5-second inter-slot gap for Dr. Track to call out the next number?

   _____ minutes

   Would it have been more efficient to hold the meeting underwater using American Sign Language, or would that choice have a prejudicial impact on the result of deliberations?

   YES _____ PROBABLY _____ WHAT? _____

First-Generation Standards:

3. In Singapore, unlicensed RFID operation is allowed in the band 923−925 MHz. Is it possible to choose a 500-kHz channel in this band for reader transmission to ensure that Class 0 tag signals are also in the band?

   YES _____ NO _____

4. A Class 0 reader using a symbol time of 25 microseconds and ID2 to singulate tags is counting tags with 128-bit IDs. How long does it take for the reader to receive a complete tag ID and error check?

   _____ microseconds

---

[2] One of many consequences of the embarrassing incident involving laser surgical modification of the cornea, a toy caboose, and Miss Cody Pendant from Twelve-Step Temporary Employment Agency.

[3] Reputed to involve indecent use of geothermal energy.

5. A Class 0 reader monitors a conveyor. Tagged boxes on the conveyor are within the read zone for 1 second; on average, boxes are spaced apart by 3 meters and move at 1 meter per second. The reader continuously reads at full speed as in problem 4 above, simply discarding reads whose CRC and ID do not agree, and stopping every 100 milliseconds to issue a **RESET** and calibration sequence for new-entering tags. If the reader 24 hours per day, and assuming no actual metaphysical intervention, how many ghost tags will it detect in a week?

   _____ tags

6. A Class 1 reader monitors the same conveyor, in Global Scroll mode. To save time, no mask bits are used in the reader command. Production worker Amon Breick carelessly leaves an extra tag on the table close to the reader antenna, so that this tag replies to every **ScrollAllID** command and its backscatter signal is much larger than that of tags on the conveyor. How many conveyor tags will be read under these conditions?

   _____ %

   To avoid catastrophic failure when Amon is not on break, the reader software is modified to issue a **Quiet** command to each tag that is successfully read, so that another tag can participate. If the reader issues a **Quiet** command, with the 96-bit ID of the tag included in the mask, each time it successfully reads a tag, and the remainder of the command (other than the mask bits) takes 1 millisecond to send, what is the impact on the peak read rate? Assume the reader transmits at 60 kbps.

   _____ reads per second with Quiet vs. _____ reads per second without

Gen 2 Protocol:

7. How big does a memory bank have to be before an EBV requires a second byte?

   _____ words

8. Here is the baseband signal a Gen 2 tag receives. The top shows a closeup of the preamble; the bottom shows the complete command except for the CRC-5 error check bits.

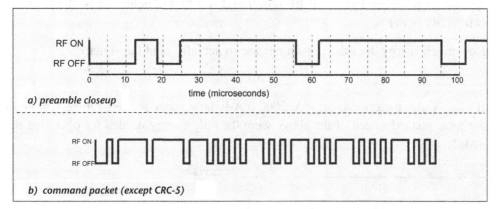

a) preamble closeup

b) command packet (except CRC-5)

Based on this packet signal:

What is Tari?

_____ microseconds

What is the average reader data rate, assuming an equal mix of 1's and 0's?

_____ bits per second

What command is being sent?

_____

What session flag does the command apply to?

_____ (0 to 3)

What tag backscatter link frequency should be used to respond (estimates are ok)?
What is the tag-to-reader data rate?

BLF:_____ kHz   data rate: _____ bits per second

If this is your reader, and you know that between 1 and 3 tags are in the field of view of the reader at any given time, what is wrong with the parameters of this command?

Active Tags:

9. If you were a sensing device, would you want to associate with a PAN named Bluetooth? Is Zigbee any better? Shouldn't the IEEE get real about these names?

   YES _____

   NO, WHY? _____

   WHICH QUESTION AM I ANSWERING? _____

10. Assume a Wi-Fi device sends a packet with 1000 bits in it at 2 Mbps. The radio transmitter produces 10 mW of RF power and is 30% efficient: that is, the RF output is 0.3 (DC power in).

    How much energy does the transmitter use to actually send the signal?

    _____ mJ

    If a 3.7-V AA battery can provide 2000 mA-h, how many packets can be sent before the battery is exhausted, if the above were the only energy needed for operating the node?

    _____ packets

At one packet per minute, how many years is that?

_____ years

Review the discussion of an active tag in chapter 5 (Figure 5.32 and text). How does the energy usage above compare to the values cited there? What can we conclude about the actual energy used in data transmission as a fraction of the total energy used by the device?

CHAPTER 9

RFID Applications

9.1 What Is It All for?

People who work on technology mostly do so because it's interesting, but to get someone to pay the bills, the technology must also satisfy a real need. To be practical, an RFID system has to identify something that's worth keeping track of, and for which wireless means are superior to other labeling approaches.

When we look at potential RFID applications, we need to not only think about whether it makes sense to use RFID — instead of a bar code, a conventional printed name, or a couple of marks with a hammer — but also what kind of RFID technology to use. Is an application suitable for inductively coupled or radiatively coupled devices? Active, battery-assisted, or passive? Is there a published standard that sets requirements for the specified application? Do we need anticollision capabilities to count when lots of tags present simultaneously? Are there requirements for security and privacy? For what period of time do we need to keep track of the object? Every application has a unique set of requirements and constraints, and the technology that works for one may be inappropriate for another. Let's look at the key issues that need to be considered in any proposed implementation.

9.1.1 Cost

The potential benefit of identifying an object is inevitably related to the cost of the object. Expensive objects are hard to replace and worth knowing about. Cheap objects are less so. The amount of money you can spend to track something is usually some fraction of the value of what's being tracked, where the fraction is determined by the benefit you hope to gain from keeping track. It's not a big deal to spend US$100 on a tag to track a shipping container with US$50,000 of items in it. Spending the same amount to identify a candy bar that costs US$1 is absurd. The cost of the item being tracked sets an upper limit on the amount of money that can be spent, and thus on the technology that can be employed. Passive tags, with a cost of a few cents to a dollar (embedded in a printed label), can be used for tracking consumer goods in the supply chain. Active tags, costing tens of dollars, are used for expensive objects with long useful lifetimes. Battery-assisted tags are in between in cost and capability, and thus in application.

The RF in RFID
© 2013 Elsevier Inc. All rights reserved.

9.1.2 Range

How far away do you want to be able to read a tag? A library checkout system needs only a few centimeters of read range to identify a book sitting on top of the checkout station, but around 0.5 m to identify items being removed from the library without being checked out. Keeping track of doctors in a hospital may requires some tens of meters of range, and the ability to read through at least a few walls. Read range is often a complex tradeoff: if you can afford a lot of readers, or if you can funnel all the items you're tracking through a few pinch points, you don't need long range. Once you determine the optimal read range, you've also constrained the technologies available. For 20–50 cm, passive HF tags are fine; passive UHF is great up to 10–20 m, battery-assisted UHF up to 100 m, and active tags are needed beyond that.

Note also that as read range grows, so does location ambiguity. Where is that tag you just read? This is an easy question to answer when the tag has to be lying on top of the reader to be detected. It is much more subtle when a reader can detect every tag in a dockyard or train yard. The problem of locating an object by radio means in the presence of reflections and obstructions is big enough to deserve a book all by itself, and many have been written on the topic. For our purposes here, it is enough to note that the question of location must be included in the tradeoff between read range, cost of readers, and cost of tags.

9.1.3 Useful Life in the Application Environment

A properly-encapsulated passive tag IC is unlikely to fail in normal use. The lifetime of the tag is generally set by the limitations of the ancillary materials, such as corrosion of the antenna, leakage of the weatherproof casing when present, or failure of the adhesive or other attachment means. Passive tags in robust containers can be considered for applications requiring years or even decades of readability, but the definition of a robust container may not be trivial. A weather-sealed plastic case may be sufficient for an application like tracking automobiles at a dealership, but quite inadequate for tracking steel pipes or being driven over by a forklift.

A tag with a battery can last between a few months and a few years; then the battery or the tag must be replaced. This may not be a significant obstacle; for example, a tag attached to measurement equipment with a regular calibration schedule can have battery maintenance during calibration. More cynically, if the current CIO installs the system and then retires, replacing the batteries is a problem for his or her successor.

9.1.4 Quantities and Rates

How fast do you need to count? A veterinarian identifying a pet dog need only read the single tag embedded therein. Tagged boxes on a conveyor moving at 1 m/s must be read in a few tens of milliseconds, in the presence of (probably) less than 10 other tags in the read zone. When a pallet of tagged items passes through a portal or reader-enabled doorway, hundreds of tags may be simultaneously present in the read zone, and must be read in one to a few seconds. The technology and protocol to be used must be chosen with these speed and anti-collision requirements in mind.

9.1.5 Security and Privacy

Users of identification technology may have an interest in preventing the information contained therein from becoming available to unauthorized people. Security and privacy considerations may not be strongly coupled to cost: a passport is not expensive, but the holder may not wish to broadcast the information contained within. Industrial users may not wish to expose their inventory levels or sales to competitors. Means for keeping information secure may be very simple − using a tag with a few centimeters of read range so that physical security is equivalent to information security − or complex, with elaborate encryption and authentication.

To provide concrete examples of the application of the general ideas above, we will first revisit the established RFID applications we discussed briefly in chapter 2, to see how successful applications have addressed each key requirement. Finally, we'll look at new and envisioned future applications, with an eye toward elucidating the important obstacles that must be surmounted, and the likely paths to commercial success.

9.2 Old Tricks

Those who have actually been reading the book in a more-or-less continuous fashion (rather than in scattered moments in medical waiting areas and airliner rest rooms) may recall that in chapter 2 we touched upon a number of existing RFID applications:

- Aircraft identification
- Railcar tracking
- Automobile tolling
- Animal tracking
- Container tracking
- Tracking of consumer goods through the supply chain.

For those readers new to the field at that time, the particular approach used for each of the applications above may have seemed random and even mysterious. We are now in a better position to understand how technologies and requirements were matched.

9.2.1 Aircraft Identification

The cost of an airplane is very high indeed, and the consequences of misplacing them can be even more expensive than the objects themselves. While it is frivolous to say that cost is no object, spending hundreds to thousands of dollars is certainly reasonable when the safety of a US$100M airplane and potentially hundreds of passengers is at issue. This is a good thing, because the other dominant issue in aircraft ID/IFF is range. It is not very helpful to track a flying airplane from 20 cm, or even 20 m, away: distance in this application must be measured in kilometers, and thus active UHF transmitters are unavoidable. Location is extremely important, as befits this long-range application, but airplanes generally fly above obstructions (except when mountains get in the way). Thus, location can be reliably established in most cases by means of time of arrival and angle of arrival of a signal.
If I know how long it took for the signal to arrive, delays here being in easily-measured tens of microseconds, and what direction I was looking in when I saw it, then I know where the transponder was when the signal was sent. (In practice, location information may be provided by detection of the backscattered signal from the airplane: signal strength is lower but accurate timing is easier.)

The lifetime of an aircraft transponder is measured in years or decades, and thus it is normally powered using the electricity generated by the aircraft engine, rather than by a stand-alone battery. The environment is the inside of an aircraft cabin: acceleration, heat, and vibration are expected, but rain, snow, and animal attacks are not. Anti-collision is really important in this application, but only in the sense of the airplanes avoiding that fate: distances are so large that even if transmissions were all simultaneous at the aircraft, it would usually be possible to separate the received signals at air traffic control. Rarely will more than a few tens of airplanes be simultaneously present within the range of a single radio receiver, and only a small fraction of them are seen at any moment by a rotating directional antenna. Computational resources are available, and full duplex wireless links can be supported as needed. Thus the identification space needed is small − four decimal digits − and simultaneous inventory is straightforward. A typical transponder is shown in Figure 9.1. The unit is big enough to permit manual entry and human-readable display of identifying codes during flight and includes an on-switch, lighting, and an indicator showing the pilot when an interrogation occurs. A radio of this type can be fabricated using conventional discrete semiconductor devices and integrated circuits, printed circuit boards, and molded parts, and may cost hundreds to thousands of dollars.

The basic transponder protocol is very simple. The transponder is interrogated at 1030 MHz and replies with 12 digital bits (4096 possible codes) at 1090 MHz. A mode C transponder adds an estimate of aircraft altitude, providing full 3D location information. The process takes several milliseconds, which is OK because there aren't that many airplanes in the sky at one time.

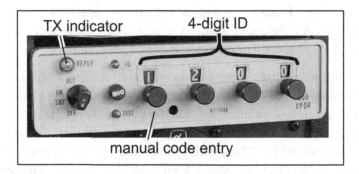

Figure 9.1
Aircraft transponder. *Image courtesy Wikipedia public domain.*

Security is not important in civilian applications – after all, anyone can look upwards and see an airplane in clear weather – but is vital for military IFF. Fortunately, the computational resources needed for encryption are available.

9.2.2 Railcar Tracking

Railcars are not cheap, but they are in general much less expensive than airplanes. Rail transport in the United States is often employed for moving massive quantities of low-cost commodity goods, so the value of even a loaded railcar may be in the thousands of dollars rather than hundreds of thousands to millions. But tens of thousands of cars are in use in the United States at any given time, and maintenance is also expensive. It is sensible to spend enough to ensure that tags will have a long life in the field, preferably several years, so that they can be installed and left alone. A railcar tag can therefore cost a few tens of dollars, must last through years of heat, cold, rain, snow, and a bit of whacking, and should not need a battery.

If we are willing to spend the money to install readers at each rail line in a yard, the read range needed is on the order of the size of track and cars – a few meters, generally unobstructed. HF technology could be used but would require relatively close tolerances, whereas typical passive UHF range of a few meters is quite sufficient with no special precautions. If a reasonably directional antenna is employed, only one railcar is likely to be visible to the reader at any given time, so only the simplest means of resolving collisions between tags are required, and the location and direction of travel of the identified object can be readily established. Railcars don't have pilots who can reset a transponder value as requested by air traffic control, so the identification space of the tags must be large enough to provide a reasonably unique number for tens to hundreds of thousands of items.

Therefore, railcar tracking is sensibly performed using passive UHF tags in weatherproof casing, robustly attached (e.g. by nuts and bolts) to railcars, positioned to be read as the car passes a stationary reader. Examples of typical readers and tags were shown in Figure 2.7, repeated here as Figure 9.2 for convenience. The tags need to have a pretty big identifying number, but it need only be written once (possibly at manufacture).

The protocol is described by the American Association of Railroads standard S-918-00. A passive tag operating in the 902−928 MHz range is used; different channel assignments are available for trackside and yard readers. The tags use a 20/40 kHz subcarrier modulation: a binary 0 is 1 cycle of 20 kHz and 2 cycles of 40 kHz; a binary 1 is 2 cycles at 40 kHz and 1 cycle at 20 kHz. Thus, each symbol takes 100 μs to send (10 kbits/s). Tags turn on when the local field exceeds 3.5 V/m, which is roughly 3 m range for a 1-watt transmitter with some antenna gain. A tag reply contains 128 bits, of which 26 are "procedural" (headers and protocol management) and 102 bits available for ID information. The data may include category of equipment, a car number, length, number of axles, and other information. The tag reply takes about 10 ms, during which the train (if moving at the maximum speed of 80 miles per hour or 40 meters per second) travels 0.4 m. So as long as tags are spaced at least 3−4 m apart, no special means are required to prevent tag replies to the same reader from colliding with each other. Nearby readers offset their transmit channels by 2 MHz, which suffices to keep the tag replies (about 130 kHz wide) to different readers distinct.

9.2.3 Automobile Tolling

Automobile tolling is a nice example emphasizing the value of tracking the object rather than the cost of the object. An automobile is quite expensive, but radio identification must

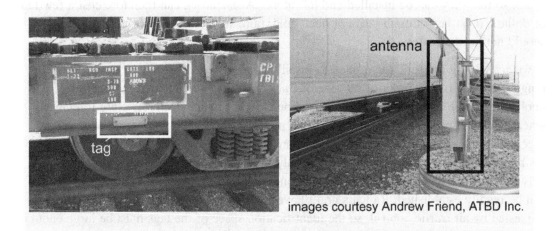

images courtesy Andrew Friend, ATBD Inc.

Figure 9.2
Typical railcar tag and reader antenna.

compete with human toll collectors and coin or currency counting machines. The cost of tags and readers must be low enough to provide a net benefit versus labor costs and alternative automation methods. Automated toll collection still coexists with the manual kind in the author's locality (northern California), though in the eastern United States many toll facilities permit only radio identification.

It was originally envisioned that readers embedded in a roadway could be used, in which case read ranges of 1 m would suffice, allowing HF technology to be used. However, in practice, reader antennas mounted above roadways are much more common, and thus the required read range is on the order of 5 m (to allow for tall trucks), so UHF technology is needed. Read reliability needs to be high: drivers get really irritated when they are given a ticket even though their toll transponder was clearly mounted on the windshield. In some applications, it is desirable to record transit at full highway speeds. The combination of relatively long range and high reliability at high speed has generally driven toll authorities to use battery-assisted tags.

A nationwide standard does not exist in the United States. Proprietary transponders are used in some regions. In California, transponders follow California Title 21. The standard specifies operation at 902–928 MHz and uses a rather more elaborate downlink than AAR S918. After a 33 ms startup pulse, the reader sends amplitude-shift-keyed Manchester data – that is, a transition from high to low amplitude signifies a binary 1, and from low to high a binary 0. The downlink data rate is 300 kbps, fast enough to allow a 3-byte header/ delimiter, 2 bytes of transaction type, 2 bytes describing the agency, and a 2-byte CRC for error checking. Uplink is frequency-shift-keyed: four cycles at 1200 kHz signify a binary 1, and two cycles at 600 kHz a binary 0. The tag is required to respond to a field greater than 0.55 V/m but to not respond to less than 0.45 V/m, and backscatter modulation must involve a cross-section change between 45 and 100 cm^2. When the tag receives a valid polling message, it responds with the same header/delimiter, 2 bytes of transaction type (all 1's) and its unique 32-bit ID. That's enough ID space to cover a few billion cars – more than even the state of California is likely to have any time soon. The reader sends an acknowledgment including the tag ID and its own ID. If the tag hears the ACK, it must ignore identical polling requests for the next 10 s. In the context of automobile tolling, this simple measure is sufficient to minimize collisions between tag responses.

We can see that the whole exchange requires some tens of milliseconds. At 100 kmh (60 mph), an automobile travels 3 m in 100 ms, so a reader configuration that provides 5–10 m of effective range is sufficient to give good read performance. Note that this application involves substantial integration challenges, particularly when tolling is to take place at full highway speed. In a typical installation, an inductive loop detects the presence of a car; a second loop 10 m later detects exit from the tolling area, and can trigger a photograph of the car's license plate. The reader is mounted on a gantry, providing a read zone about 3—4 meters long on the road surface. After the read is completed, the reader must communicate the tag ID to the authentication database fast enough to capture a photograph of the relevant license plate upon exit if the account is not validated.

The battery-assisted tag we encountered in Figure 2.20 is depicted in more detail in Figure 9.3. The tag is about 9.5 cm wide, so it isn't big enough for a simple straight half-wave dipole antenna. It is contained in a robust plastic case, quite sufficient for service when mounted inside an automobile. Within is a conventional circuit board, with a large round lithium battery in the center. Two bent-dipole antennas are visible, one for backscatter transmission and one for receive. A surface-acoustic-wave filter is used to remove most out-of-band radio signals from the receive antenna. A high-impedance amplifier is used to detect when the received power reaches the power level specified by the standard. A crystal reference is needed to provide a clock for the control logic and to ensure accurate anti-collision timing. A PIN diode is used to modulate the impedance of the backscatter antenna. (This is a diode in which the middle, between the two contacts, has very few free electrons or holes when no current is flowing. By changing the current flowing through the diode, its conductivity can be modulated. PIN diodes are often used in microwave applications where a variable resistor is required.) The tag also has a beeper, used to let the driver confirm that the tag has been read.

The lithium battery by itself costs around US$6 in quantities of a few thousand. The SAW filter is around US$0.50 in similar quantities. The Schottky diode is about US$0.30. Standard 32 kHz crystals are around US$1; beepers are a bit cheaper (couldn't resist).

Figure 9.3
Title 21 toll tag with major components marked.

So we're already at a bill-of-materials cost of almost US$9 without the board, plastic package, minor components, and the cost of assembly and test. Quantity purchases may reduce some of these individual costs, but one of the problems with the automobile market is that the quantities are not very large on the scale of electronic products: tens to perhaps hundreds of thousands per year. That isn't enough leverage to secure greatly reduced prices. The vendor has to be able to charge around US$20 to make any profit on this sort of item, which corresponds to historical pricing for these tags. There is no way to make this object for US$1. This relatively expensive approach is still practical, because it is marking a large expensive object (an automobile).

The lithium battery has about a 5-year lifetime. When the battery runs out, the transponder is no longer readable, and a relatively expensive replacement process (in which a new transponder must be provided by mail to the subscriber) is needed.

An application like this has challenges that rail or airplane tracking do not. The cooperation of the user is not assured. For example, if the tag has a removable battery, the driver may take the battery out, place the tag in plain sight, and then argue that it was present and not read. The same problem arises in the other direction when a battery becomes exhausted, and the disgruntled driver receives a ticket when they had (as far as they knew) obeyed all the rules. Drivers have an incentive to clone someone else's tag information, so that they don't have to pay their own tolls. Battery-assisted tags are more than 99% reliable when properly mounted and operating with a fresh battery in good weather, but 95% read reliability can result from poor mounting, weak batteries, and nearby metal objects.

9.2.4 Animal Tracking

One of the motivations for the early work of Koelle and coworkers, described in Chapter 2 (Figure 2.6), was to provide a means of identifying cattle that did not involve branding, which ruins a significant area of the hide for use as leather. Thus, RFID for animal tracking has a long history. Cost constraints vary tremendously depending on the species in question. Cows are huge animals that weigh up to 1000 kg, a significant portion of which can be converted to food and other saleable products; a single head of cattle may sell for US$1000. It is not unreasonable to spend up to several dollars to track an individual animal, but US$100 would be excessive. Other common food animals are smaller, less valuable as individuals, and thus require less expensive tracking means. Individual hogs sell for a few hundred dollars: even a US$10 tag is a significant cost adder. Individual chickens sell for a few dollars; tagging them with the same technology used to track a cow is impractical.

Mature cows are more than a meter high and 2 m long; tracking them without individual human intervention requires at least 1 m read range. Pigs and chickens are often raised in physically constrained conditions and could be identified from close range, but at the cost of numerous readers that must be installed, powered, and maintained. Read range of at least a few meters is useful to minimize the required number of readers. If the low-end range of 1 m is used, it is likely that one can arrange for only a single tag to be present in the read zone, and no elaborate medium allocation is required. If a longer-range technology is used, collision avoidance or resolution will be required. All animals are basically bags of salty water; the range of an LF tag will be unaffected by the animal to which it is attached, but UHF tag reads may be blocked by the body of the animal.

Cows can live up to 25 years, but cattle raised for beef are likely to be slaughtered at 20 months. Pigs have a similar 1–2 year lifespan in food production. Any tag should have at least a few years utility.

In summary, animal tracking requires moderate to long read range, cost of a few dollars, and useful tag lifetimes of 1–2 years. Depending on the technology and range chosen, anything from one or two to tens of tags, might be present simultaneously. Passive tags are obviously the preferable approach, but one can consider both LF/HF and UHF technologies. Security is generally not a major issue, since anyone who can read the tags can see the animals.

Work in the late 1980s and early 1990s focused on LF approaches, and resulted in ISO 11785 and the later 14223. These standards use an LF tag and a very simple protocol with minimal provisions for collision resolution. The reader simply sends a 134 kHz continuous signal for 50 ms, with 3 ms quiet windows. A tag either replies during the ON time or waits until the quiet window. Tags that reply when the reader is active use differential phase modulation at 4.2 kbps (1/32 of the carrier frequency). Tags that reply during the quiet time send a frequency-shift-keyed signal in which 134 kHz signifies a binary 0 and 124 kHz a binary 1. The default ID, described by ISO 11784, uses 1 byte to describe the application, a flag, 9 bits for a country code, and 32 bits for a unique ID. The ID space of a few billion values is enough to cover all the animals likely to be marked at any given time, but will roll over after a few years in a large jurisdiction. ISO 14223 describes a more elaborate but backward-compatible tag that can accept commands and support a larger memory.

An LF tag with a 1-m range requires a coil with lots of turns; practical ranges are often measured in the tens of centimeters. Animal tags are typically implemented as large coils encapsulated in thick plastic. With cattle or pigs, the tag may be attached to the animal's ear, implanted under the skin, or swallowed and trapped within the stomach. At 130 kHz, the induced currents in water have little effect, so tag reads are not blocked by the marked animal's tissues or those of its neighbors. However, the limited range even with a large antenna requires that the animals be induced to travel near the reader in something like single file, which is not always easy, particularly for herding animals like cattle or sheep. In recent years, with the wide use and consequent falling price of ISO18000-6C (Gen 2) tags and readers, UHF has been reevaluated for these applications. Figure 9.4 depicts a typical UHF ear tag. Note that a human-readable label is combined with the radio device. Weather-sealed plastic packaging is used. Even after packaging for multiyear field lifetimes, UHF tags are cost-competitive with older LF tags. Great improvements in sensitivity in the last decade have improved the read reliability of UHF tags in the presence of animal tissues. It is considerably more convenient to perform inventory with read ranges of 3–5 m. At the time of this writing (early 2012) several countries, including Brazil, Paraguay, and South Korea, have mandated cattle tracking using UHF technology. Others, including Canada, have required the use of LF technology.

Note that tracking of pets is also an interesting application but one with rather different constraints. The goal is generally to identify a lost pet for return to its owner, so the animal is handled individually anyway and long read range is not needed. The tag should be permanent and need only contain an ID. Passive LF tags encapsulated in glass, which can be placed under the skin of a pet, are often used for this application. The combination of no battery and resilient external materials ensures long life in the field. A typical tag is 3–4 mm in diameter and 15 mm long. Read range is only a few centimeters, but sufficient for the requirement, unless the object of inspection is unusually hostile to strangers.

9.2.5 Container Tracking

As noted in chapter 2, multimode shipping containers are widely used throughout the world and accumulate in substantial numbers in typical port facilities, rail yards, and distribution centers. Shipping containers are big objects (typically 12 m long and 2.5 m high), so you wouldn't think they would be hard to find — until you go to a major seaport and see the

Figure 9.4
A UHF ear tag used for cattle identification. *Image courtesy http://rfid.net.*

containers stacked there (Figure 9.5). Their contents are often worth tens of thousands of dollars, and may be time-sensitive, so the value of being able to find them is substantial; a purchase cost of US$100 for tracking is acceptable. The storage area in a large seaport may span kilometers; the read range must be at least hundreds of meters. As a consequence, tens or hundreds of containers may be in range of a single reader, and some means of allocating the wireless medium must be employed. In this context, merely detecting the tag is of little use; it is necessary to provide a reasonably accurate location as well. The lifetime of a container is up to several years; if battery-powered devices are to be used, at least several months of service must be provided on a charge, and it must be possible to replace the battery or tag while in service. Tag read security is not of major consequence, but if the tag contains information on the contents of the container, that information must be secured when necessary. It may also be of interest to combine the tag with a door seal that records when the container door is opened.

We can conclude that the appropriate technology for this problem is a battery-powered active tag, combined with multiple reader antennas to provide location information. An example protocol is ANSI 371/ISO 24730. The protocol specifies a minimum range of 300 m, with 120 beacons per second and 3 m location accuracy. The protocol provides for magnetically coupled exciters with unique IDs that can trigger beacons and provide supplemental location information. Section 1 of the standard describes transmission in the 2.4–2.483 GHz ISM band. Transmissions consist of a 511-chip binary-phase-shift-keyed pseudorandom sequence, transmitted at 30.5 Mbps, occupying a 60-MHz bandwidth

Figure 9.5
Shipping containers in transit at a seaport. *Photo: NOAA.*

(most of the ISM band). The "blink interval" (time between transmissions from a single tag) is at least 5 s, with 0.6 s of randomization. Since each beacon is about 17 μs long, this leaves room for several tens of thousands of timeslots. A reasonably offered traffic level of 25% of the available slots provides room for around 10,000 tags with modest chance of a collision; thus, no other provisions for medium allocation are needed or provided. The standard supports timed beacons, beacons triggered by proximity to an exciter, and beacons triggered by an external event.

Recalling that light travels about 1 m in 3 ns, a single chip of the beacon is about 10 m long. Thus, if the packet can be timed to within about 1/3 of a chip, its distance from a given reader can be estimated to within around 3 m. With three independent readings of absolute time, or four readings of relative delay, the location of the tag can be unambiguously determined – if the measurement isn't too contaminated by reflections and diffraction.

An example of such a tag was shown in Figure 2.22. The tag is a few centimeters in a side and contains an expensive battery and multiple circuit boards. Tags of this type sell for on the order of US$75–100 in small quantities, and have lifetimes of months to years in the field.

The problem of tracking containers throughout their travels must also be considered. Containers on trucks entering a port can be tracked using the same methods used for finding them in the yard, but with the proviso that the shipper has placed the tag and registered the load with the port authority. A pile of containers on a ship can in principle be tracked through the ship's manifest, but it is also possible to support a separate, independent system of tracking by satellite links to the topmost containers, which will also be able to receive GPS signals, and thus determine their own location, combined with *ad hoc* networking with containers that don't have a clear view of the sky. Fortunately, the wide applicability of GPS has made high-performance, low-cost receivers widely available.

The uplink is a different story. Using the Friis equation introduced in chapter 3, we can estimate what is needed to talk to a satellite. If we send a 1-watt signal at 6 GHz to a satellite in geosynchronous orbit, about 33,000 km up, using, e.g. a 13-dB-gain antenna,[1] and assume the satellite has a 40-dB-gain receive antenna, the signal it gets is about −115 dBm. If we go rather slowly – say, a few tens of kbps, so that the bandwidth of the receiver is thus some tens of kHz – this signal can be received above the noise in the receiver. So we're going to be on for at least some tens of milliseconds, and then there's the problem of receiving an acknowledgment: it takes about 0.1 s for our signal to even reach the satellite, and thus at least 0.1 s more to get an ACK for our transmission.

---

[1] OK, we're cheating here too. Who is going to point the antenna at the satellite? If we use an adaptive antenna array, we need a lot of computational power to figure out where the satellite is, and we need a fairly elaborate set of phase shifters or multiple RF transmitters. If we skip it and use a 3-dB dipole, we have to come up with another order of magnitude of link budget somehow: go slower, or use a 10-W transmitter.

Our tag needs to be on for at least 1/4 second each time it tries to connect to the satellite, if we want a confirmed link. During the time the transmitter is on, it will use several watts. Thus, we will spend on the order of 1 J per transmission, if careful power management practices are employed in the transmitter and receiver. A Joule per transmission allows for thousands to a few tens of thousands of transmissions from a reasonable-sized lithium-ion battery. If our tag updates once per hour, it could last for a few years, granting that beacons are rare enough that we don't need to provide for medium allocation. (That is, we hope that each ship on the ocean has few enough tags sending data that we can just send at random and usually get through.) Additional energy is needed to pass the GPS information on to other tags via Wi-Fi or some other link technology, and we haven't accounted for any additional energy required to make the link secure, a possible customer requirement. On the other side, a 40-dBi antenna at 33,000 km distance will cover around 1000 km; multiple antennas are needed to see significant portions of the earth. So you can see that, while active tracking of containers via self-contained geosynchronous communications is not impossible, it is not simple.

An alternative approach is to employ low-earth satellite links. For example, Globalstar operates a network of satellites active near the equator at an altitude of about 1400 km. In this case, the required transmit power is reduced substantially. With a 3-dBi-gain transmit antenna (i.e. a dipole with a decent view of the sky) and a reasonable 20-dB-gain receive antenna at 1600 MHz, we find a path loss of about −130 dB. If we start with a modest 20 dBm, the satellite sees about −110 dBm of signal, sufficient for a low-data-rate link if some coding, spreading, or repetition is used. The power during transmission is less than 1 watt, so it is reasonable to again expect thousands of transmissions. It is no longer necessary to have a directional antenna that is either pointed by hand or adaptively locates the desired receiver (an energetically expensive task).

At the time of this writing (early 2012) commercial solutions are becoming available for global tracking and security. An example commercial product uses low-earth satellite uplink and sends a 9-byte message in about 1 s using direct-sequence techniques, providing for a few thousand transmissions from a compact 300-g transmitter. A photo of the tag interior is presented in Figure 9.6. A complex board with several RF-shielded regions (the shields having been partially removed to expose the circuitry) can be seen. A separate enclosure contains the battery. This elaborate active tag has sophisticated capabilities, but it isn't going to be cheap in small quantities. (Recall that 100,000 tags per year is small volume for electronics manufacturing.)

Once a container reaches a distribution yard, typically as a trailer on wheels, the problem is a bit easier. A large yard can contain on the order of 1000 parking slots, all at the same vertical level − it is not normal practice to stack wheeled vehicles on top of each other, as happens at a seaport. The same active beacon tracking described above can be used in

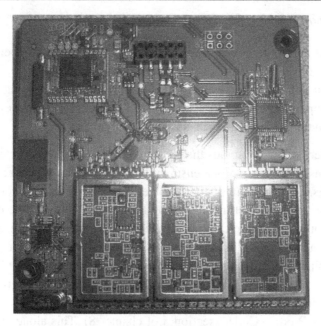

Figure 9.6
Example of satellite transponder tag. *FCC report, TWV-SXL1FLEX.*

this context, with the benefit that containers are no longer buried at the bottom of a stack, so the line of sight to the reader antennas is less likely to be obstructed. Magnetic exciters can be mounted on the yard truck that moves the trailers around, so that when a trailer is moved, the tag beacons regularly, while not doing much when the trailer is stationary and no information is conveyed anyway.

A challenge to this approach for distribution yards is economic: the infrastructure required, which includes multiple reader receivers mounted high above the yard, with provisions for timing accurate to a few nanoseconds between the receivers, is expensive. It might cost US$100,000 or more to outfit a facility. This is acceptable for a large shipyard but expensive for a distribution center. An alternative approach that provides sparser information but is much less expensive to implement is to use passive tags placed on each trailer, and place a reader in the yard truck that moves trailers around. As the yard truck moves through the yard, each tag is read. If the location of the yard truck is monitored, the location of each tag can be reliably inferred to within a few meters. The tag cost is reduced to a few dollars (a cheap UHF passive tag in a weatherproof plastic housing), and each yard truck requires a few thousand dollars of readers and location equipment. Visibility is not provided when an outside driver first brings a trailer into the yard, but if yard trucks traverse the whole facility several times a day, the delay in locating a given load will be a few hours at most.

9.2.6 Supply Chain Tracking for Consumer Goods

Tracking individual consumer goods in the supply chain confers a number of potential benefits, but the cost of the solution must not exceed the value of the benefits obtained. It's really important to keep the investment manageable and the recurring cost of the tags low. Low cost is important. You can't spend a lot of money saving money. Oh, yes, and did I mention that everything needs to be cheap? Don't forget about cost.

Hopefully the reader is now conscious that cost is of overriding importance in supply-chain-tracking applications. This factor alone ensures that the majority of such applications employ passive tags. Since consumer goods sometimes require up to a couple of years from manufacture to sale, the long lifetime of passive tags is also helpful. The other inevitable consideration in supply-chain applications is the identification space. Where the airplanes visible to a single radar station may be numbered in the tens or (rarely) hundreds, the number of nominally identical objects sold by a major consumer-goods company in a year may number in the billions. At least 64 bits, and preferably 96, are required to provide a convenient space in which to identify the manufacturer, model, and serialized object (see "The Electronic Product Code," section 5 of chapter 8). This alone excludes slow LF tags from consideration. Read range of a few meters allows the use of readers located at dock doors or other pinch points to read most or all tags on a pallet full of (large) items and permits high-speed reads in conveyorized facilities without the need to orient the items to be read. Thus, in many supply-chain scenarios, UHF tags are favored over HF tags.

You might only have one bicycle at home, but a retail store has tens and a warehouse or distribution center hundreds or thousands in close proximity. Making money requires that thousands or tens of thousands of items of each type are in roughly the same place at the same time. Robust medium allocation and collision resolution is indispensable for any practical supply-chain-tracking system. (Closely spaced tags also scatter radio waves and cause problems in the physical layer, as we alluded to in section 4 of chapter 7.)

So a consumer goods tracking technology must provide range of several meters, hundreds or thousands of reads per second in the presence of hundreds of competing tags, and at least about 100 bits of memory: a UHF passive tag implementing the Gen 2 protocol. That wouldn't be so hard except for the first four or five requirements, which are in order, low cost, low cost, low cost, low cost, low cost, and did we mention low cost? The integrated circuit (if one is used) must be manufactured, tested, and assembled for a few pennies. If embedded in a printed human-readable and bar-coded label, the total cost may rise to ten or twenty cents. That's a large cost adder for a bin of lettuce that the produce company may sell to a retailer or distributor for a few dollars, or a television set manufactured in China and sold at a few percent gross margin to a US retailer. Consumer goods tags must employ specialized technologies to achieve high volume at low cost, as described in more detail in chapter 5.

In addition to the direct cost of labels, sensible application can be challenging. Figure 9.7 shows a pallet of lettuce containers, each with a human-readable, bar-coded, RFID-enabled label attached. These labels must be applied as the lettuce is harvested, so a network-enabled smart label printer must be provided in a farming area that may have spotty or nonexistent coverage by any high-speed wireless network. The photograph depicts a pleasant sunny day, but field work proceeds in the rain and mud, where keeping labels clean and sticky is a challenge. Even when everything goes smoothly, label application represents an additional operation to the laborious tasks of cutting heads of lettuce, inspecting them, placing them on the movable truck-mounted conveyor for packing, packing heads in bins, and stacking the bins for shipment back to the storage facility where the lettuce is washed and chilled. Each hour's delay between picking and cold storage reduces expected useful life of the product by about 1 day; additional steps are painful.

When a pallet like that shown in Figure 9.7 enters a facility, it will typically encounter a reader with antennas looking into a loading dock area, often known as a ***portal reader***. Ideally, the reader should read every tag on the pallet as the pallet passes through the doorway on the way to appropriate storage. This can be challenging, due to the properties of the materials contained in the boxes — for example, lettuce and other food products are mostly water, and thus reflect and absorb strongly at UHF frequencies. Various tricks can be used to improve read percentage. The pallet can be placed on a rotating table during transport; the combination of Gen 2's slotted-Aloha anticollision capability and a variety of reads from multiple angles will often suffice to ensure that all or nearly all tags are acquired. Location in this context becomes the ability to read tags in one doorway but not

Figure 9.7
Pallet of lettuce containers, each with a field-applied human-readable RFID label. *Image courtesy Stefan van der Bijl, Tanimura, and Antle, used with permission.*

in adjacent doorways. Selectivity can be improved by selecting differing circular polarization for leftward- and rightward-looking antennas; software rules are also helpful to infer with good confidence the location and direction of motion of a given tag when read multiple times by different readers.

Data security in supply chain tracking is usually ensured by keeping all the important information on secure servers: the tag ID is just a pointer into a database, so reading it doesn't reveal much if you can't access the database. Even in this context, some minimal physical security precautions should be taken if, e.g. tag data that a vendor depends upon can be erased or overwritten. This is why the Gen 2 protocol provides password protection for locking or killing a tag. But any security approach comes up against the inevitable monster of cost. Even if the computational requirements on the tag are modest, good security requires fairly elaborate provisions for keeping private information private but also distributing it where it needs to go. For example, if you decide to give each groups of tags a separate password (rather than e.g. using a common password for all products from a single vendor), you have to have some sort of infrastructure that tracks which products got which password, and makes that information readily available to anyone who needs to access the tag information, while making it not available to hackers, criminals, and disgruntled employees who wish to mess with the tags. The more complex the procedure, the more likely it is that honest but harried employees will develop workarounds that degrade security but make their daily lives tolerable.

The problem of security is even more challenging in industries where different legitimate participants have conflicting interests. The pharmaceutical industry in the United States provides an illuminating example. A few large manufacturers of drugs sell their products to a handful of major distributors, who in turn provide supplies to a broad mixture of large and small clients (hospitals, clinics, retail pharmacies, and so on). The manufacturers often sell the same product at different prices to serve the needs of large customers while maximizing revenue from small customers, a common practice in any industry. Since many sales are indirect through distribution, this creates an opportunity for arbitrage profits within the supply chain: that is, an item is obtained at a low price from a manufacturer and then sold not to the target, large customer, but to a small customer at a much higher price, providing a wonderful profit to the intermediary. The problem is amplified in the United States by the very high price of many pharmaceuticals, and the opaque pricing of most participants in the health care business. Such practices would become much more difficult if every container of medicine were tracked from manufacture to use. Thus, item tracking is in the financial interest of the manufacturer but not of the distributors.

Where a secure physical-layer data exchange is required, the computational needs of implementing any real encryption algorithm will likely override other considerations and force the use of HF (13.56 MHz) tags. When an inductive tag is in close proximity to the reader antenna, quite substantial power (measured in milliwatts rather than microwatts) can be provided, allowing nontrivial digital circuitry to operate. Secure HF tags implementing the Advanced Encryption Standard (AES), and SHA-1, are commercially available, at some cost in read range.

9.3 Plus Ç'est La Même Chose

Like just about every other technology in the modern world, the effects of RFID are overstated in the short term and underestimated in the long term. The pronouncement by Wal-Mart executives that radio waves shall travel through boxes of metallic items did not make them do so, but the continued labor of scientists, engineers, production workers, and business managers has produced something approximating the desired result. Real progress is made as procedures and equipment come together to make technologies do what they are capable of.

Let's first look at large-volume applications like consumer goods. The glorious world of 5 billion UHF tags per year envisioned by the early MIT researchers around the turn of the century is about in reach at the time of this writing (early 2012). To understand why volume matters, a bit of review of semiconductor economics is in order. Recall from section 5 of chapter 5 that a wafer of 1/2-mm-square RFID chips might produce around 40,000 pieces. To produce 5 billion tags, one requires about 125,000 wafers, give or take a factor of 3 — chip designers and manufacturers are constantly improving density, and then giving the area savings away in increased capability. That is about 10,000 wafers per month: a sizable wafer fab capacity. No one vendor makes all those chips, but cumulatively the number of wafers is substantial enough to justify actual expenditures on improving their manufacture, something a fab operator cannot afford to do for a mere niche business. In the same vein, software operators can afford real investments when customer sites number in the tens of thousands instead of the tens, and when only one type of product (in this case, EPC Gen 2/ ISO 18000-6) must be supported. We can anticipate substantial reductions in the cost of RFID IC's, and increases in overall system capability, over the next decade, as the real benefits of economies of scale become important.

When coupled with favorable economics, the availability of technologies that work creates a virtuous cycle of implementation. At the moment, tracking of apparel is rapidly expanding. Clothing (particularly for the ladies, speaking as an engineer whose choices in clothing archetypally define lack of fashion sense) is relatively expensive, and in cases where fashion matters, its value may be transitory. If an item doesn't sell in season it may never sell at all. Tagging items may cost tens of cents, and still be worthwhile if a substantial improvement in inventory turns can be achieved. Furthermore, most apparel is RF-transparent, so it is fairly easy to count all the tags in a box, or all the items on a rack, in a single operation.

Improvements in inventory management have been shown to be substantial simply because it is so much easier to count tagged items that it actually gets done. Even very inexpensive items may justify the cost of tracking, simply because of the inventory labor savings. Without a convenient counting method, retail workers often would open a box, note that the first item was (say) a Medium size, and presume that the remaining items were, too, creating errors in inventory that would propagate through the rest of the retail operation. With a handheld reader and tagged items, a box can be accurately inventoried in the same time previously used for an erroneous estimate. Since each item is uniquely identified, accidental multiple reads — for example, from neighboring boxes not yet examined — can be properly accounted for. Similar improvements are obtained in management of display shelves and racks, where clothing misplaced by consumers can be readily found and put back where it belongs, and stock levels of each size and type can be accurately matched to the mixture of potential purchasers. And the time required for floor inventory is greatly reduced, reducing the cost of this necessary but unproductive operation and making the store more available for sales to customers. Key challenges that remain for UHF technology are not very relevant in this application: missing 1 or 2% of tags on a rack is not an important obstacle, and handheld readers with intermediate range can be used, avoiding the need to accurately locate an individual tag by any sophisticated radio means.

In contrast, the oft-cited but never implemented ability to check out a shopping cart of items from, e.g. a grocery market, in a single read faces almost exactly the inverse situation. The economics is very challenging, and the requirements are unachievable. To use RFID to do simultaneous checkout, all the items in the store must be marked. If a checker must search through a pile of stuff in someone's cart for the scattered items that must be bar-code-scanned, much of the time benefit is lost, and if the checker doesn't bother, the grocer is likely to lose all their profit margin in items not charged. (Groceries are a famously low-margin business.) This means tags that cost a few pennies must be appended to items that cost a few tens thereof: a losing proposition.

Furthermore, to be successful, such a checkout must have a very low frequency of false positives or false negatives. As noted above, a grocer cannot afford to have consumers go home with products that are not paid for. Every tagged product in the basket must be read and the corresponding price paid, at about the 0.1−0.01% level. In the other direction, customers get very ticked off when they pay for something they didn't buy. An offended customer can easily be lost to a specific retailer for years, giving away thousands of dollars of future sales. So a reader must never read an item in a neighboring cart as being in the current assay, again at about the 0.1−0.01% level. And all this has to happen in a randomly oriented jumble of items containing metallized packaging, foods loaded with salty water, and other aqueous items, scattering and absorbing incident waves while also interacting with antennas in a positively nefarious fashion. Near-field tags that tolerate water don't have enough range for a whole cart, and are just as sensitive to metallized items as far-field tags. Existing UHF technology would be hard put to achieve 1% total false positives and false negatives in such an environment, 10−100 times poorer than what is needed.

Proposed alternatives are involved, e.g. having goods read as they are placed in the cart by a cart-mounted reader. This approach is again spectacularly challenging to implement. Instead of a few readers at checkout lines, every cart must be equipped, adding costs for installation, power, and maintenance. The same challenges of false positive and false negative accuracy apply, and are made worse by the lack of mutual supervision. A consumer has an incentive to place an item in the cart without it being read, and is in most cases out of sight of an employee in doing so. Simple tricks like laying the active sides of two meat products, or two metallized containers, against each other would be readily learned and implemented.

And finally, someone still has to bag the purchased goods to make it possible for the consumer to carry them home, limiting the possible labor savings even if the approach can be made to work.

On the other hand, it is plausible to imagine that RFID tags may soon appear to supplement bar codes in the existing checkout model. Using improvements in near-field and evanescent antennas, short-range readers can be collocated with the existing bar-code scanners, to obviate in many cases the need to orient, stretch, clean, and otherwise manipulate items to obtain a successful bar code read. It is easy to see, especially while waiting in line at the grocery store that such improvements can increase the throughput of a typical grocery checking operation by 10−30%. There is no requirement that every item be tagged, just items that are likely to fail initial bar code reads. The financial benefits are measured in improved productivity at the checkout, which directly affects customer satisfaction. These benefits must be balanced against the painful cost of an additional and redundant marking of products.

Various niche consumer-goods markets exist where RFID application is particularly beneficial. Jewelry is small but valuable, easy to lose and worth the expense of tracking. Shelf antennas and either HF or UHF tags can be used to monitor the presence or absence of each item in a display case; an example is shown in Figure 9.8. An HF antenna is intrinsically short range. If UHF tags are used, a good solution is to employ a near-field antenna (section 5 of chapter 6). A segmented loop (Figure 6.35) antenna can be used for this purpose but requires that the tag antennas have a loop or other magnetically coupled structure. An unshielded (leaky) transmission line, possibly meandered, may work well for a rectangular shelf area (Figure 6.36). A nonradiating antenna minimizes spurious reads at long distances (on other shelves) while providing good performance near the antenna.

We've discussed above the travails of tracking fresh produce (Figure 9.6). Merely recording what bin went where is of modest benefit and marginal for justifying additional cost.

Figure 9.8
Tagged watches are placed on a flat evanescent-field antenna for tracking.
Image courtesy http://rfid.net.

The temperature history of fresh foods, such as produce, fresh meat, and fresh fish, has a powerful influence on their taste, perceived quality, and safety. The combination of a battery-assisted tag with temperature sensing and logging enables instant access to the temperature history of a given batch. Because battery-assisted tags have much better sensitivity than pure passive tags, they can be embedded within the food (for good thermal contact) and still be reliably read from meters away. An appropriately packaged tag can then be cleaned and recycled back to the farm for repeated use. This approach enables distributors and retailers to instantly evaluate the status and likely shelf life of batches of product. Distributors and retailers can operate under a first-to-expire rule – shipping the product that has the shortest remaining shelf life first, thus optimizing overall product value – rather than first-in/first-out. And in the unlikely but painful event of a major product recall, lot tracing becomes rapid and automated, greatly reducing the cost, time, and risk of insolvency.

To secure the full benefits of such an approach may also require changes in a retailer's business model: for example, providing consumer access to the measured storage history of products may enable a vendor to charge a higher price. If measurements are needed over the whole product history, the pricing benefits need to be shared between producers, distributors, and retailers in order to provide an incentive for all participants to do their part. Otherwise, temperature loggers will continue to conveniently malfunction or mysteriously disappear when truck's air conditioners fail.

Tracking more valuable items has more demanding requirements but the potential for greater returns. Passive tags can be used for specialized requirements, where storage or access requirements ensure that short range is sufficient for identification and location. Small but high-value objects, like the stents inserted into arteries and veins for mechanical support, can be tagged and stored in shelves equipped with built-in HF or UHF readers, providing continuous inventory visibility and expiration tracking. Passive tags can also be conveniently used to mark an event for a more capable information-gathering device, often a high-resolution video camera. If valuable objects are tagged, and can only enter and exit a given facility through a finite set of regions, each equipped with a passive reader and the whole monitored by video surveillance, the moment at which a given object entered or left the area of interest is readily provided by one or more tag reads. These provide a time stamp to immediately direct investigation to the relevant moment in a long video record, which otherwise might be extremely laborious. The video record, in turn, enables the use of the powerful pattern-recognition and context awareness of a human observer, which is difficult to replace with an RFID tag.

Another area where passive tags can be preferable is in the tracking of long-lived assets. Many physical objects can outlive the recordkeeping that supports them. Examples are airplane parts in service, utility poles, and supply plumbing for water or natural gas. In these cases, a local record-keeping — means a passive tag — can in principle ensure that the history of the object is available where it is needed, regardless of changes in the organizational structure or ownership of the object. Passive tags also send no signals when not being queried, a major benefit in interference-sensitive environments; the Federal Aviation Administration has specifically forbidden the use of active or battery-assisted tags on airplanes in the United States.

The multiple-decade service life of such objects exceeds any reasonable battery-powered solution. Although a passive tag doesn't require a battery, it must have enough memory to provide a useful historical record. The resulting increase in chip area is acceptable for high-value applications like this, but the total volume is modest relative to the cost of custom semiconductor components. Retention lifetime for memory must also be comparable to service life, despite possible extremes in temperature: the tag is unhelpful if it has forgotten what it was told a few years ago. In addition, the tag must be packaged and protected so that it can survive the expected operating conditions for decades. Power poles are outdoors, exposed to rain, snow, heat, cold, vermin, mold, and even lightning. Airplane components are mostly shielded from precipitation, but exposed to wide temperature ranges and sustained vibration. Tag packaging is arguably more critical than tag performance in outdoor sustained tracking applications.

Intermediate-sized, mobile objects, like medical ultrasound or infusion stations and other movable instruments, require a solution that provides long read range and location information, and does so in the obstructed environment of a hospital or clinic building. Battery-powered active tags are generally necessary in this type of application, typically using a short-range low-power protocol: 802.15.4 or low-power 802.11 (Wi-Fi). Identification badges for physicians in hospitals can also be radio-equipped, enabling critical people to be found when they are needed. The location problem in this context is not trivial or ignorable. Various approaches are used. Comparison of signal strength from multiple access points has been shown to provide robust location indoors for 802.11

devices; similar techniques can be used with 802.15.4. An alternative physical layer, defined by 802.15.4a, is specifically designed to support time-of-flight location. It provides 500-MHz-wide channels with 2-ns symbols, along with orthogonal preamble symbols to enable accurate timing in the presence of interference and multipath propagation. This makes it possible to time the arrival of a signal to a few nanoseconds, equivalent to location accuracy of 1 m or less.

Workers at large facilities, like refineries or generation plans, can be equipped with radio-linked identifying badges, so that in the event of an emergency all personnel can be rapidly accounted for, and rescue arranged where needed. This may require specialized longer-range technologies, like those used for container identification as discussed above, depending on the size and configuration of the facility in question.

It is worth noting that 802.15.4/Zigbee and 802.11 used for identification and tracking are not always distinguished from sensor network applications of the same technologies. The identity of an object is just one of many aspects of the physical universe that can be acquired by a sensing interface thereto.

9.4 New Capabilities

Technological change comes in two forms: completely novel ideas that cannot be predicted and straightforward application of existing technologies to achieve new capabilities. Completing a book is quixotic enough with attempting additional impossibilities; therefore, we shall content ourselves here with a few words on the latter topic.

Commercial implementation of the improved ISO-180006 battery-assisted tag standard described in the previous chapter has led to BAT products with read ranges exceeding 100 m in an open area. The reader will recall (section 7 of chapter 3) that ranges of this order are expected for a diode detector combined with a low-noise reader receiver. Absolute range will be greatly reduced in the presence of obstructions, but battery-assisted tags will provide greatly improved readability at short range in the presence of scatterers and absorbers. As noted above, this capability can be exploited for produce tracking, where the tag can simply be placed atop the water-containing products and still be read.

We have stated previously that monostatic antennas reflect part of the incident carrier back to the receiver, increasing its effective noise floor. Adaptive cancellation of the carrier, a technique well known in radar, can be used to minimize the noise contribution of the reflected signal. Improvements of 20–40 dB are achievable, resulting in monostatic antenna performance comparable to that of a bistatic arrangement. Improved receiver noise is particularly important for long-range battery-assisted tags, whose advantage in receive sensitivity is vitiated if their reply cannot be heard.

Adaptive antennas can be constructed using an array of radiators, elaborated versions of the simple two-element array introduced in chapter 6, where (at least) the relative phase of each radiating element can be adjusted. Changes in phase, and optionally amplitude, can be used to steer the beam either continuously, or in a discrete set of directions. Adaptive arrays have been used in defense applications for decades, and techniques of construction and control are well known.

An adaptive array can be used to locate tags by angle of arrival (really angle-of-tag read): in the absence of reflections, the tag must be along the line the beam was pointing at when the tag was read. Combining beam direction with a *received signal strength indication* (RSSI) provides three-dimensional location information, again in the absence of reflection or scattering. In order to provide accurate angular information, high antenna gain is needed; recall from section 2 of chapter 6 that beamwidth is inversely proportional to the square root of the directive gain. In order to simultaneously provide coverage of a large area, the beam must be stepped or swept, so that over the course of a reasonable time, all available directions are inventoried in succession. By noting changes in position of specific tags on successive sweeps, the direction of motion of a tag can be ascertained. Accuracies on the order of 1 m or less have been achieved in large open areas, making this approach very useful in tracking tag motion at, e.g., a distribution center door. Location information minimizes confusion about which tags are at what station. Location and direction of motion together establish fairly conclusively whether a tagged item is moving into the distribution center or being placed onto a truck for transport elsewhere. However, in obstructed environments where reflections are present, such as storage areas with metallic shelves, location becomes ambiguous at best.

Recall that an antenna beam's angular width is inverse in the antenna gain, which depends in part on the antenna's physical size. It is not possible to get a very narrow transmitted beam without a very large antenna, practically difficult for wavelengths of 30 cm. However, a null (an angle at which essentially no energy is transmitted) can be very narrow, even for a modest antenna size. An array antenna can exploit this fact to locate a tag fairly accurately even with a small antenna, as described by Hansen and Oristaglio. The simplest arrangement uses a two-element array, like that discussed in Chapter 6. The normal reader signal is sent in phase from both antennas and adds in the usual fashion to form a beam. A pseudorandom scrambling signal of similar total power is also transmitted from the array, but in this case with the two elements 180° out of phase. Thus, along the forward direction, the signals from the two antennas cancel: a null is present. Within the null, tags hear only the normal reader signal and can respond. Outside the null region, tags hear both the reader signal and the scrambling signal, and are thus unable to decipher the reader's commands. The technique has been shown to provide effective read zones of a few centimeters width at >1 m distance, much smaller than achievable with a conventional small array antenna.

Ultrawideband (UWB) signaling is defined as transmissions using >500 MHz bandwidth. UWB transmission is permitted, with rather stringent limitations on transmitted power, for frequencies less than 1 GHz, or in the 3−10 GHz band. Traditional UWB signals are just very short pulses; recall that a pulse of 1 ns long will have a bandwidth of about 1 GHz. Short pulses are easy to time accurately, so UWB signals have long been used to provide ranging and thus location, subject to the usual constraint that the line-of-sight signal can be distinguished from any reflected or scattered signals present in the environment. Pulses are usually shaped and combined to minimize "DC" content and manage their spectral width. Information is added using pulse-position modulation, often combined with polarity inversion. Mutually orthogonal pulse trains enable the receiver to extract accurate timing, distinguish multiple reflections, and reject interferers.

UWB tags are useful for providing accurate location in large areas. The use of wide bandwidth signals is helpful in penetrating walls and other nonopaque obstacles. For long-range applications, active beaconing tags are used, but passive tags designed for UWB operation have also been demonstrated. These tags achieve read range on the order of 20 m and provide the same benefits of accurate location, tolerance of environmental reflections, and wall penetration.

Orthogonal frequency-division multiplexing has also been used for UWB communications. This approach is computationally intensive and more appropriate for powered short-range high-rate links than for RFID.

9.5 Things to Keep in Mind: Makers

Making something work once in the lab is not the same as shipping a product for revenue. Each type of product and each business area are likely to have special requirements that must be met to sell a product or service at all, or to do so profitably.

9.5.1 Radios

All intentional radiators must be certificated by the Federal Communications Commission to be sold in the United States, even if they operate in unlicensed bands. Most jurisdictions throughout the world have similar requirements. Certification usually involves testing of the product − for example, an RFID reader − at a certified compliance testing laboratory. Testing includes both radiated and conducted emissions (that is, signals that might sneak out through the power cord or other connections to the device). Testing includes a wide range of frequencies, often far from the band in which operation is intended and thus subjected to little scrutiny during product development. Compliance testing takes a couple of days and a few thousand dollars even when everything goes smoothly. Ferreting out the source of a fourth-harmonic mixing product that happens to lie in a protected band can be laborious and painful to optimistic delivery schedules. It is a good practice to do preliminary characterization, either with a compliance laboratory or in-house with a good broadband spectrum analyzer, before committing to a date for approval.

The problem is, of course, greatly complicated for products intended for sale into international markets, since approval must generally be gained for each jurisdiction. Differing countries also impose different specific restrictions on which bands must be protected, what uses are licensed or unlicensed, and how testing is done. Experienced compliance laboratories can be very helpful in dealing with these problems, but they don't work for free.

9.5.2 Safety

Specific markets have specific safety requirements. Hospitals have traditionally been reluctant to adopt too many wireless devices for fear of interference with life-critical instruments. Medical devices must in general be approved by the Food and Drug Administration in the United States and equivalent authorities in most other jurisdictions. Wireless devices intended for use on airplanes must satisfy the Federal Aviation Administration. Consumer products may require independent safety approval, often provided by Underwriter's Laboratory (UL) in the United States.

9.5.3 Environment

The International Electrotechnical Commission has promulgated standards for environmental exposure (IEC 60529); in the United States, the National Electronics Manufacturers Association publishes a similar set of requirements. Equipment intended for outdoor use will generally need to at least be IP68-compliant, certifying that dust and water cannot enter the enclosure. Additional application-specific requirements often exist. Tags may be squashed or driven over; readers may need to work at subfreezing temperatures.

9.5.4 Documentation and Support

It's obvious how it works, right? No, it isn't. It if was, you wouldn't have read this book. Good products need good documentation. Don't assume the user knows everything you know; in fact, don't assume the user knows anything about RFID. You may have lived and breathed your product for the last 2 years, but your customers have many problems and your product is likely only one of their assignments. Don't expect them to remember what you told them 3 weeks or 3 months ago. Work with the people who need to use the products to see what confuses them. Keep documentation simple and as graphical as possible. Get the users started so they can learn for themselves what works for their application. The better a job you do designing the interface, the less documentation you need, and the better the documentation, the less time you and your colleagues will spend on the phone or at the customer site fixing trivial problems.

9.6 Things to Keep in Mind: Users

If you are applying RFID technology to solve a specific problem, first you need to clearly understand the problem you're trying to solve.

If you are counting on RFID implementation to improve process efficiency, you need to first analyze the process you're trying to improve. Procedures that have arisen organically may not be documented; documentation that does exist may have little correspondence to what actually happens. Once you know what is actually being done, you may find that opportunities for improving operations through facile identification differ from your original intentions.

When you know what you want to do, consider how you will obtain the cooperation of the people who have to do it. Shrinkage (products going missing in the supply chain) can result from people walking off with what is not theirs, but it can also happen due to innocent errors, such as the misidentification of a product size or type, as noted briefly in section 3 above. If people are stealing products, they are not very likely to cooperate with the introduction of technology that will catch them at it; but there are other less-reprehensible reasons for resisting change. Successful technology adoption is greatly aided when the people involved have incentives to change: if their job becomes easier and faster or they get paid for increased productivity. People whose responsibilities are threatened by a different approach to a problem may not be helpful. You need to convince the fellow with a clipboard and a pencil that his job will run more smoothly if he learns to use a handheld with a reader.

If your implementation involves fixed equipment, remember that it is likely to require both network access and mains power. In many industrial environments, running 110 V AC power costs more than the reader it powers; power over Ethernet often provides significant savings as well as simplicity.

RFID is a radio technology, typically operating in an unlicensed band. It is subject to interference. Interferers can be in-band: for example, many conventional cordless telephone handsets operate in the 900-MHz unlicensed band. Other RFID readers, including those you control and neighbors that you don't, can interfere with your tag; Gen 2's Dense Reader option can mitigate this interference when you control the reader in question. Out-of-band interferers can also be important. A cell phone tower with an antenna that looks directly at your intended operating space may be radiating enough power in the US cell band (875−894 MHz) to make operation difficult at 902 MHz; similar problems exist in other jurisdictions. Broadband interferers include poorly maintained spark plugs on forklifts, electrical insect zappers, and any other equipment that generates an arc discharge in air.

The radio performance you need is intimately tied to the way the application is structured. Parts moving on a conveyor require modest read range, but high read speed and reliability. Avoiding false positives from neighboring conveyors, parts on forklifts, and tags carried by people walking by is also potentially challenging. Valuable objects kept in a stockroom can be tracked at the pinch points (entry and exit doors), but if this is done with HF tags, the region needs to be small so that the tags are reliably within the read range of large loop antennas. UHF tags provide better read range but are more subject to blocking by people's bodies. UHF tag reads are also sensitive to orientation; even if circularly polarized antennas are employed, orientations exist where a given tag is invisible to a given antenna. In tagging specific products, optimal locations on a container ("hotspots") for tags can often be established.

All your work identifying an object is of little utility if the resulting data is not available and intelligible where it is needed. Networking and database management are the larger part of most real implementations.

9.7 Silly But Fun

If you stick a UHF tag on the outside of a diaper (safely located so as not to chafe), you'll be able to read the tag when the diaper is dry. When it disappears, time for a change! Saves parents poking their kids' pants all the time, at the spectacularly impractical cost of carrying around a portable reader.

Well, I said it was silly.

9.8 Capsule Summary

Potential applications have become real applications when the technology can meet the relevant requirements for the application, typically including cost, read range, environmental tolerance, useful life, speed, and security. These requirements dictate which technologies are used for a given purpose. Aircraft use large, expensive active transponders; railcars use simple passive tags with robust packaging. Automobile tolling requires battery-assisted or active devices to achieve compliant-read performance. Animal tracking can use LF or UHF passive tags, and the best approach is still not established. Shipping container tracking uses battery-powered active transmitters.

Future growth in RFID applications will be driven less by technology and more by value added, as the technology becomes more capable and less visible. Passive tags are becoming cheap and ubiquitous, enabling their use in applications that could not support a technology on their own. Improved readers will provide accurate location information (at least in open areas) and long range from compact antennas. UWB approaches will provide similar capabilities in dense scattering environments. Wireless identification using active and battery-assisted networked devices will merge with sensing applications.

If you make an RFID product, you need to consider regulatory approval, safety, and how to help your customers use the product. If you want to use RFID technology to improve your activities, you need to consider how to fit it into what you want to do, and how to make it easy for the people involved to adopt a new way of working.

Further Reading

Online Information

A number of web sites focus on news and resources related to RFID. RFID Journal, www.RFIDjournal.com, provides news and information, some free and some with subscriptions. RFID Journal is clearly an advocacy site, and not a place for serious technical information. The RFID Network, www.rfid.net, is oriented toward providing more specific product information and evaluations, complemented by interviews with folks in the business (including, on rare occasions, the author of this book). RFID 24-7, www.rfid24-7.com, similarly provides RFID-related news.

Location

"Direction sensing RFID reader for mobile robot navigation," M. Kim and N. Chong, IEEE Transactions on Automation Science and Engineering, vol. 6, p. 44 (2009)

"Method for controlling the angular extent of interrogation zones in RFID," T. Hansen and M. Oristaglio, IEEE Antennas and Wireless Propagation Letters, vol. 5, p. 134 (2006)

"3D passive tag localization schemes for indoor RFID applications," A. Almaaitah, K. Ali, H. Hassanein, and M. Ibnkahla, IEEE International Conference on Communications (2010)

UWB Tags

Ultra-Wideband Radio Frequency Identification Systems, F. Nekoogar and F. Dowla, ISBN 1441997008, Springer, 2011.

"Remote monitoring and tracking of UF6 cylinders using long-range passive ultra-wideband (UWB) RFID tags," F. Nekoogar and F. Dowla, UCRL-CONF-231891, Lawrence Livermore National Laboratory, June 2007

Cold Chain Tracking

"A sub-μW embedded CMOS temperature sensor for RFID food monitoring application," M. Law, A. Bermak, and H. Luong, IEEE Journal of Solid-State Circuits, vol. 45, p. 1246 (2010)

"Temperature management and the cold supply chain", WP-05-0811, Intelleflex Corporation, www.intelleflex.com.

Implementation

"A process- and knowledge-based model to identify and to evaluate the potential of RFID Applications." N. Vojdani, J. Spitznagel, and S. Resch, World Automation Congress 2006

"RFID applications in automotive assembly line equipped with friction drive conveyors." D. Tang, R. Zhu, W. Gu, and K. Zheng, Proceedings of the 15th International Conference on Computer Supported Cooperative Work in Design (2011)

"Evolution of RFID applications and its implications: Standardization perspective," B. Lee, Y. Kim, H. Kim, PICMET 2007

Exercises

1. Libraries use RFID for four separate functions:
    * associating a book or other item with a user when it is checked out;
    * verifying that all items leaving the library have been properly checked out;
    * noting returned items and allocating them to the proper location for reshelving;
    * performing shelf inventory.

    For each of these functions, examine the following questions:
    * What read range is needed? What assumptions are you making?
    * What environmental conditions are the tags exposed to?
    * How many tags must be read at once? This may depend on your checkout model (how many items a user may check out, and how many operations are needed), and interacts with the read range you assumed for inventory.
    * What security precautions are required?
    * What value is gained relative to a bar code?

    Then answer the following basic questions:

    Type of tag: UHF, HF, or LF?   _____

                      Passive, BAT, or Active? _____

    How much memory does the tag need? _____ (bytes)

    How are tags packaged and attached?

    _____

    Where are readers to be located?

    _____

2. A semiconductor fabrication facility is a complex place with three floors: a top floor filled with air handling equipment and toxic gas scrubbers, a main floor with semiconductor processing tools and wafer movement and management, and a bottom floor with vacuum pumps, gas supplies, and other support equipment. There are a lot of people involved (as well as a lot of robots), little visibility, and many toxic gases and liquids, as well as high voltages and powerful radio frequency generators. The management would like to be able to locate all the personnel at any time in case of emergency. Two models are under consideration:

    1. Divide the facility into many small areas with entry by card keys (which are themselves passive 13.56 MHz tags); track location through swipes.

2.   Use a personnel badge that can be detected and monitored at any location in the facility.

Examine the same issues as in exercise 1, for each of the two scenarios above:
* What read range is needed? What assumptions are you making?
* What environmental conditions are the tags exposed to?
* How many tags must be read at once?
* What security precautions are required?

Then answer the similar set of basic questions:

Type of tag: UWB, UHF, HF, or LF?   _____

                       Passive, BAT, or Active? _____

How much memory does the tag need? _____ (bytes)

How are tags packaged and attached?

_____

Where are readers to be located?

_____

If you use a tag with a battery, how are batteries to be monitored and serviced?

_____

3.   Buy some tags and a reader. Plug the reader in and follow the instructions to read a tag. Create a web site describing yourself as an RFID-implementation expert. Get an associated Twitter account so you can let everyone know how it worked out.

Did it help that you read the book first? _____YES _____NO

Which book?

[ ] This one

[ ] A good one

[ ] A funny one

[ ] All of the above

Afterword

Every book faces the limits of the ability of the author and the patience of the reader. This book has focused almost exclusively on physical-layer operation of tags and readers using radiative coupling. We have only touched on the variety of inductively-coupled RFID technologies and protocols that are more appropriate for many important applications. We have left for others to illuminate the vital topic of how the voluminous data collected by a network of readers is converted into useful knowledge for the enterprise they serve. We have left unexamined, save by citation, the important social issues surrounding the effect of radio frequency identification on privacy, and personal security as well as the security of corporate data.

The reader may wish to remedy these omissions by taking advantage of the many existing resources in the physical library and on the web, some of which have been listed in the current tome. In doing so, the reader may proceed with confidence that what was absent yesterday will be available tomorrow: A search on the keyword "RFID" in the IEEE's digital library in late 2003 produced only about 100 citations; in early 2007, only three years later, over 900 documents now result from the same request.

I hope you've found this book informative, and perhaps on occasion as entertaining as a technical tome can hope to be. I'd like to offer thanks again to the many people who helped, and accept responsibility for the inevitable errors of commission and omission. Comments, corrections, and criticism may be addressed to me at dan@enigmatic-consulting.com

Daniel M. Dobkin
Sunnyvale, CA
March 19, 2007

Five years have passed. "RFID" now turns up 7,827 results on the IEEE's digital library, more than I or any other single person can ever hope to read and absorb. I am pleased to note that some of those publications cite the first edition of this book as a reference. I would also like to thank the people around the world who took the time to write with compliments for the first edition, as well as suggesting corrections and improvements, many of which they will find incorporated here. There is no more gratifying experience for an author than to learn that readers have benefited from his work. I hope those who read the first edition will have found this second one at least modestly improved, and that new readers will have found it sufficiently current to be useful.

And finally, because it absolutely does not belong here, here is my only mechanical engineering joke:

> What did the stone beam say to the column?
> "If a rock can do it, why can't a lever?"

Think about it.

Daniel M. Dobkin
Sunnyvale, CA
March 10, 2012

Appendix 1: Radio Regulations

A1.1 Couldn't Wait for Global Warming

In April of 1912 the great passenger linear Titanic glanced off an iceberg on its maiden voyage from Great Britain to the United States. The Titanic was equipped with the most advanced wireless communications available in its day, but so were many of the ships that plied the north Atlantic in those days. While cold winter weather, arrogance, optimistic planning, and a lost pair of binoculars were the proximate causes of the disaster, the lack of regulations regarding radio usage and interference avoidance made a contribution to its severity.

This much-publicized tragedy catalyzed the end of the anarchy in the use of wireless at sea and the beginning of regulation of broadcasting in general. In August of 1912, a US law gave the Secretary of Commerce the responsibility of administering licensing of radio stations within the United States. Broadcast radio in the United States arose around 1921, and the initial legal framework recognized *de facto* possession of spectrum as constituting a property right: the first broadcaster in any jurisdiction to broadcast in a given frequency band owned the spectrum they transmitted in. However, confusion and contrary court decisions caused this simple guideline to falter, leading to broadcast anarchy and interference with popular stations in major broadcast markets. In 1927 the US Congress passed the Radio Act, which established a Federal Radio Agency with responsibility for stewardship of spectrum; seven years later the Congress replaced the FRA with the Federal Communications Commission (FCC). The FCC through its early years followed a strict command-and-control regulatory model, in which licenses for spectrum were granted only for specific uses in specific locations, with no rights to trade, modify, or assign the license.

Almost all other nations in the world followed suit in establishing a national agency to regulate the use of the radio and microwave spectrum. Coordination between these agencies is today provided by the **International Telecommunications Union** (ITU), which operates under the auspices of the United Nations. The ITU gathers the national regulatory agencies once every three years at the **World Radiocommunications Conference** (WRC), which like most such exercises generates only slow and painful progress towards world consensus on the regulation of radio technology.

Regulatory attitudes began to change in the 1980's, as the mantra of deregulation slowly spread through American government and politics. In 1985, the FCC approved the use of **Industrial, Scientific, and Medical** (ISM) spectrum at 900 MHz and 2.4 GHz (and a bit around 5 GHz) for unlicensed use, with limitations on power, allowed modulation techniques, and antenna gain to minimize consequent interference. Additional spectrum around 5 GHz, known as the **Unlicensed National Information Infrastructure** (UNII) band was made available in 1997. Similar actions were being taken by other regulatory agencies around the world. The manifest success of these actions in promoting innovative use of spectrum induced the FCC to make additional 5 GHz spectrum available in 2003 (Figure 3.15), as well as to allow ultrawideband radios to operate without licenses in spectrum previously reserved for other uses (Figure 3.27).

The proper model for regulation of spectrum usage is a topic of active debate within and outside of governments. The success of unlicensed spectrum has demonstrated the feasibility of a 'commons' model in which neither detailed use regulations nor property rights are applied; the success of the auctioned spectrum used for cellular telephony has demonstrated the utility of a market in spectrum rights. Both models are arguably superior to the command-and-control model practiced for most of the 20th century.

It is important to recall the physics that frames the changing debate. The 1912 Act in the US dealt with radio wavelengths in the range of 10's of meters to over 1600 meters (that is, from 10's of KHz to around 10 MHz). Buildings and even small mountains represent sub-wavelength obstacles at these frequencies. Diffraction is easy, absorption is small, and very high powers were often used. Direct ranges of tens to hundreds of kilometers are typical. The lower frequencies are below the plasma frequency of the ionosphere (the layer of partially-ionized gas that lies above the stratosphere), and thus are reflected by it, so that they can hop across hundreds or thousands of kilometers. That is, early radio stations broadcast everywhere, limited only by radiated power, and interference was intrinsically a public issue.

By contrast, the transmissions of RFID readers and tags are much easier to corral. For indoor networks, low radiated power, combined with good antenna management and a helpful building (thick concrete or wooden walls and conductive-coated windows) can keep most of the radiated power within a single structure. In many locations, foliage provides a strong limitation on outdoor propagation distance except for line-of-sight applications. Management of interference becomes much more akin to a local property right, administered by property owners for indoor networks and perhaps by local authorities (city or county governments) for outdoor networks. There is less need and less justification for government intervention in radio operation in these circumstances.

As communications evolves towards the use of still higher frequencies, with more intelligent radios and antennas, it can be expected that the laws and codes regulating the use of radio will continue to evolve as well. It is to be hoped that further innovation will produce more ways of identifying objects at lower cost as a result. That's the fun part; the remainder of this appendix is necessarily best suited for curing insomnia, though still less so than the original documents. Only the brave need continue.

A1.2 FCC Part 15

Use of unlicensed spectrum in the United States and other countries is not unregulated. Limitations are placed on total power, modulation approaches, antennas, and conditions of operation, in an attempt to minimize interference within the intended band. Further limitation of radiated and conducted emissions (that is, stuff that sneaks back into the AC power system through the equipment power cord) is imposed to minimize interference outside of the intended bands.

The regulations of the FCC are contained in Title 47 of the US code of Federal regulations. Part 15 deals with operation of unlicensed radio transmitters. Equipment must generally be approved prior to marketing, and labeled as such when sold (15.19); the main thrust of the approval process is to ensure that the regulations regarding interference are met by the equipment. Users are responsible to see that the limitations in the regulations are met during operation. A table summarizing some of the key aspects relevant to RFID applications is provided below, based on the revision of these regulations released August 14, 2006. Some of the choices in modulation and architecture made in the various protocols may become clearer on review of the relevant regulations. Of particular interest are ISM-band regulations (15.247). This table is meant to help the reader understand what the law is so as to remain within it; when in doubt, review the original text (available at the Government Printing Office website, http://www.access.gpo.gov), or seek legal counsel if appropriate.

For equipment installer, it is important to consider specific restrictions and loopholes. The intent of the regulatory apparatus for unlicensed radios is stated in 15.5, paragraph (b): an unlicensed radio must not create harmful interference, and must accept any interference it encounters. The basic framework followed to ensure this result is to require certification of all unlicensed equipment before sale (15.201). Furthermore, the rules seek to allow the use of only those antennas and power amplifiers which have been certified with a particular transmitter (15.203, 15.204). However, several loopholes are provided in this regulatory fence. First, 15.23 specifically allows 'home-built' radios that are produced in small quantities and are not sold. In addition, 15.203 implies non-certified antennas can be used when systems are 'professionally installed', and 15.204 specifically allows antennas of the same type as the antenna with which the system was certified, so long as directive gain is equal to or lower than the certified antenna.

Designers and manufacturers of commercial radio equipment are specifically permitted to develop and test their hardware, but specifically forbidden to offer it for sale on the open market prior to certification. Certification is a complex rule-bound process, best relegated to the many testing laboratories that exist to perform this function.

Table A1.1 provides a summary of the sections of part 15 which seem most relevant to unlicensed RFID operation. Some of the regulations are framed in terms of field strength. Field strength in dBμV/m at 3 meters, can be converted to EIRP in dBm by subtracting 95.2 dB.

Table A1.1: Summary of Code of Federal Regulations Title 47 Part 15 as of August 2006.
Unquoted text is the author's brief summary; quoted text is taken from the regulations

Section	Topic	Summary, excerpts, remarks
15.1	Scope	This section sets out the conditions under which unlicensed operation is permitted
15.5	General unlicensed operation	"Operation of an intentional, unintentional, or incidental radiator is subject to the conditions that no harmful interference is caused and that interference must be accepted that may be caused by the operation of an authorized radio station, by another intentional or unintentional radiator, by industrial, scientific and medical (ISM) equipment, or by an incidental radiator."
15.19	Labeling requirements	Certificated devices must be labeled in specific ways; in particular, the unique FCC ID number must be present.
15.21	Information to user	The user's manual must note that modifying the unit may void the user's authority to operate without a license.
15.31	Compliance measurement procedures	Read 'em and weep. This is why you pay a testing laboratory.
15.33	Measured frequency range	Frequency range for radiated measurements. Start at the lowest frequency generated by the device, and (for devices operating at <10 GHz) to the tenth harmonic or 40 GHz, whichever is lower. Thus for 900 MHz devices the limit will be 9 GHz; 2.45 GHz devices must be checked to 24.5 GHz, and UNII-band (5.2–5.8 GHz) to 40 GHz.
15.35	Measurement detector functions and bandwidth	Specifies how measurement tools are to be configured; another section hopefully delegated to your testing laboratory.
15.201	Certification	All intentional radiators (except home-builts as in 15.23) should be certified under part 2 subpart J (a separate chapter of the regs, not covered in this table) prior to marketing.
15.203	External antennas	All devices must either use built-in antennas or non-standard connectors not easily available to the public [maybe true years ago but hardly today], except those that are required to be 'professionally-installed'. In the latter case, the installer is responsible for ensuring that

(Continued)

Table A1.1: (Continued)

Section	Topic	Summary, excerpts, remarks
		the limits of radiation are not exceeded. Not entirely consistent with 15.204 below.
15.204	External power amplifiers and antenna modifications	An antenna can be marketed for use with an intentional radiator if it is of the same type as that used for certification testing, and of equal or lower gain. A power amplifier can be marketed separately but only for ISM/UNII band devices with which it has been tested.
15.205	Spurious emissions only	Lists bands in which only spurious emissions are allowed. Bands of interest for RFID are: 960–1240 MHz 2310–2390 MHz 2483.5–2500 MHz 4500–5150 MHz 5350–5460 MHz Limits for emission are stated in 15.209. Note there are limits on conducted emissions to the AC power line, and low-frequency (< 1 GHz) emissions which are not specifically cited in this summary, though emissions may exist due e.g. to LO frequencies, and must be measured in certification.
15.209	Radiated emission limits, general requirements (mainly spurious radiation for ISM systems)	For f < 900 MHz, field strength shall be less than 200 μvolts/meter measured at 3 meters; for f > 900 MHz, field strength shall be less than 500 μvolts/meter (except as provided in 15.217 through 15.255 for specific frequency bands) Note that these field strengths correspond to an EIRP of about −49 and −41 dBm, respectively, as per the conversion cited above. Spurs must be less than the intended fundamental. Measurements should use an averaging detector above 1 GHz. Measurements at distances other than 3 meters are converted using procedures from 15.31,33,35.
15.240	433.5 to 434.5 MHz RFID band	This band is specifically directed to asset tracking in ports and shipping areas. Band-specific radiated field limits are provided. These devices are forbidden from operating within 40 km of certain air force bases. Users must inform the FCC of the location of the devices and notify the FCC if they are moved.
15.247	ISM band regulations	This one is important and long. Summary presented separately after the table.
15.401	UNII general	
15.407	UNII band regulations	The equivalent of 15.247 for the UNII band.
15.501	UWB general	Ultrawideband operation — that is, radiators that are allowed to intentionally radiate into bands normally reserved for other uses.

(Continued)

Table A1.1: (Continued)

Section	Topic	Summary, excerpts, remarks
15.503	UWB band definitions	fH,fL are the 10-dB-down upper and lower frequencies fC is avg of fH, fL fractional BW = (fH − fL)/fC UWB = fractional BW > 0.2 OR BW > 500 MHz Several sections cover requirements for special applications of UWB (e.g. through-wall imaging); they are not summarized here.
15.517	Indoor UWB technical requirements	Must be only capable of operation indoors; e.g. mains-powered, not intentionally directed outside the building, no outdoor-mounted antennas. UWB bandwidth must be contained between 3,100 and 10,600 MHz; radiated emissions at or below 960 MHz must obey 15.209. There are special restrictions on emission into certain bands. Additionally, in 50 MHz centered on frequency of maximum radiation, less than 0 dBm EIRP.
15.519	Handheld devices	Technical requirements are similar, though these devices may of course be used outdoors.
15.521	Other UWB uses	UWB may not be used for toys, on board aircraft, ship, or satellites. The frequency of maximum radiated emission must be within the UWB bandwidth.

15.247 in Detail

Frequency bands covered: 902−928 MHz, 2400−2483.5 MHz, 5725−5850 MHz

"Frequency hopping systems shall have hopping channel carrier frequencies separated by a minimum of 25 kHz or the 20 dB bandwidth of the hopping channel, whichever is greater... The system shall hop to channel frequencies that are selected at the system hopping rate from a pseudorandomly ordered list of hopping frequencies. Each frequency must be used equally on the average by each transmitter. The system receivers shall have input bandwidths that match the hopping channel bandwidths of their corresponding transmitters and shall shift frequencies in synchronization with the transmitted signals."

Frequency-hopping systems in the 902−928 MHz band: at least 50 hopping frequencies if the bandwidth is less than 250 kHz; at least 25 hopping frequencies if the bandwidth is 250−500 kHz. Average time of occupancy at any frequency shall not exceed 0.4 seconds in a 10 second period. Bandwidth >500 kHz not permitted. Systems employing less than 50 hopping channels are limited to 0.25 W conducted power; systems using at least 50 channels may transmit at 1 Watt.

Frequency-hopping systems in the 2400−2483 MHz band: at least 15 non-overlapping channels must be used. Occupancy in any channel less than 0.4 seconds in a period of

(0.4•(# of channels)). Systems that use 75 channels are allowed 1 Watt transmit power; any other FH system is limited to 1/8 watt.

Digital modulation techniques are allowed in 902–928 MHz, 2400–2483 MHZ and 5725–5850 MHz; 6 dB bandwidth shall be >500 KHz, with 1 Watt maximum transmit power. [Note that there is no longer a 'spread-spectrum' requirement, which was present in the original incarnation of the regulations.]

The power limits above assume antennas with no more than 6 dBi of gain. Power must be reduced 1 dB for each dB of antenna gain above 6 dBi (that is, the regulated quantity is actually the EIRP of the system).

All systems must protect the public from harmful RF energy; see Sec. 1.1307(b)(1).

OUT OF BAND: in any 100 KHz slice outside of the target band, the radiated energy must be at least 20 dB less than that found in the 100 KHz slice with the highest intentional power (typically center of the radiated band or near the center). Radiations in the restricted bands defined in 15.205 must be within the limits of 15.209.

COORDINATION: an individual transmitter can adjust hops to avoid interference, but coordination between multiple transmitters is not allowed. [This appears to be a precaution to avoid people simply having a bunch of transmitters all hopping, but the net effect of which is to fill all the band and exceed the limits on radiated power.]

A1.3 European Standards

Individual national regulatory bodies continue to operate within European states, but the nations of the European Union generally seek to harmonize their individual regulatory actions, under the supervision of the *European Conference of Postal and Telecommunications Administrations* (CEPT, the acronym being derived from the French). Until 2001, and the *European Radiocommunications Committee* (ERC) was responsible for spectrum engineering, frequency management, and other technical regulations; in 2001 the ERC was reorganized and renamed the *Electronic Communications Committee* (ECC). Under the ECC/ERC, the *European Radiocommunications Office* (ERO) issues recommendations on spectrum allocation and regulation. The *European Telecommunications Standards Institute* (ETSI) is responsible for issuing recommended standards for radio protocols, testing, and operation within the EC.

ETSI recommendations must generally be followed by all EC nations. However, each country can and often does choose their own sub-bands, and each country may create some "interface" standards that can supplement but not contradict the ETSI requirements.

For RFID use in Europe, three documents are of particular interest:

- **ERC Recommendation 70-03**: this document covers spectrum allocation for a number of low-power applications, including RFID; it also gives some channel- and band-naming conventions that are used in other documents.
- **ETSI 302 208**: regulations for the operation of RFID in the 865−868 MHz bands at power up to 2 Watts.
- **ETSI 300 220**: overall regulations for short-range radio devices. Some RFID systems have been sold and operated under these requirements, though in general 302 208 is preferred.

At the time of this writing, ETSI documents are available for free download (after registration) from the ETSI web site, http://www.etsi.org. Brief summaries of some of the key portions of the relevant documents are provided below.

ERC 70-03

The original document is dated 1997; this discussion is based on the October 2005 update. Annex 11 of this document defines three sub-bands in the 865−868 MHz range that can be used for RFID. The sub-bands differ mainly in the radiated power allowed, which is expressed in terms of *effective radiated power*(ERP), which is like EIRP but referenced to a standard dipole antenna rather than an isotropic antenna. Thus ERP = transmit power (dBm) + antenna gain (dBi) −2.2 dB; 36 dBm EIRP is about 33.8 dBm ERP.

Sub-band name	Frequency range (MHz)	Allowed power (ERP)
b1	865−868	0.1 W
b2	865.6−867.6	2 W
b3	865.6−868	0.5 W

That is, in the region 865−865.6 MHz, transmitters are limited to 100 mW; from 865.6 to 867.6, 2 W ERP are allowed, and from 867.6 to 868 MHz, 500 mW are permitted.

Annex 1 provides differing names for some of the same bands and some additional bandwidth, with much lower power limits and limits on duty cycles and modulation schemes.

Sub-band name	Frequency range (MHz)	Allowed power (ERP)
g	863−870	25 mW
g1	868−868.6	25 mW
g2	868.7−869.2	25 mW
g3	869.4−869.65	0.5 W, <10% duty cycle
g4	869.7−870	5 mW

ETSI 302-208 "Radio Frequency Identification Equipment operating in the band 865–868 MHz with power levels up to 2 W" (version 1.3.1, 2010)

ETSI 302 208 is specifically directed to RFID applications using the 865–868 MHz band and the b1/b2/b3 nomenclature. Interrogators must employ 200 kHz channels, with stringent spectral mask requirements. This means that maximum reader data rates will be roughly 40% of those obtained in the US.

Readers may use one of four channels, each 200 kHz wide. The bottom channel is at 865.7 MHz, with the three higher channels spaced at 600 kHz intervals from this one. Dense interrogator mode is to be used, so that the tag replies are located between these "high-power" channels, in "low-power" 200 kHz channels. Transmitted power shall not exceed 33 dBm e.r.p., where the reader will recall that e.r.p. refers to the transmitted power relative to a standard dipole antenna. Thus, for a 6 dBi patch antenna, you have to subtract $6 - 2.3 = 3.7$ dB from the allowed power. Thus, this requirement is about 5 dB more restrictive than the 36 dBi EIRP requirement used in the United States.

ETSI 300-220 "Radio equipment to be used in the 25 MHz to 1 000 MHz frequency range with power levels ranging up to 500 mW" (version 2.1.1, 4/05)

EN 300-220 is not RFID-specific but creates some general categories for operating short-range, low-power devices in the same general bands addressed by 302 208. The bands addressed are the g-g4 bands described in ERC 70-03. Power is strictly limited, and duty cycle (the percent of time the transmitter can be on) is limited unless the transmitter implements LBT, except for band g4, where only very low power operation is allowed.

In the past, some handheld readers have been able to operate in band g3 under 300-220, since it can be expected that a handheld reader will not be on most of the time, and RFID printers have been allowed in band g4: since the tags are on a roll passing right over the reader antenna, actual power radiated from the printer is very small. It is expected that as countries adopt 302 208, RFID operation will be required to meet 302-208 requirements and 300-220 will no longer be allowed, although the anticipated revisions of 302-208 may delay adoption.

A1.4 Those Other Few Billion Folks

The rest of the world is hardly less important than the US or Europe, but is rather more fragmented and opaque from a regulatory point of view. The situation was graphically summarized in Figure 2.13 in chapter 2. From time to time EPCglobal publishes a very useful summary of regulatory status worldwide; the latest at the time of this writing is "Regulatory status for using RFID in the UHF spectrum, 3 March 2006". Generally speaking most nations of the world follow either the FCC or ETSI approach, or some combination of the two. However, in most countries providing FCC-like operation, only a small subset of the 26 MHz available in the US is provided. This is in part due to the prevalence of the GSM cellular telephony standard in many regions of the world, which uses some of the bandwidth allocated to the ISM band in the US.

A few words on some of the key nations seem appropriate:

In Japan, the *Ministry of Posts and Telecommunications* regulates the use of the radio spectrum. The MBT appears to delegate the creation of 'voluntary' Japanese standards to the *Association of Radio Industries and Businesses* (ARIB). Equipment certification is performed by the *Telecom Engineering Center* (TELEC).

The *Ministry of Information Industry* regulates radio operations in the People's Republic of China (the mainland). The regulations do not appear to be readily available in English.

Radio operation in Singapore is regulated by the Infocomm Development Authority (IDA). Regulations are generally available in English. RFID operation is allowed in the 866−869 MHz band without a license, and in the 923−925 MHz band at less than 0.5 W ERP. At power levels from 0.5 to 2 W ERP, a site license is required. See http://www.ida.gov.sg/Infocomm%20Adoption/20061002182022.aspx.

The regulations for unlicensed operation worldwide continue to evolve with the technology. By the time this book gets into your hands, some of what is described here will have changed (even when I got it right in the first place!). There's no substitute for checking the source materials. The APEC Telecommunications and Information Working Group web page http://www.apectelwg.org/apec/alos/osite_1.html provides a very useful listing of telecommunications regulatory links and sites for many countries in Asia and elsewhere. Other relevant web sites are listed below:

International Telecommunications Union, Radio sector: http://www.itu.int/ITU-R/

ITU-R World Radio Conference site: http://www.itu.int/ITU-R/conferences/wrc

US Federal Communications Commission: http://www.fcc.org

US Code of Federal Regulations, Title 47:

http://www.access.gpo.gov/nara/cfr/waisidx_03/47cfr15_03.html

European Telecommunications Standards Institute: http://www.etsi.org;

European Radiocommunications Committee http://www.ero.dk/documentation/docs

Japan Ministry of Posts and Telecommunications radio-related laws: http://www.soumu.go.jp/joho_tsusin/eng/laws.html

Japan Association of Radio Industries and Businesses (ARIB): http://www.arib.or.jp/english/index.html

Japan Telecommunications Engineering Center (TELEC): http://www.telec.or.jp/ENG/index_e.htm

National Radio Spectrum Management Center, People's Republic of China: http://www.srrc.gov.cn/ [in Chinese]

Singapore Infocomm Development Authority: http://www.ida.gov.sg

Appendix 2: Harmonic Functions

A2.1 Sines and Cosines

The archetype of a smooth periodic signal is the sinusoid (Figure A2.1), typically written as the product of the angular frequency ω and time t. The two closely related functions sine and cosine, abbreviated $\sin(x)$ and $\cos(x)$, where the arguments of the functions are here expressed as *radians*. The argument can also be expressed in degrees. There are 2π radians in a circle, so one radian $= (180/\pi) \approx 57$ degrees.

Both functions are periodic with a period of 2π radians, so if we write the sine as $\sin(2\pi f t)$, where t is time, then $f = 1/\text{period} = \text{frequency}$. We often use the *angular frequency* $\omega = 2\pi f$, in which case the sine becomes $\sin(\omega t)$. Frequency is measured in *Hertz* (abbreviated Hz); 1 Hz is one full cycle of the function per second. Thus when the frequency is 900 MHz = 900 000 000 Hz, the angular frequency is about 5.65 billion radians per second.Both of these functions alternate between a maximum value of 1 and minimum value of -1; cosine starts at $+1$, and sine starts at 0, when the argument is

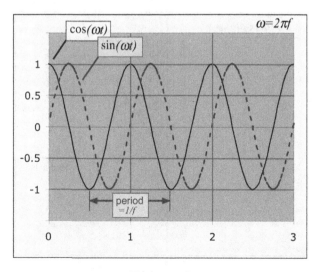

Figure A2.1
Sines and Cosines as a Function of Time.

499

zero. We can see that cosines and sines are identical except for an offset in the argument (the *phase*):

$$\cos(\omega t) = \sin\left(\omega t + \frac{\pi}{2}\right) \tag{A2.1}$$

We say that the sine lags the cosine by $\pi/2$ radians or 90°.

A2.2 Complex Numbers and Complex Exponentials

Let us now digress briefly to discuss complex numbers, for reasons that will become clear in a page or two. *Imaginary* numbers, the reader will recall, are introduced to provide square roots of negative reals; the unit is $i = \sqrt{(1)}$. A *complex* number is the sum of a real number and an imaginary number, often written as e.g. $z = a + bi$, where bi indicates the product of the real number b and the imaginary unit i. Electrical engineers often use; j instead of i, so as to use i to represent an AC current; we shall use that convention here. The *complex conjugate* z^* is found by changing the sign of the imaginary part: $z^* = a - bj$. This is a useful operation, since the quantity $z + z^*$ is then a real number.

Complex numbers can be depicted in a plane by using the real part as the coordinate on the x- (real) axis, and the imaginary part for the y- (imaginary) axis (Figure A2.2).

Operations on complex numbers proceed more or less the same way as they do in algebra, save that one must remember to keep track of the real and imaginary parts. Thus, the sum of two complex numbers can be constructed algebraically by

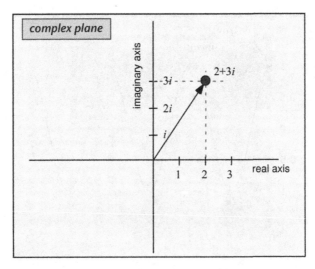

Figure A2.2
A Complex Number Depicted as a Point in the Plane.

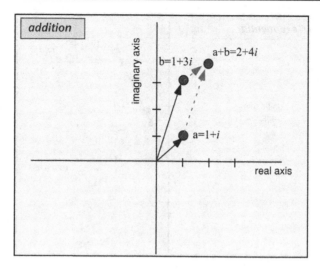

Figure A2.3
Addition of Complex Numbers.

$$(a + b\,j) + (c + d\,j) = [a + c] + [b + d]j \tag{A2.2}$$

and geometrically by regarding the two numbers as vectors forming two sides of a parallelogram, the diagonal of which is their sum (Figure A2.3).

Multiplication can be treated in a similar fashion, but it is much simpler to envision if we first define the length (also known as the *modulus*) and angle of a complex number. We define a complex number of length 1 and angle θ to be equal to an exponential with an imaginary argument equal to the angle (Figure A2.4). Recall the number e is about 2.718, and is the base for the *natural logarithm*. Any complex number (for example, b in the figure) can then be represented as the product of the modulus and an imaginary exponential whose argument is equal to the angle of the complex number in radians.

By writing a complex number as an exponential, multiplication of complex numbers becomes simple, once we recall that the product of two exponentials is an exponential with the sum of the arguments:

$$e^a \cdot e^b = e^{[a+b]}. \tag{A2.3}$$

The product of two complex numbers is then constructed by multiplying their moduli, and adding their angles (Figure A2.5),

$$(\rho_1 e^{j\theta_1}) \cdot (\rho_2 e^{j\theta_2}) = [\rho_1 \rho_2] e^{j[\theta_1 + \theta_2]}. \tag{A2.4}$$

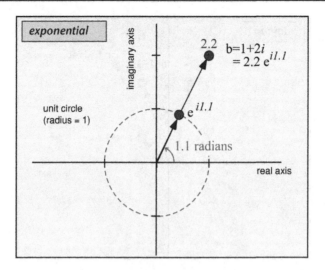

Figure A2.4
Complex Numbers Written Using Imaginary Exponential Functions.

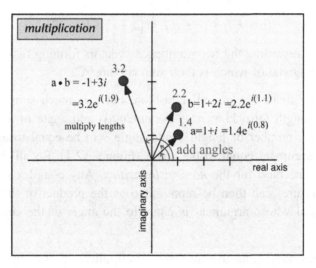

Figure A2.5
Multiplying Complex Numbers by Multiplying Lengths and Adding Angles.

The imaginary unit j is an exponential with an argument of $\pi/2$, so $1/j$ is an exponential with an argument of $-\pi/2$—that is:

$$\left(e^{j\frac{\pi}{2}}\right) \cdot \left(e^{-j\frac{\pi}{2}}\right) = 1 \rightarrow \frac{1}{j} = -j. \tag{A2.5}$$

We have taken the trouble to introduce all these unreal quantities because they provide a particularly convenient way to represent harmonic signals. Since the *x*- and *y*-components of a unit vector at angle θ are just the cosine and sine, respectively, of the angle, our definition of an exponential with imaginary argument implies:

$$e^{j\theta} = \cos(\theta) + j\ \sin(\theta). \tag{A2.6}$$

Thus, if we use for the angle a linear function of time, we obtain a very general but simultaneously compact expression for a harmonic signal:

$$\begin{aligned} e^{j(\omega t + \varphi)} \quad &= \cos(\omega t + \varphi) + j\sin(\omega t + \varphi) \\ &= [\cos(\omega t) + j\sin(\omega t)] \cdot [\cos(\varphi) + j\sin(\varphi)]. \end{aligned} \tag{A2.7}$$

In this notation, the signal may be imagined as a vector of constant length rotating in time, with its projections on the real and imaginary axes forming the familiar sines and cosines (Figure A2.6). The phase offset ϕ represents the angle of the vector at $t = 0$.

In some cases we wish to use an exponential as an intermediate calculation tool to simplify phase shifts and other operations, converting to a real-valued function at the end by either simply taking only the real part, or adding together exponentials of positive and negative frequency. (The reader may wish to verify, using equation (A2.6) and (A2.7), that the sum of exponentials of positive and negative frequencies forms a purely real or purely imaginary sinusoid.) However, in radio practice, a real harmonic signal $\cos(\omega t + \phi)$ may also be regarded as being the product of a real carrier $\cos(\omega t)$ and a complex number $I + jQ = [\cos(\phi) - j\sin(\phi)]/2$, where the imaginary part is obtained through multiplication with $\sin(\omega t)$ followed by filtering. (Here I and Q denote 'in-phase' and 'quadrature'—that is, 90° out of

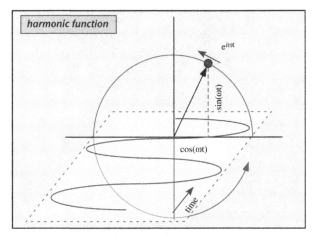

Figure A2.6
An Imaginary Exponential can Represent Sinusoidal Voltages or Currents.

phase—respectively.) This formal decomposition is carried out in practice using two mixers excited by local oscillator signals in quadrature; see Chapter 4 for examples of this configuration.

We can use the multiplicative properties of exponentials (equation (A2.3)) to derive some very useful identities relating trigonometric functions:

$$
\begin{aligned}
e^{i\theta} e^{i\varphi} &= e^{i(\theta+\varphi)} \rightarrow (\cos(\theta) + i\sin(\theta))(\cos(\varphi) + i\sin(\varphi)) \\
&= (\cos(\theta + \varphi) + i\sin(\theta + \varphi))
\end{aligned}
$$

equating real and imaginary parts:

$$
\begin{aligned}
\cos(\theta)\cos(\varphi) - \sin(\theta)\sin(\varphi) &= \cos(\theta + \varphi) \\
\sin(\theta)\cos(\varphi) + \cos(\theta)\sin(\varphi) &= \sin(\theta + \varphi).
\end{aligned}
\tag{A2.8}
$$

So we can express cosines and sines of a sum or difference as products of cosines and sines of the individual values. We can readily derive expressions for products of cosines and sines, and double- and triple-angle relationships.

Finally, we note one other uniquely convenient feature of exponentials: differentiation and integration of an exponential with a linear argument simply multiplies or divides the original function by the constant slope of the argument:

$$
\frac{d}{dx}(e^{ax}) = ae^{ax}; \quad \int e^{ax}\, dx = \frac{e^{ax}}{a}.
\tag{A2.9}
$$

That is, exponentials convert calculus to algebra! Electrical engineering makes extensive use of this remarkable fact in analyzing networks containing capacitors and inductors, as we discuss in Appendix 3.

Appendix 3: Resistance, Impedance and Switching

A3.1 Electric Company Detective Sherlock Ohms

Electric current is the flow of electrically charged particles along a wire. (The particles that flow are usually electrons, which by convention have a negative charge as a consequence of Benjamin Franklin guessing wrong; therefore the direction of current flow is usually opposite the direction in which the particles actually move. Sorry.) A resistor is a device that doesn't store electrical charges, but merely resists their flow. Electrical engineering starts with Ohm's law: the voltage across a resistor is proportional to the current flowing through it. In mathematical terms:

$$V = I \cdot R \tag{A3.1}$$

where I is the current, R the resistance, and V the voltage. Electrical current is usually carried in thin wires, and generally returns to where it started to form a *circuit*. We can draw a simplified picture of an electrical circuit, using line segments for wires, a circle to symbolize a voltage source (like a battery or generator), and a squiggly line to indicate a resistor. This is known as a *schematic diagram*, or just schematic for short. The schematic for a simple circuit containing only a resistor and a voltage source is shown in Figure A3.1. The figure also shows a *ground* symbol. Ground is just the location in the circuit where the voltage is defined to be equal to 0. The voltage along a wire is constant, so the voltage on all the wires directly connected to ground is 0. The voltage source creates a voltage difference V_1 between the top and bottom connections. The wire carries this voltage to the left side of the resistor. The right side of the resistor is connected by a wire to ground and so is at zero voltage. By Ohm's law, the current that flows through the resistor is the ratio of voltage to resistance:

$$I = \frac{V_1}{R}. \tag{A3.2}$$

The current is constant all around the circuit, because there's nowhere else for it to go (assuming the electrons are all stuck on the wire): this statement can be formalized as one of *Kirchoff's laws*.

When the same current flows through two resistors in sequence, the resistors are said to be connected in *series*. Resistors can also be connected in *parallel*, providing multiple paths

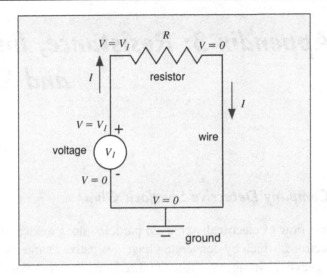

Figure A3.1
Example Schematic Diagram.

for the current to flow in. These alternatives are illustrated in Figure A3.2. When the resistors are connected in series, the voltage at the top of resistor R_1 is found from Ohm's law as IR_1. This is also the voltage at the right side of resistor R_2. Since the voltage across R_2—the difference between the voltages at the left and right—is similarly IR_2, the total voltage across the two resistors is the sum of the individual voltages. That is, the two resistors in series act just like a single resistor whose value is the sum of the two individual resistors:

$$V_1 = IR_1 + IR_2 = I(R_1 + R_2) = IR_{\text{ser}} \tag{A3.3}$$

where $R_{\text{ser}} \equiv R_1 + R_2$.

When resistors are connected in parallel, the total current splits between the two branches. We can again find a single resistance equivalent to the two resistors, but the calculation is a bit more complex. We can find the current through each resistor individually using Ohm's law since the voltages on the left and right sides of the two resistors are the same:

$$I_1 = \frac{V_1}{R_1}; \quad I_2 = \frac{V_1}{R_2}. \tag{A3.4}$$

We then impose the condition that the two resistor currents sum to the total current:

$$I_1 \quad + I_2 = I = \frac{V_1}{R_1} + \frac{V_1}{R_2} = V_1 \left(\frac{1}{R_1} + \frac{1}{R_2} \right) \rightarrow$$

$$I \quad = \frac{V_1}{R_{\text{par}}} \text{ where } R_{\text{par}} \equiv \frac{1}{\dfrac{1}{R_1} + \dfrac{1}{R_2}} = \frac{R_1 R_2}{R_1 + R_2} \tag{A3.5}$$

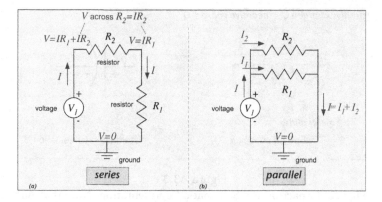

Figure A3.2
a) Two Resistors Connected in Series; b) Two Resistors Connected in Parallel.

That is, the two resistor in parallel look like a single resistor whose value is a somewhat messy formula. If the two resistors are of equal value, the parallel resistance is 1/2 of the resistance of either one. If one resistor is much smaller than the other, the parallel resistance is nearly equal to that of the smaller-valued resistor and the other has little effect.

Sometimes people like to introduce the *conductance* $G = 1/R$. The conductance is the ratio of current to voltage. The conductance of resistors in parallel simply adds together:

$$G_{\text{par}} = \frac{1}{R_1} + \frac{1}{R_2} = G_1 + G_2. \tag{A3.6}$$

Any circuit can be decomposed into series and parallel combinations of components. Thus, any combination of resistors can be converted into a single effective resistor value, though the computation may be laborious for a complex circuit.

A3.2 Resistance is Useless?

No—just insufficient.

A resistor is a *passive* component: it does not add power to a circuit but merely converts a current into a voltage. Two other very common passive components are *capacitors* and *inductors.*

A capacitor is a device that stores charge. It is constructed by placing two conductive plates very close to one another (Figure A3.3). Since like charges repel, if some negative charge accumulates on one plate, electrons are driven away from the other plate, producing a net

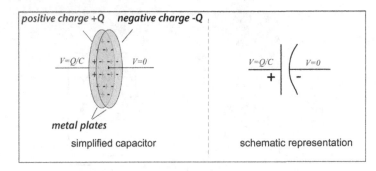

Figure A3.3
Simplified Physical Capacitor and Schematic Representation. Note that Although the
Schematic Symbol is Often Asymmetric, In Fact Most Capacitors Can Have Either Polarity
Applied to Either Plate.

positive charge. The charge is proportional to the voltage between the plates; the constant
of proportionality is the *capacitance C*. That is:

$$Q = C \cdot V \tag{A3.7}$$

where Q is the charge on the capacitor. It would be nice to use an analog of Ohm's law to
analyze circuits with capacitors and resistors, but there is a problem: the voltage is
expressed not in terms of a current but of a charge. Since current is the flow of electric
charge, the total charge is the time integral of the current:

$$Q = \int I dt. \tag{A3.8}$$

So instead of a simple linear relationship between current and voltage, analysis of a
capacitor leads to an integral equation:

$$\int I dt = C \cdot V. \tag{A3.9}$$

It is possible to treat circuits by writing out and solving the appropriate integral equation,
but fortunately there's a way to make the process look a lot more like the analysis we used
for resistors. The trick is to assume that all the circuit currents and voltages are the real part
of a complex harmonic function (these were introduced in Appendix 2)—that is:

$$V(t) = V_0 e^{j\omega t} \quad I(t) = I_0 e^{j\omega t}. \tag{A3.10}$$

The coefficients V_0 and I_0 are complex numbers but are not time-dependent. The reason this
is useful is that the time integral of the current is obtained by dividing the current by a
constant. Equation (A3.9) becomes:

$$\frac{I_0 e^{j\omega t}}{j\omega} = C V_0 e^{j\omega t} \rightarrow V_0 = \frac{1}{j\omega C} I_0 = -jX_c I_0. \tag{A3.11}$$

That is, if we use (complex) amplitudes, we can once again express the voltage as the product of the current and a quantity analogous to the resistance, the *capacitive reactance* X_c. The voltage is imaginary if the current is real: physically, this means that the voltage across a capacitor is not in phase with the current, but lags the current by 90°. This is the direct result of the fact that it takes a while to build up charge on the capacitor: at the moment when current is maximum, the charge accumulated in the previous cycle has gone to 0, and by the time the charge is finished accumulating (corresponding to a maximum value of voltage) the current has returned to 0. The capacitive reactance is inversely proportional to the capacitance: large capacitors offer little impediment to the flow of (AC) current, whereas small capacitors require large voltages for significant current flow.

Armed with reactances, we can analyze circuits containing resistors and capacitors in more or less the same way we analyzed resistive circuits. However, in this case, the constant of proportionality between overall current and voltage will be a complex number, the *impedance*, usually written as Z. For example, the impedance of a series combination of a resistor and a capacitor is the sum of the reactance and resistance:

$$V_0 = I_0 \left(R + \frac{1}{j\omega C} \right) = I_0 \cdot Z \tag{A3.12}$$

as illustrated in Figure A3.4. The impedance plays exactly the role of the resistance in Ohm's law, but it is a complex number, and its value is dependent on the frequency of the currents and voltages.

The other important passive component is the inductor (Figure A3.5). An inductor is a length of wire arranged so that the magnetic interaction of the currents flowing on the wire is significant. This magnetic interaction can be regarded as an extra contribution to the momentum of the electrons in the wire: current in the inductor doesn't like to get started and once it does get started it doesn't like to stop. These propensities can be summarized mathematically by saying that the voltage across an inductor is proportional to the rate of change of the current, that is to the time derivative of the current:

$$V = L\frac{dI}{dt} \tag{A3.13}$$

where the constant of proportionality is the *inductance L*. By assuming once again that the voltages and currents have a harmonic dependence, we obtain an expression for the inductive reactance:

$$j\omega L I_0 e^{j\omega t} = V_0 e^{j\omega t} \rightarrow V_0 = j\omega L I_0 = jX_L I_0. \tag{A3.14}$$

The inductive reactance is also complex but of the opposite sign to that of a capacitor: the voltage on an inductor leads the current by 90°. The reactance of an inductor increases with

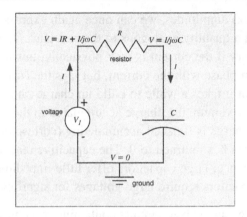

Figure A3.4
Series Circuit Containing a Resistor and Capacitor, Analyzed Using Complex Impedance.

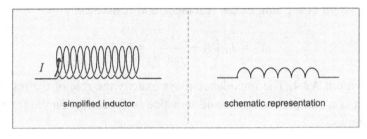

Figure A3.5
Simplified Physical Inductor and Schematic Representation.

increasing frequency and larger values of inductance. Inductors are the electrical opposite of capacitors. The impedance of a series resistor—inductor circuit is (Figure A3.6):

$$V_0 = I_0(R + j\omega L) = I_0 \cdot Z. \qquad (A3.15)$$

It is informative now to combine all the elements we have studied by finding the impedance of a series combination of a resistor, capacitor, and inductor. (In this and subsequent expressions, we omit the 0 suffix, assuming that all the voltages and currents are time-independent complex amplitudes.) Adding the reactances we obtain:

$$Z_{LCR} = \left(R + j\omega L - \frac{j}{\omega C}\right). \qquad (A3.16)$$

The interesting property of this expression is that the capacitance and inductance contribute with opposite signs, so for any given values of L and C, there is some frequency where the reactances exactly cancel, and the circuit just looks like a resistor as far as the overall

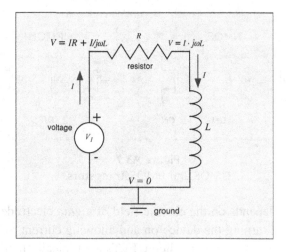

Figure A3.6
Series Circuit Containing a Resistor and Inductor, Analyzed Using Complex Impedance.

voltage and current are concerned. This condition is known as a *series resonance.*
The resonant frequency occurs when the two reactances are equal in magnitude:

$$\omega_{\text{res}}L = \frac{1}{\omega_{\text{res}}C} \rightarrow \omega_{\text{res}} = \sqrt{\frac{1}{LC}}. \tag{A3.17}$$

If the resistance is small compared to the magnitude of the inductive and capacitive reactances, the current flowing through the circuit will be much larger at resonance than at frequencies where the two reactive elements no longer cancel. The ratio of the reactance to the resistance of the circuit is known as the *quality factor, Q.*

$$Q = \frac{\omega_{\text{res}}L}{R} = \frac{1}{\omega_{\text{res}}CR} \tag{A3.18}$$

The bandwidth of the circuit at resonance is inversely proportional to the quality factor.

Parallel circuits of resistors, capacitors, and inductors can also be analyzed in the same fashion, but the resulting expressions are more difficult to work with because of the presence of complex denominators. A parallel combination of an inductor and a capacitor also exhibits resonant behavior, but instead of the current reaching a maximum at resonance, it is minimized.

A3.3 Switching

Active components can amplify signals and make new ones. The common active components are transistors, of which the most popular type today is the field-effect

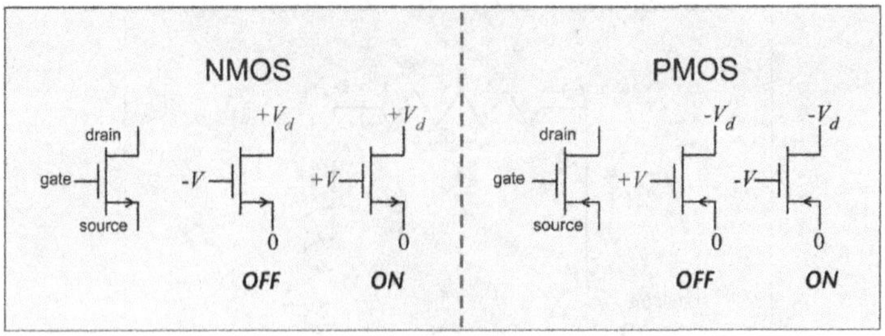

Figure A3.7
NMOS and PMOS Transistors.

transistor (FET), which depends on the electric field of a *gate* electrode to attract charge carriers into the *channel*, turning the device on and allowing current to flow between the *source* and *drain*. When constructed on silicon, FETs are typically fabricated using a thin layer of silicon dioxide to separate the gate and the channel and can either use electrons (NMOS) or holes (PMOS) as charge carriers. To operate an NMOS device, a positive voltage is applied to the drain; a positive voltage on the gate will allow electrons to flow from source to drain (with conventional current flowing in the other direction). A PMOS device uses opposite polarities on all the contacts and is thus turned on by a negative voltage applied to the gate, with conventional current flow in agreement with the flow of holes from source to drain. These conventions are illustrated in Figure A3.7. A FET can be used as a diode by connecting the gate and drain together. For example, an NMOS device connected in this fashion will conduct current readily when a positive voltage is applied to the drain (and gate), since the gate is positive with respect to the source but will turn off when a negative voltage is applied to the common drain/gate contact.

Appendix 4: Reflection and Matching

A4.1 Reflection Coefficients

In most of our discussion, we have assumed that radios, tags, and antennas are all well-matched, so that any power coming from one goes into the other. What happens when this is not the case? How do we measure the deviation from ideality, and what can we do about it? In this appendix we provide a very brief introduction into reflection coefficients and impedance matching.

In Microwave Land, a *port* is a connection from one microwave environment to another—for example, from a cable to an antenna. The cable, or any other signal-carrying electrical connection whose properties are well-defined and don't change along its length, is often known as a *transmission line*. A signal traveling along a transmission line ideally doesn't change its shape as it moves down the line, but only its phase. Transmission lines generally have a *characteristic impedance*, the ratio of the voltage due to a current traveling along the line to the current, that is a real resistance: 50-Ω transmission lines are very commonly encountered in microwave applications. The signal traveling along the transmission line to an antenna or other component may be partially reflected if the impedances of the antenna and the line do not match (Figure A4.1).

The ratio of the reflected signal to the incident signal is the reflection coefficient:

$$\Gamma \equiv \frac{\nu_{\text{ref}}}{\nu_{\text{inc}}}. \tag{A4.1}$$

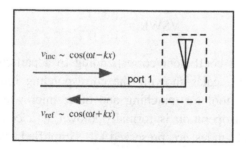

Figure A4.1
An Incident and Reflected Signal.

513

The reflection coefficient is in general complex, since the phase of the incident and reflected waves may not be the same. It can be shown that the reflection coefficient is related to the impedance seen by a wave at port 1:

$$\Gamma = \frac{Z_1 - Z_c}{Z_1 + Z_c} \tag{A4.2}$$

where Z_1 is the (generally complex) impedance of the port and Z_c is the characteristic impedance of the transmission line, typically 50 Ω. Recall that the impedance of a capacitor C with frequency f is $Z_C = 1/(j\omega C)$, and an inductor has an impedance of $Z_L = j\omega L$, with $\omega = 2\pi f$.

The reflection coefficient must always have a magnitude less than 1 (if port 1 has only passive circuits inside of it) and varies from $+1$ for an open (infinite load impedance) to -1 for a short (zero load impedance), as can be easily verified from (A4.2). The magnitude of the reflection coefficient in dB is often known as the *return loss*; the terminology implies that this is the loss suffered by the incident signal making a return trip to the sending instrument.

The phase of the reflection coefficient changes if we measure it at a different location along the cable, because the incident wave gains phase as we move to the right, and the reflected wave gains phase as we move to the left. Since the ratio of the voltages takes the difference of the phases (see Appendix 2), the phase of Γ changes by 4π each time the measurement plane moves one wavelength. Since the total voltage at any location is the sum of the incident and reflected voltages, the total voltage will also vary with position along the cable. The ratio between the largest and smallest magnitude of voltage is known as the *voltage standing wave ratio*, often abbreviated VSWR. (The importance of this somewhat funky parameter is partially historical; in the days when phases of microwaves were very difficult to measure it was relatively easy to move a pickup along a waveguide and measure the difference in power received.) By reference to equation (A4.2) it is easy to see that VSWR can be expressed in terms of the reflection coefficient:

$$\text{VSWR} = \frac{1 + |\Gamma|}{1 - |\Gamma|}. \tag{A4.3}$$

If we display the reflection coefficient corresponding to a particular complex impedance in the complex plane, with scales showing the corresponding impedance, we get an extremely useful graphical tool for matching and other microwave circuit operations, the *Smith Chart*. (Such an operation is formally known as a *conformal map*; circles map to circles and angles are preserved.) A simplified chart is shown in Figure A4.2.

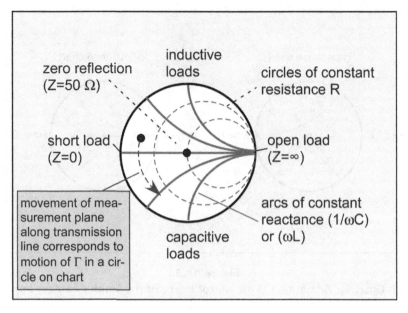

Figure A4.2
The Smith Chart.

The Smith chart maps the infinite impedance right half-plane (with positive values of the resistance) into a finite region (a circle of radius 1). It provides a very useful visual summary of what adding any element to the circuit of port 1 does to the consequent reflection. Moving along the transmission line towards the load moves the reflection coefficient counterclockwise on a circle around the point $\Gamma = 0$ (as shown in Figure A4.2); for each half-wavelength distance, the impedance makes a complete circle around the chart. Adding a capacitance or inductance moves the reflection coefficient on a circle of constant resistance; adding a (series) resistor moves Γ on an arc of constant reactance. Another nifty property of the Smith chart is that the picture for admittances (the reciprocal of an impedance, corresponding to elements added in parallel instead of in series) is just the same chart but reflected through the y-axis (Figure A4.3).

When used for design purposes, the Smith chart is usually displayed with a large number of circles of constant resistance and arcs of constant reactance, each labeled with either the corresponding value in ohms, or the normalized value (that is, the resistance or reactance divided by the characteristic impedance of the transmission line, Z_c). An example of this type of practical design chart is shown in Figure A4.4.

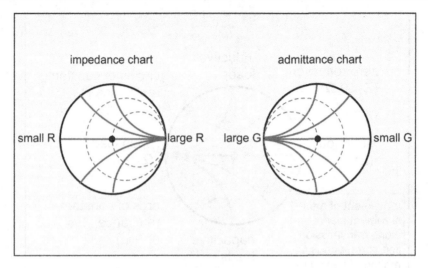

Figure A4.3
The Smith Chart for Admittance is the Mirror Image of the Smith Chart for Impedance.

A4.2 A Simple (But Relevant) Matching Example

Let us consider the problem of *matching* a short dipole antenna to a tag IC: that is, adjusting the impedance of the IC so that all the power that the antenna delivers is absorbed by the IC. Two conditions must be met to ensure best power transfer:

1. The real part of the load impedance must be equal to the real part of the source impedance.
2. The imaginary part of the load impedance must be—(imaginary part of the source impedance): that is, the load impedance is the complex conjugate of the source impedance, so that when we add them together to get the total the imaginary parts cancel, leaving only the resistances.

Plausible values for the equivalent circuits for these two objects are shown in Figure A4.5. Here, the antenna is represented by the resistor, capacitor, and inductor in series; the component values have been adjusted to be reasonably representative of a dipole antenna a bit shorter than resonance. The load is a lossy capacitance like an IC input, though the values have been adjusted to make the load a bit easier to see. The challenge is to find a way to convert the actual load impedance, a lossy capacitance at the bottom right part of the chart, to the matched load impedance, a rather more lossy inductance at the top right.

One possible first step is to use a large series inductor to both resonate the IC capacitance and that of the antenna. However, this requires a rather large inductor and will not solve the problem of the mismatch between the 50-Ω source and 15-Ω load. A better approach is to

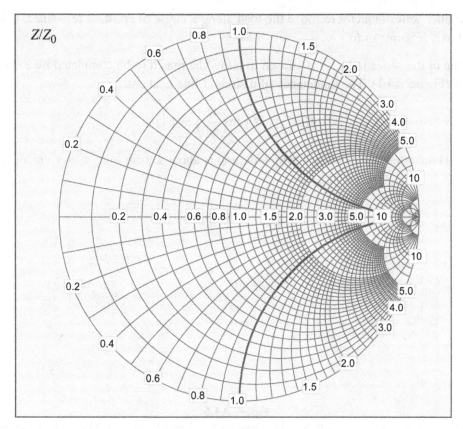

Figure A4.4
Practical Design Smith Chart, Normalized to the Characteristic Impedance of the Line.

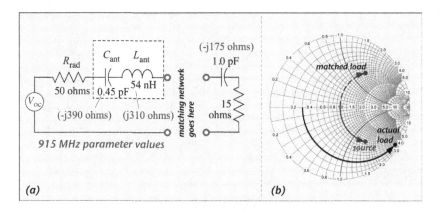

Figure A4.5
Example Matching Problem. a) Equivalent Circuit Model for Antenna and IC b) Corresponding
Loads on the Smith Chart. Component Values Adapted for Visibility.

use a smaller series inductor to move the load along a circle of constant resistance, towards the real axis (Figure A4.6).

The value of the series inductor is chosen to allow the match to be completed by a shunt inductor (Figure A4.7). The inductance of a straight line is about:

$$\ell(\text{nH}) = 2L\left(\ln\left(\frac{4L}{w}\right) - 1\right), \tag{A4.4}$$

so a 21 nH inductor fabricated in 1 mm wide line is about 2.5 cm long, a convenient size.

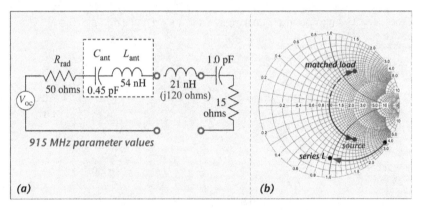

Figure A4.6
a) A Series Inductor is Placed Before the Load b) The Series Reactance Moves the Load on a Circle of Constant Resistance Towards the Real Axis.

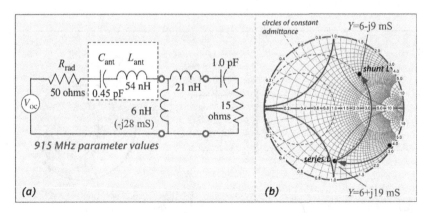

Figure A4.7
a) A Shunt Inductor is Placed Across the Load to Complete the Match b) The Shunt Susceptance Moves the Load on a Circle of Constant Conductance. Note that a Simplified Admittance Chart has been Overlaid on the Impedance Smith Chart.

The shunt inductor can move the load along a circle of constant conductance. By choosing the proper value for the series inductance, we can arrange for this circle to intersect the desired conjugately matched load. The 21 nH and 6 nH inductors are more readily realized with short strips of line than the 42 nH inductor that would be needed to match using a single element. More importantly, the use of two matching elements allows us to transform the real part of the load to match the real part of the source. A single inductor would resonate out the series capacitances but leave us with a load resistance of 15 Ω, implying poor power transfer from the antenna—most of the power is scattered away. The use of the series-shunt inductors allows us to make the load appear like a 50 Ω resistor to the source, achieving optimal power transfer.

The series-shunt matching procedure shown here is particularly convenient for tag antennas, since the requisite inductors are realized as short lengths of feed line connecting the integrated circuit to the antenna. Many other matching topologies are possible, employing series and shunt capacitances and lengths of transmission line.

Index

Note: Page numbers followed by "*f*" and "*t*" refer to figures and tables, respectively.

Printed in the United States
By Bookmasters